Safety for Future Transport and Mobility

Hans-Leo Ross

Safety for Future Transport and Mobility

Springer

Hans-Leo Ross
Lorsch, Hessen, Germany

ISBN 978-3-030-54882-7 ISBN 978-3-030-54883-4 (eBook)
https://doi.org/10.1007/978-3-030-54883-4

The work was first published in 2014 by Carl Hanser Verlag München Wien with the following title: *Funktionale Sicherheit im Automobil: ISO 26262, Systemengineering auf Basis eines Sicherheitslebenszyklus und bewährten Managementsystemen*

This Springer imprint is published by the registered company Springer Nature Switzerland AG
The registered company address is: Gewerbestrasse 11, 6330 Cham, Switzerland

Preface

During the last pages of this book, my father died after a long illness at the age of 89 years. Of course, he had a great influence on my thinking and my professional development. He was born in the years between the two world wars, and his childhood was marked by the aftermath of the First World War and people mourned the Imperial Era.

At that time, Göttingen was still considered the technical centre of science, and the language of science was still German.

At the very beginning, I would like to apologise to all readers for thinking like a German when writing this book. I live my daily life in German, with my children and my wife I only speak German. Most of the conversations about the topics in this book I have conducted in German. I continue to try regular contact with China, France, England and Scandinavia and with companions from the USA, but they all confirm that I think typically German. With many of these people, I speak English, but I still think the complex matters in German thoughts. It is an insight of this book that the language context is a major influence on the understanding of the content. Norbert Wiener and the scholars of these days also knew that the linguistic context has an essential influence on what and how you understand something. The mathematicians had an intensive exchange with the universities and professors in China, and by reflection and communication the common understanding was formed.

The parents of Norbert Wiener had decided to emigrate to America, as so many Jewish families had to do. He himself not only conducted an intensive exchange with the scholars in Germany, but also with France, England and Russia, forming networks such as the philosophical network of the Vienna Circle (Wiener Kreis). However, the new rising power in Asia was China, which already offered interesting aspects to many scholars. China offered a good ground for the scholars and the exchange with Japan soon proved to be just as difficult because the harbingers of the Second World War were in the offing. My father was able to attend a gymnasium for a few more years, but after the war, the focus was on feeding the family and educating the younger siblings. After the Second World War, Germany was divided, we in the West oriented ourselves more towards France and the USA;

the East oriented itself towards Russia and a few made trips to communist China. Aeronautics and space travel now developed in the English language, and in the development of cybernetics, the Russians were more likely to stand on their own feet. German was lost as a scientific language.

My father became first a truck driver, later a CNC miller and machinist in a chemical plant. Later on, he was very active in a large enterprise for the unions and the works council. All this influenced my own professional career. Even as a student, I worked in chemical plants. In one plant, hydrogen was also produced by electrolysis and used for hydrogen peroxide and other bleaching agents. Some chemicals from the plants were used for catalysts. During my professional education, I learned how to maintain and install telecommunication systems. At that time, telecommunication exchange and switching centres still existed in large buildings with relay technology. We were allowed to look into the future during our apprenticeship and practised the first programs on a Sinclair computer with a Z80 processor. During my basic studies, IT technology was still called "electronic data processing", and we learned the basics of algorithms and embedded control on UNIX machines or with assembler on 8-bit controller. During my studies of communication engineering, the iron curtain opened and Germany was reunited. As part of my diploma thesis, I built and programmed a three-phase converter to test the performance of IGBTs and investigate their behaviour under massive EMC interference. A special challenge was a robust software that could control the inverter with very precise pulse widths in the audible frequency band using 8-bit controllers.

After my studies of electrical engineering, I first had enough and did not want to fill my head with more wisdom. But as a pensioner, I wanted to study philosophy again. My father followed my professional career, but now he mostly just listened to me or told me about his experiences and observations when the first cars were developed into a Volkswagen or sporty racing cars. The one or other story from his time as a truck driver was also always present. He did not know the scientific basics of all process engineering topics, but he always said, a manager can say what he wants, a physical law cannot be overridden. With the law of conservation of energy, nobody could contradict him, with everything else we developed the ambition to sound out the limits of physics.

My first job as an engineer was in plant construction at an engineering contractor with a focus on liquid–gas process engineering. My first own project was an automatic rail tank wagon loading system in a refinery. My boss said it is not a big project because it was only about the release for loading and automation of the transfer locomotive. About this mistake, I got to know the safety engineering. It was actually only a matter of a few signals within the nominal function, but the safety analyses and measures taken to ensure safety were enormous. Therefore, I quickly became a safety expert in the automation projects. The experienced engineers preferred to keep their distance from the safety technology. Another interesting project was the planning and implementation of the first safety concept for an LPG liquid–gas tanker according to the IMCO Gas Code. At that time, the current machinery directive still stated that software-based safety functions were not permitted. All shutdown systems consisted of relay technology. Here, of course, the connection between reliability and safety was quickly learned. The one or other redundancy naturally helped to design the safety technology in such a way that the gas system on the ship was not permanently in a safe state and was therefore not available for the intended function. At times, I also headed the service department for the gas tankers, so in addition to the product liability aspects, I also became familiar with the lifecycle issues and the business aspects. Here, the availability of transportability was the decisive factor, so investments in maintenance and service were made with foresight. Afterwards, I changed to the sales department of the still leading manufacturer of safety-related safety control systems. I was responsible for the UK, Northern and Eastern Europe. In particular, the offshore and pipeline business was an essential aspect of my sales area. Among other things, I was involved in the automation and safeguarding of Europipe 2 or in projects for the remote control of unmanned oil platforms or shuttle tankers. Here in Norway, satellite technology or the IT networks of the oil companies were also used to implement safety functions. During my time in sales, I studied business administration with a focus on marketing on a part-time basis at the University of Basel, because I wanted to better understand the basics of business relationships. Moreover, I was able to put legal topics, innovation and technology management, etc., into practice very quickly. In Norway and the UK, the new emerging IEC 61508 was discussed very intensively. I had active contact with the fathers of this standard. In my company, there were many people who also developed the German derivation of the standard, the VDE 803. The then current safety control system is still the basis for the architecture metrics in this standard today. Since I returned from my sales trips permanently with new safety requirements, I took over as head of product management. Here, new safety concepts were developed and the topic of safe Ethernet was systematically implemented in products. Thus, we developed the safety technology from the relay-based approach of that time into safe network technology.

In the period around the turn of the millennium, electronics were also used more and more intensively in automotive engineering and electronics were now also needed as a protection and safety mechanism. I changed to the automotive industry and was now responsible for setting up the organisational structures for the development of safety technology at Continental. In my first working week, I was already involved in customer discussions for an active front steering system, a by-wire steering system, attended my first FAKRA meeting and was involved in the escalations around EHB, a by-wire brake system. FAKRA was a department of the German Automobile Association (VDA) that developed the vehicle safety standards. Later, I took over the leadership of the German mirror committee, which developed the ISO 26262. After the merger with Siemens VDO, I also took over the management of the corporate engineering process, methods and tool department. In 2011, my twins were born and in 2013 my daughter saw the light of day. While the boys are gathering their first experiences as citizens of the earth, I wrote my first book about the experiences and insights into functional safety. In the context of electromobility, it quickly became apparent that the regulations for high voltage and touch protection showed that legal regulations were necessary in order to give electromobility the necessary safety. The SimTD field test showed the first challenges for automated driving. It soon became clear that a functional safety standard would not be sufficient for the targets of automated driving. As in the case of E-mobility, the homologation regulations had to be adapted. However, all traffic regulations and the basis of the homologation were based on the Vienna Convention. There, the driver or the guide of draught animals was named as the person responsible for behaviour in road traffic. Now the vehicle should be guided by a system in public road traffic. It was necessary to initiate worldwide working groups to obtain a possible revision of this worldwide basis. In 2016, a change of the Vienna Convention came into force, which allows system driving in public roads. However, that was just the starting point; the driver still plays a significant role. The approval regulation UN ECE R79 was considered the basis for the approval of vehicles, especially the appendix 6 allows an override by electronic systems, which was already used for the by-wire systems several times, had to be revised. Furthermore, the regulation did limit the allowed steering assist power and defined several speed limits for electronic steering functions. The driving was only allowed hands-on, and hands-free driving was forbidden in most national homologation standards. Working groups consisting of functional safety and homologation experts start with the proposals for modification, and under the umbrella of UN ECE the basics are still under development. National implementations and harmonisations with existing regulations start. The outcome of currently pending and future proceedings cannot be predicted with any certainty, so that reliable regulations are not mature enough for design of acceptable safety functions yet.

In my second book, I published in English language, and here I already contributed more about software security than what was described in ISO 26262. At that time, we did not think that we needed such complex safety standards as the aviation industry, but more than what was written in ISO 26262 was certainly necessary.

When the standard was published, I wanted to prove to myself that safety engineering is not just processes and methods. I switched to a Korean supplier who was about to set up his new development centre for Europe. Here I was primarily responsible for cross-functional development. This task included setting up the entire development infrastructure for chassis systems and setting up corresponding development teams. A team had to be set up that could develop base software for electronic braking systems in accordance with the safety regulations, standards and AUTOSAR compatible. Furthermore, projects were acquired for electronic parking brake systems that were integrated into Bosch or Continental Brake Systems. A basic development and a development for electronic series products naturally required a completely different process landscape. At the beginning, I also took over the management of ESP development until the main responsible person for ESP development in Europe could be taken on board. After the successful acquisition of several series projects and joint basic developments with the German vehicle manufacturers, the Koreans decided to transfer the basic development back to Korea. I found a new challenge at Bosch Engineering a safe German port also for my family. Service for safety was already an interesting aspect during my time in the plant construction business.

During my time in sales, I had already acquired the sport pilot licence (PPLA), so I was pleased to take over the coordination of the safety activities of Bosch General Aviation Technology. Applying a vehicle motor management system for an aviation propulsion system taught me what safe availability really means. In addition to the flight area, I also coordinated the security activities for the service area automotive engineering and for the connectivity activities. On this topic, the consulting activity and the development of security concepts for Automated Valet Parking was a very valuable experience. In these areas of application, the legal safety aspects again became the focus of my work. Not only the extremely legal-oriented processes in the aviation industry, but also the massive differences between the legal regulations in Europe and North America posed a challenge. In the field of electromobility, people suddenly looked at China to see with astonishment what high standards were being established there. Many aspects of standards were transformed into laws or legal recommendations in China. My third book was in German again, where the German road traffic act was at the beginning of all considerations. I already wrote a lot about control engineering and the challenge that the vehicle control systems had to change to make the dream of automated driving come true. I received a further boost from the fact that I was able to support a lecture series at the Heilbronn University on the subject of "functional safety". I really enjoyed my role as a lecturer. This gave me a closer connection to science, and I was able to write some good theses, especially on the topics of control engineering and software engineering, and I got to know great experts and promising students. This also reawakened my interest in cybernetics. During my studies, the term cybernetics came up again and again, but I always pushed it too far into the branch of philosophy, which I only wanted to deal with later. In a Christmas lecture, I discussed the benefits of E2E machine learning with the students, it was great what aspects the students brought to the topic.

While 10 years ago, it was thought that automation was an evaluative process resulting from driver assistance, it has now become clear that automation is more of a disruptive development.

Especially, the relation to the requirements of the operational safety concept showed essential insights into automated driving. In the context of vehicle safety, several customer projects were tackled where start-ups could be helped to enter the field of E-mobility. Series concepts for the E-powertrain with range extender were also developed for established customers. A special experience was the activities in the field of fuel cell technology. Safety concepts for the propulsion, charging and storage of hydrogen in trucks were developed for an American company. On my many trips to the West Coast of the USA, I met many old friends of safety technology and also new people who are dealing with completely new ideas of vehicle technology and its automation. Whether it was technology that originated in graphics cards for PCs or car manufacturers who wanted to create entire cars from the 3D printer on customer demand, all this broadened my horizon immensely. With the 2018 revision of ISO 26262, not only commercial vehicles came in the focus of the standard, also ASICs and FPGAs, etc. Here I was able to adapt to support the new software and safety processes to the new technology and integrating new microcontroller concepts such as CPUs according to the ARM specification. This was applied to the development of lidar technology as well as to the use of new hardware concepts such as many cores or hardware accelerators for artificial intelligence. At the beginning of 2019, I moved to the central engineering department and, in addition to vehicle safety, was involved in new areas of application for safety technology. In the course of this activity, one can deal very intensively with the state of science and technology, not only for reasons of product liability. The idea that this book brought the subject of cybernetics beyond the legal limits, and thus the limits of social acceptance, arose from my disposition to search for the ultimate answer or origin. In my search for possible Ph.D. fathers for a topic on how to relate the behaviour of the vehicle to the people acting in road traffic, I was repeatedly reminded of the fundamentals of cybernetics. But communication technology, control engineering and the state of the art in safety engineering did not provide an adequate answer. Nevertheless, the discussions with the professors and the thesis on risk analysis provided me with a good framework to consolidate my image of automation in mobility.

Besides the law of conservation of energy, we also know the law of conservation of momentum from physics. Today, many people fear the conservation of information theorem. We have to deal with all these conservation laws and learn to control the risks; otherwise, we will not be able to get a hold of the technology. All these conservation laws can be put into a different context again and again, how one then adds the necessary energy to a system can happen in many different ways, where one can let off steam as an engineer. Nevertheless, such fundamentals of laws, like the Vienna Convention, cannot be completely abolished or overridden. It is a way of preserving information, which has been in people's minds for generations, and on which people's sense of justice is based. However, the better we understand such principles, the more likely we are to find solutions that we can introduce for the benefit of humankind and whose risks we can assess appropriately.

My father saw his task in providing a good home for the family, and we did not have to experience a lack of essential things. If he recognised a lack, he often only showed us a way how we ourselves could acquire the necessary means or the necessary knowledge. My mother has always been the organising hand, who also took care of my father until the very last day. She kept the household budget together and taught us how to use the available resources sensibly.

My father died in the midst of the turmoil of the corona pandemic. In the end, it was hard for me to make him understand all these connections. In his last days, he continued to ask me when we would finally bring electric Volkswagens on the road and how long it would take with the fuel cell. Unfortunately, I was no more able to answer him these questions.

Words of Thanks

Unfortunately, I could not thank to particular persons, because I would not be able to give them a weighting and I would have forgotten an essential person anyway. Beside my parents, my children and wife, I had so many exciting experiences with friends, colleges, professors, students who gave me a lot of reflexion to build up my mind for this book. Especially, the digital media are impressive, so that you could get in contact with people all over the world and what kind of feedback you got from the social media. However, the personal contact is during many rounds, in workshops, at conferences, or when we met just for discussions in a restaurant. For example, I met people from all over the world in a beautiful restaurant in my hometown Lorsch, and we discussed in this beautiful small town about our future ideas in regards to mobility and communication. During the corona lockdown, we established even a digital exchange with experts from aviation and automotive industry. Some of the experts did enjoy their beer during that fruitful exchange, since they did not need to return by car afterwards.

My memories are with my father and my younger brother, who are now reunited in heaven. In my thoughts, they are also united with the many great experts from whom I have learned so much. The feedback from the deceased I must now collect in my imagination. What might they have thought about this book?

Lorsch, Germany Hans-Leo Ross

Introduction

My first book did focus more on the standards for safety and addresses methods how to develop safety mechanisms for automotive electronics. The historical and philosophical background has often been reflected upon in previous books. In this book, special emphasis is placed on cybernetics, which is considered the mother of control engineering, communications engineering and automation technology. The derivations of cybernetics into sociology give us many indications of how we can evaluate human behaviour. The origin of many methods from today's system engineering comes from people who have made a name for themselves in the further development of cybernetics. The methods and principles have established themselves in other industries and fields of application. Such methods and principles cannot be transferred directly into a new context, but they certainly offer different perspectives and approaches to thinking about a problem in a completely different way. Cybernetics has also given us ideas for predicting the future or even for travelling into the future or past. Although fortune telling is not a proven means of safety, there have always been warlords in history who listened to oracles, fortune-tellers and shamans more than to their strategists. Systematic use of predictive methods, however, must be applied to the intended functionality as well as to the possible behaviour in case of failure; otherwise, we are not able to build a safety argumentation. Today, strategists have learned to include such evaluations and uncertainties in their strategic planning. Unfortunately, populism is still heard despite today's clarification. Based on systematic methods, the populism could be exposed as charlatanism.

Due to corona and the diesel gate, the society seems to change their mobility preferences. At the beginning of the new millennium, the idea of driver assistance systems was to be developed into autonomous vehicles. These were intended to make the driver's essential tasks of driving easier at first and to reduce them later on. This was to be achieved through evolutionary growth based on existing technology. In recent years, we became more and more aware that the idea of a sustainable technology had to become more and more a technological revolution in order to actually make the new functions suitable for humans. Even more, the society and consequently the authorities did not accept that autonomous vehicles provide safer road traffic just by automation of driver functions. The engineers realised that a human driver does much more things during driving, as the current systems based on today's technology could ever realise. At the beginning, it was thought that it would be sufficient to realise the human brain and its functions by increasing the computing power for technical systems. The microcontrollers were enhanced by microprocessors from server technology, and the technology of graphics cards was used to provide additional functionality. Automotive microcontrollers and microprocessors usually still work with program memory, as John von Neumann had suggested a hundred years ago. On such controllers with von Neumann architectures, there are many new ideas about artificial intelligence or there are many offers for machine learning software solutions. Most of the vehicles are using current control units with traditional software on their actuators especially for brake and steering systems. The EE architecture for electrical energy and for the communication base is still on the platforms for motor vehicles with combustion engines. Whenever various parties implement new safety or security features, the communication gets less reliable due to timing latencies and logical inconstancies. Network technology has not really found its way into cars yet, most communication nodes are still very similar to CAN bus architectures. Especially, the existing ESP systems or steering assistance systems have not yet become remote controlled or by-wire braking or steering systems. Even the "active front steering systems" developed 20 years ago are only used to a small extent in the vehicle, mostly there are still three-phase power steering assist systems where the driver has to be held responsible for certain errors as a fallback level. Unfortunately, it must be said that today's system approaches by far do not meet the requirements to actually automate such human capabilities. The ability to orientate oneself in road traffic, to make the right decisions in situations, or even to avoid a child who suddenly runs onto the road in front of a school like a reflection, cannot be provided by today's vehicle systems. There is a lot of talk about uncertainty, imponderability and high complexity, but the sensors, the algorithms for object or environment perception as well as the actuators, cannot fulfil the required functions by far. Human perception, the human nervous system and intuitive reaction do not need 18 years of training for nothing before a human driver is ready to drive a car in traffic. Although national traffic and safety authorities motivate industry to find appropriate solutions, they require at least the same level of safety as for traffic with human drivers. Accident statistics are repeatedly shown to indicate that systems can control a vehicle far more reliably than the average motorist. However, research into the causes of these accidents does not necessarily show cognitive misconduct as the cause of the many traffic accidents, but speeding, risky overtaking, stress, alcohol, fatigue and so on are often cited as the primary causes. This is often countered by the fact that systems always behave deterministically, as the developers intended, and therefore, all these factors are no longer considered as causes of accidents in a system-controlled vehicle. In other words, one tries to argue that the automation system should be allowed to have the same accident statistics as the human driver. Even if one or the other state order is inclined to such arguments, society is questioning whether such positive risk balance is acceptable. For years, the automotive industry has been telling us that speed, more power, more interior space, higher seating position and increased comfort were the non-plus-ultra of the automotive industry. The fast and powerful SUVs had to be safe and the driver allowed himself riskier driving manoeuvres. Trucks with a speed limit of 80km/h were not accepted as well as a sports car that was locked below the 250km/h limit. Now such limits are being massively questioned by consumer protectionists, environmental associations and also traffic experts. Is 30km/h sufficient in city centres? Will such measures reduce particulate matter? Some manufacturers hope to increase their position in the market with environmentally conscious vehicles, others have already buried the economic viability of individual mobility. Can we only save our cities from collapse by reducing traffic or will this turn our inner cities into ghost towns and we have to get used to images like Detroit in Europe? Perhaps the renaturation of American cities will also become a positive example for Europe. Where is society's idea of safety heading? In Asia, more and more programs are being set up to bring together large numbers of people in a socially acceptable way in a small area to create a feel-good community. Entire residential areas are methodically planned and implemented on the basis of human-centric design. Disabled, elderly people, children should remain safe and mobile without impairment or adverse effects to the mobility and freedom of others. Express roads are planned to provide an optimal traffic flow, partly underground, so that pedestrians and cyclists have their own traffic space on the surface. This means that such expressways can be adapted more easily for automated driving functions. The car traffic moves in completely different lanes and levels, thus the risk of accidents between road users of different strengths is clearly reduced. In the Arabic countries, entire underground cities are planned; football stadiums will also be accessible from the underground and get air-conditioned. Each type of transport gets its own route. Will this also be a solution for Europe? Will our society accept such huge construction projects? Probably not in Europe. Is Europe therefore falling increasingly behind in terms of future mobility concepts? But why only think about tunnels, why not build upwards? We know bridges over bridges, motorway junctions over several levels from the USA. Does one feel comfortable in such cities? Not really. But there are new solutions to this problem, why build bridges over bridges, why not fly directly? The fuel cell can also be built-in aeroplanes, already more than a hundred years ago zeppelins flew with hydrogen. There is enough sun, then let us use the sun's energy to electrolyse hydrogen. Then air taxis will fly from one skyscraper to another. What a great show, the whole sky full of silent air taxis powered by electric motors. The noise of the rotor blades will be reduced to a minimum with artificial intelligence, noise will not be a problem. There will be no accidents, air traffic is already safer than individual road traffic. Maybe hydrogen will also be used in silent rocket propulsion systems to bridge larger distances between continents. Such rocket engines only need to be used to shoot the transport capsule into the atmosphere, after which it can glide to earth at the desired point without external energy. Will we love cities like this, where the big aeroplanes fly in swarms around our ears? Well, in the Arab cities, it will be too hot to feel comfortable on the surface anyway, so the swarms of aircraft would not seriously disturb anyone and in principle would not endanger anyone. In the deserts, solar energy is enough and people will never really feel comfortable in such areas. So why not use such areas to advance the basics for new technologies? These aspects would have been considered cynical in 2019, but many misjudgements and over-motivated forecasts, especially for the automotive industry, will become more transparent due to the corona crisis. Of course, the individual citizen lacks an orientation, the scientists and scholars are behaving cautiously, and the politicians try to save what can be saved. Some very populist and some with slogans of hope. Oren Harari made us already a bridge from the last industrial revolution transparent: "The electric light did not come from the continuous improvement of candles". At the moment, people like to hear sentences like: "Every crisis is also an opportunity!".

One can of course follow such slogans as: "Every business is a software business now", but Dean Leffingwell did not mean you can solve any problem with software. Thomas Watson forecasted 50 years ago: "I think there is a market for maybe five computers worldwide". He never expected that people use the opportunity to develop their own market and that they saw in the computer the chance for progress and new business. In Chap. 3 of the book, I try to bring some first experience with a networked nation from Russia, the idea of a digital nation—Cybertonia. As early as the 1950s, ideas began to be developed to network the whole country. What was an advantage for central control was also an opportunity for productive local economic cells. This means that these local cells could have developed their own intelligence. From an economic point of view, Peter Bartels coined the phrase: "Anyone who believes that they can master digitization in isolation will find it difficult in a networked world. Networking includes competition".

Also, in China, they see digitization as a great opportunity for developing the nation. They never even tried to attack the Western autoindustry with their own combustion engine concept. Instead, the future was seen in the electric motor, which initially relied solely on battery electric vehicles and now also on hydrogen, and the fuel cell technology makes significant progress. In addition to this, China has also secured access to raw materials such as rare earths, lithium and cobalt. worldwide. The Chinese communication companies are world leaders, and the whole world is thinking about how to implement a communication strategy in the nations that is independent of Chinese technology. Chinese wisdom goes well with this: "When the wind of change blows, some people build walls, others build windmills". Of course, everyone thought that with the industrialisation of course China primarily thinks of building the Great Wall of China higher and higher, but here we were wrong. In the digital age, you will not be able to make progress through walls, this is only being attempted at the border with Mexico. In the digital age, you will have to learn that progress can only be achieved with maximum freedom, but limited by the ability to control it through the organisation. This goes with the quote from Melvin E. Conway: "Organizations which design systems are constrained to produce designs which are copies of the communication structures of these organizations", which we can still attribute to the EE architectures of our cars today. But that also means that with today's organisational structures, we can only try to stick to the old principles and cannot make any real progress. But what will our mobility look like in the future? How do you reconcile digitization and mobility? To make money with the new technology one day, a market has to be created again. The organisational structures have to be adapted to the needs of the market. The need for internal combustion engines will no longer excite anyone. Nevertheless, today's car will have to meet our mobility needs for at least another 10 years. But innovations are no longer expected in this area, so the fuel-powered car will only be able to prolong its lifecycle somewhat through hybridization. But there is a discrepancy between digitization and mobility. The radio technology enables us to overcome distances very well and communicate virtually, but we will not be able to exchange affection and feelings. Apparently, there is an essential aspect behind all future products and values, which could be formulated as follows: "In a digital world, what will be most valuable one day that cannot be digitized".

What are the chances of finding the new markets? One of the often-demanded needs is "safety". In German as well as in Chinese language, there is only one word for safety and security. How the two terms are distinguished from each other is not elucidated in any legislation, in the different social classes, professions and in the different languages. On the one hand, we speak of social security, on the other hand we speak of safety in the sense of accident prevention. Here in the book, the term safety is used as a generic term under the aspect that cybersecurity is an essential prerequisite for safety. In the technical sense, the term "dependability" is also often used as an umbrella term, but this term has not yet found a clear definition in the automotive world. Safety and security are demanded, requested and propagated by many parties, beneficiaries or so-called stakeholders in the mobility society. The need for safety behind these demands is quite heterogenic. Especially through populism and lobbying, primary needs often take a back seat, and the press also has a strong influence on opinion making. The safety need of the individual is in a different relationship than the safety need of society. Investment security, contract security, location security and trade security are terms used to describe hygienic factors to safeguard business. But also, terms like trust, reliability, loyalty, etc., are demands for good business partners. Threats, dangers and hazards are used to articulate the opposite, to say what one does not want. Here we experience again and again that the feeling of safety and security often contains a contradiction in itself. A sheltered environment is often a preferred safety feature. The sheep are protected from the wolf. A person who feels permanently observed becomes more and more insecure in his behaviour. An environment that is too closely monitored restricts the freedom of the individual, and when organised on a large scale, it restricts the freedom of an entire society. Here one finds another term, which is defined very differently, the term protection. How does protection relate to safety? Is protection a safety mechanism? What is the generic term, how do the terms differ? In addition to the fuzziness of the term safety, the relation to risk is correspondingly fuzzy. The perceived risk of the individual and the socially accepted risk are very different parameters, which are often understood very differently in the development of technical systems and are applied very differently in the argumentations. When developing a new product, one must be guided by the socially accepted risk.

Many of these questions and definitions have not been clarified for society and industry, and many initiatives are needed to find our way into the new mobility. In this book, aspects of how to grasp the challenges and make them transparent will be highlighted. The book does not describe solutions but only gives hints on the right way to go in order to find solutions. Often a solution is simply offered for discussion. These are many questions about the automobile market or the mobility market of the future as such. The question arises immediately: What does this have to do with safety? Many safety guidelines have their origins in trade legislation. Essential aspects, which particularly address the domain of norms and standards, consider state of the art or state of science and technology in the context of product liability. There are quite a number of initiatives to reduce such trade barriers, which will of course be rekindled by the discussions in the post-corona period. But essentially due to the risks of the corona pandemic, society has developed a more intense sense of safety and security. Why are the individual nations discussing IT security laws? What is the object to be protected? Are state-mandated Trojans an effective means against terrorism and sabotage? How much transparency is necessary and which transparency is dangerous because too much information about security, safety and protection mechanisms is disclosed.

Chapter 1 tries to approach the topic from the legal side. It is based on the hypothesis that safety is seen as the absence of intolerable risks, whereby the state authority prescribes the minimum level of safety requirements within the framework of its laws. I would like to warn in advance that the interpretations attempted here will indeed lead to the desired justice in court "safely". Each case must be considered in its specific context. The state, legislators and regulatory authorities must be seen as key stakeholders in future mobility. It ensures safe roads, it prescribes the traffic rules, it monitors traffic, it ensures justice in court, and the public prosecutor brings charges in the case of capital crimes. But also, insurance companies, various lobbying organisations and consumer protection groups introduce quite a lot of safety requirements, which often also relate to the state of the art in case of damage. This legal context is a major driver for the architecture and design of means of transport and systems that control such means of transport.

Chapter 2 is about making the risk of mobility comprehensible. What is the relationship between risk and safety in relation to the laws and the various standards? What are the main differences in the various risk management systems? Which risks are relevant, how can the risks be assessed? What is the risk for which party in the development of a complex networked transport system or its means of transport? What is the relationship between intended use and unintended use? What is misuse, foreseeable misuse? How are the risks of cybersecurity related to the risks of accidents? Do the same standards apply here for the extent of damage and probability of occurrence of the risk? How can the infrastructure share the risk with the means of transport? How do other road users share the risk with the vehicle for individual mobility? All these aspects and questions are examined, and the chapter tries to show different points of view and outline possible aspects for solutions.

Chapter 3 makes a digression into cybernetics. Cybernetics is considered the basis of control engineering and communications engineering. But many aspects of behavioural physiology have also been derived from cybernetics. How can we describe safe and correct behaviour in public road traffic? What help does cybernetics provide? In this chapter, I also try to give an argumentation why one should not speak of autonomous vehicles. From some aspects of cybernetics, I try to derive the essential basics for the automation of individual mobility. What does a driver do when he drives a car? What does he observe, what does he have to pay attention to? To what and which aspects must he react, to which aspects he must not react? What does all this have to do with his reflexes?

Another important aspect will be shown in Chap. 3: What are the functions that a person has to perform in the mobility context and what skills do they need? The term ability is an analogy or contradiction to what are functional requirements and what is necessary, what skills the corresponding system must have in order to fulfil such requirements. What motivated the fathers of cybernetics to do all this research and postulated all the many principles that are used in technology today? What would a technical system look like that reflects the abilities of humans? Does the human being function at all like a single control unit or is the human being a networked system that forms its consciousness through many feedback loops and learns continuously? How does the idea of bionics differ from our current idea of networked automation? Does "automation" really only mean transferring human functions into technical systems? Is it not an essential task of the automation engineer to find better and more appropriate implementations for technical systems? Isn't it the benefit of the user of the system that is at the centre of the automation task? What benefit does the automation task bring? The chapter does not deal with the commercial benefit, but rather methodically to find the bridge to necessity and the relation to applicability. What else can we learn from the functions and the "human" system? What is the task of the individual functional elements and what distinguishes them when they work together? Which means does the human being use for perception, why is it important to recognise oneself in one's own context? Which control loops are used for perception? How are the control cascades hierarchised? Which closed control loops does the driver use when driving a car? There have already been many approaches to robots, what connects the tasks of robotics with automation technology? Many of these topics have to do with the human being at the centre of technology. What chances do we have of finding a new and significant customer benefit in the digital millennium from these more than 100-year-old findings?

In Chap. 4, systems engineering is presented as a central solution approach. How to put a system into the right context, how important is the context for a correctly functioning system? What do the different views and perspectives mean for the system? Why is it necessary to describe a system hierarchically structured in different levels of abstraction? Chapter 4 creates an activity matrix, which also includes the planning of the basic operational functions. It is shown that in principle the activities are similar, but they change due to their new context. Based on a V-model approach, the classical safety lifecycle could be extended to close the gap for safe operation concepts. In the operational domain, all the necessary intended functions derive from the intended behaviour and necessary protection mechanisms could be identified. The operational domain found the basis for all active safety function, not only for system protection also for occupancy, cyclist and pedestrian protection systems. All the performance and timing constraints derive from the operational domain. Another traffic situation, different weather or road condition leads to other needed system responses and consequently to other behaviour demands of the road vehicles. During sunny days and on perfect roads with low traffic, the risk is significantly lower than in a continuously changing context. Why does the same system behave differently in different environments or contexts? Why does a human being react differently in different situations, even though human–machine interfaces would allow for uniform handling? What is the difference between function and system development? How does software development relate to this? Is a software development process always the same? What is the purpose of software in a system network? Are we talking about a software safety or security mechanism, is it necessary to implement a protection mechanism or is it about the realisation of an initial intended functionality? What has to be considered in this context? Why is a basic software and the development or integration of an operating system, a completely different task, requires a different process than to realise an application software? How could we realise safe communication systems for future mobilisation demands? How can we control the necessary performance? System engineering needs to balance technical requirements like performance, timing, availability, safety and security. Not even the safest system from one point of view means a safer solution from another viewpoint, is the more secure system safer than the high available system? All those questions and how to balance depend on the context of the intended use of the system. This Chap. 4 tries to provide several viewpoints and perspective in order to assess the necessary measures to develop safe systems.

Chapter 5 provides perspectives on systems and their context from a organisational viewpoint. Many causes of risk derive from organisational inadequacies. Several methods identify causes of organisational risk. Are the cooperation models still the same for the future mobility? The role of owner, driver, manufacturer and supplier will change. What will be the role of an operator for transportation services? Do we find structure like in the railway or in the aviation business? How do TIERs reach the end-user, just via the aftermarket or could they sell new services and safe functions during the operation of the vehicle?

Chapter 6 provides solutions for automated driving. Solutions are considered which are derived from the human system or solutions which demonstrate safe applications by means of machine learning. The limits of artificial intelligence are also shown. In order to drive automatically, the tasks that were previously assigned to the driver must now all be completely assigned to one or more systems. This of course is in addition to the functions that the vehicle had to fulfil so far. But how the distribution looks like, what the driver or the system has to perceive while driving and what he has to bring along to be able to drive safely, will probably change. We will probably have to advance our systems, because the systems today are far from mature in terms of perception and their translation into adequate and timely responses.

Some examples derive from other industry or technology areas and are mapped onto new mobility solutions. Especially, the aerospace industry has many solutions ready which can also make road transport safer. Even though aviation is still far away from air taxis, drones for the transport of passengers, aerospace technology offers very ideas, interesting approaches and safe solutions. Some predict that autonomous flying drones are faster in the sky than autonomous vehicles on the road. The risks are different in every context, so the measures are not transferable. However, comparisons and experiences in different industries can help us to better assess the possible solutions and their risks.

I conclude the book with a personal outlook on what questions can lead to safe mobility and what safety means for the benefit of humankind. The book does not contain any recipes that can be simply worked through; it is primarily intended to help people think more intensively about safety issues.

Will it be safer to fly with flying robots from one meeting to the next or simply to use digital communication? the reader himself should judge this.

Contents

Chapter 1
Safety the Basis for Future Mobility

Mobility often means a compromise between people having advantages of being mobile and people see a personal afflict of mobility. Today's traffic situation shows that noise, smell, exhaust gases and the spatial of roads, air corridors, but also railway routes, is doubtless accepted by society. Today's discussions about air taxies and press releases already about autonomous drones show the demand for finding a new contract between society and mobile service providers. How fast how many people are supposed to be moved from one place to another location is since Jules Verne part of every science fiction novel. Science fiction has always provided an imagination of future technology.

The nature of human being makes people sceptic if it comes to new technology or any creature that behave differently than the known before. Could you trust the news? Does not anybody try to misuse the new technology?

Biology, chemistry and nucleonic brought a lot of benefits for human beings, but all technology provided also new cruelty in the way how we fight against people or how we could make war. Many accidents have made us aware that we need to control the new technology in order to prevent accidents.

People need to experience and get trust in that new technology. The need for global principals of use with biological and chemical technology had been leading to the foundation of organisations like the United Nations and many others. Some of the organisations focus on.

- how to use the technology during wars,
- what are the ethical limits to impact humans (like cloning people),
- common agreements on the range of technical usage (what is generally acceptable for people, animals, etc.),
- how we manage, monitor and control the limits,
- how we deal and prevent accidents, and many more aspects.

Basic understandings and agreements had been defined that safeguards the society to beware of extensive usage of the new technology.

H.-L. Ross, *Safety for Future Transport and Mobility*,
https://doi.org/10.1007/978-3-030-54883-4_1

It seems to be the challenge for the new millennium to do that with the "Digital Technology".

What are our moral principles, concepts, the rules and limits how do we develop the society with the new "Digital Millennium"? Can we control that development? On the other hand, do we need to make the same experiences as we did in the last millennium with ABC? We approaching A (Atomic), B (Biological), C (Chemical), D (Digital), we need to find a harmonised approach how to make use and coexistence of the technology with freedom of the society, piece and benevolent of humans and their environment.

In a military context people discussing the impact of remote-controlled drones or missiles and that, the soldiers do not feel any more the cruelty of their weapons. People feel concerned that the threshold of accepted collateral damage increases if they got aware that such weapons kill civilians or even hit schools and hospitals. People fear new weapon systems. These have one aspect that they offer special protection for the users. Another aspect appeals to the fears which are poked if the weapons are not applied or how many human lives are saved by the use of such weapon systems. From it there originates fast the question, need we to diminish the new technology also around the risks in the traffic?

In civil life, people feel more and more impacted by cybersecurity or any other kind of data criminality. The misuse of vehicles to attack people becomes more and more credible scenarios. Today it will be no special risk for the road users if a hacker cracks data from traffic offenders, vehicle data bank or driving licence holder. Nevertheless, there are many publications how by hacker's attacks also traffic systems can become dangerous weapons of terrorists.

The new millennium provides us with a further challenge on how to deal with ABCD because the A seems to fail as an energy source. The Diesel-gate provides an enormous acceleration for the electro-mobility. How to generate on a green-basis sufficient electrical energy and how to store that energy on the moving vehicle are still with many question marks.

Would be hydrogen the answer? Only water coming out of our exhaust is a perfect image.

Do people need any more a driving licence, if the car makes the decisions in the road traffic? What is the reason even to own a car? Is it just the availability in case of demand?

We pay money to park a car, to drive on roads.

Vehicles occupy mayor parts of our land. Millions of vehicles are on the motorways every day in a traffic jam with just one person inside.

1.1 Safety is Much More Than Safety Engineering

The German words "Sicherheit" and "Fehler" are widely used terms. "Sicherheit" in the German understanding means also a protected environment, no fear, welfare, social security and feeling comfortable.

"Fehler" means an error, bugs, mistakes, deviation from expectations, but in many cases not a failure. In German language, failure is often translated as.

"Ausfall"—termination of fulfilling a function (e.g. athlete could not attend due to illness) or.

"Versagen"—fail in performance (e.g. athlete does not win as expected).

The new ideas of mobility require obviously a new understanding of all those terms and at least new or further interpretations of the current understanding of current society rules.

Due to the digitalisation of our life, the border between people who are afraid of the new technology and those who are enthusiastic about it must be redefined.

When in 2011 the ISO 26262 was published as an international standard, certain electronic systems were already very far further developed in the vehicle. The homologation laws even already demanded some of such systems like the airbag or the antilock braking systems. At the beginning of the twenty-first century as the first German working group in the VDA (association of the automotive industry) started with Functional Safety of EE systems, a slender standard for electronic products or systems was the target.

During the following 10 years before the final publication of the ISO 26262 in 2011, these working groups 10 of volumes with roughly 1000 requirements compiled. Even if a lot of appropriate knowledge, methods and attempts were discussed in the course of the years, one found only one fraction of it in the ISO 26262. Some information was taken up as footnotes; some disappeared completely, to the final publication of the first standard.

Worldwide different standardisation projects were initiated to translate the standard into the respective national language; into German, the standard was never translated.

The ISO 26262 should never serve as a guide but formulate requirements for activities and methods with the help of a lifecycle model.

Then in 2018, the first reworking of the standard was published. Consciously the extent of the standard became in the essentials only around commercial vehicles like commercial trucks and buses as well as two-wheeled vehicles (or tricycles) complements (part 12). Besides some editorial changes and informative changes were tried to maintain the structure. Part 11 was complemented which deals with the functional safety of semiconductor, and in Part 10, some examples about safe availability are given.

In parallel with the reworking of ISO 26262, there was a massive recovery of the electro-mobility and the terms of the "automated driving" or even "autonomous driving" stamped the public discussion to the future mobility.

The aspects to the safety-in-use, high-voltage or touch-protection and active safety for electro-mobility became a separate standard ISO 6469 "Electrically propelled road vehicles—Safety specifications". It seems to be the basis or with similar contents also to legal requirements in the various countries.

All around the subject "automated driving" developed an intensive discussion, it in a standardisation project flowed under the name: "Road vehicles—Safety of the intended functionality" which as ISO/PRF PAS 21,448 (Publicly Available Specification) was published. The demand derived from the exception in the first ISO 26262, that safety of the nominal performance was exempted and functional or technical inadequacy not even addressed in the first edition of ISO 26262.

Furthermore, safety-in-use, safety impact due to non-EE-systems or any impact from the environment of the system under consideration, all aspects of misuse or security, privacy, etc., had been not addressed and legal requirements, protection measures are still not systematically addressed in current automotive safety standards.

Key characteristics of occupancy safety systems derives from legal standards; they specify their functionality and protection targets as well their safety-related special characteristics. During the last years, those requirements increased significantly due to consumer protection initiatives from New Car Assessment Program (NCAP) or insurance companies, etc. Beside occupancies also pedestrian and cyclists getting more and more a target for road traffic safety. Especially cyclists and pedestrians are increasing identified as weaker road users that need to be protected.

This second book considers road traffic safety under a new wider perspective and put the known sources of risk in the focus and does not focus on particular standards. We need to accept that risk and safety provide different requirements only in case the context changed.

Regulations, requirements, directives, etc., are always controversial without understanding with which context or from which perspective one considers this.

1.2 Safety as a Social Right

In each country, people have a degree of protection guaranteed by the constitution of their country. Law regulates the access to the countries, and the inhabitants and citizens of the respective countries have the right with their taxes to finance this protection. Law regulates even trade of goods.

The basis for all constitutions of most countries in the world is the human right laws from the United Nations (UN).

They provide the following information in the Internet:

What Are Human Rights?

Human rights are rights inherent to all human beings, regardless of race, sex, nationality, ethnicity, language, religion, or any other status. Human rights include the right to life and liberty, freedom from slavery and torture, freedom of opinion and expression, the right to work and education, and many more. Everyone is entitled to these rights, without discrimination.

International Human Rights Law

International human rights law lays down the obligations of Governments to act in certain ways or to refrain from certain acts, in order to promote and protect human rights and fundamental freedoms of individuals or groups.

One of the great achievements of the United Nations is the creation of a comprehensive body of human rights law—a universal and internationally protected code to which all nations can subscribe and all people aspire. The United Nations has defined a broad range of internationally accepted rights, including civil, cultural, economic, political and social rights. It has also established mechanisms to promote and protect these rights and to assist states in carrying out their responsibilities.

United Nations, https://www.un.org/en/sections/issues-depth/human-rights/.

The foundations of this body of law are the Charter of the United Nations and the Universal Declaration of Human Rights, adopted by the General Assembly in 1945 and 1948, respectively. Since then, the United Nations has gradually expanded human rights law to encompass specific standards for women, children, persons with disabilities, minorities and other vulnerable groups, who now possess rights that protect them from discrimination that had long been common in many societies.

In Europe, many of the aspects derive to regulations and guidelines from the European Union and their member have to implement those in country-specific legislations.

In Germany is the understanding that fundamental rights indicate the following common characteristics:

- They oblige primarily the state, namely no matter whether it concerns executive, legislative powers or judicative, alliance, country or local authority district. Obliged means that the fundamental rights must be followed.
- They It is not distinguished whether the state operates in the way of the immediate or indirect state management or whether it becomes under private law or public law or becomes active by means of legal entities of the private law.
- The fundamental right authorisation is a characteristic of the fundamental rights property. Comparably with the civil law "legal capacity", the fundamental right bearer must be able to hold the fundamental right.

One also speaks of "each fundamental rights" whose bearer is every person, hence, also the terms "human rights"; at least the German way of understanding.

Human dignity is in general view, on the one hand, the value that is ascribed to all people equally and regardless of her differentiation characteristics like origin, gender, age or status; and on the other hand, the value with which the person positions himself as a kind about all the other living beings and things.

In the antiquity, one has derived the fundamental rights from the reason (for example, Aristoteles) of the stronger in each case; these principles have in particular also flowed in onto the traffic right.

Thus also applies to the road traffic that every person who moves on public roads and a certain protection of the place stands.

In all life situations and also therefore in all branches where companies provide products or offer services, it is a matter of providing the safety or social security of the people to a certain degree. With it linked commitment protection is for the people of him in each case parties acted of implementing, so those dangers are excluded and impediments only to such an extent must be accepted which is valid as in the respective context as reasonable.

Therefore, these fundamental rights derive in all areas where products and services are offered. This has of course from that point of view massive effects on the laws and regulations in the different ranges of application and branches. Beside human rights and the product liability are foundations also of the road traffic regulations.

In the Germany, the "Basic Law for the Federal Republic of Germany" defines the protection of basic rights as.follow:

> Article 23 [European Union—Protection of basic rights—Principle of subsidiarity] (1) With a view to establishing a united Europe, the Federal Republic of Germany shall participate in the development of the European Union that is committed to democratic, social and federal principles, to the rule of law and to the principle of subsidiarity and that guarantees a level of protection of basic rights essentially comparable to that afforded by this Basic Law. To this end the Federation may transfer sovereign powers by a law with the consent of the Bundesrat. The establishment of the European Union, as well as changes in its treaty foundations and comparable regulations that amend or supplement this Basic Law or make such amendments or supplements possible, shall be subject to paragraphs (2) and (3) of Article 79.
>
> *Source* Basic Law for the Federal Republic of Germany, 23 May 1949, Last Amended on 13 July 2017.

It shows that Germany derives basic principles from the European Union.

1.3 Legal Basis for "Auto-Mobility"

Road traffic follows historic evolutionary developed principles. Even in the Stone Age, people use proven ways to go from one place to another. At that time, it was simply easier to use the same path because in sumps, in the jungle or rocky area provide a quicker movement and less risk for the travellers. At least the Romans begin to build roads that are preferred by cars, wagons or carriages. The Via APIA to Rome is one of the most famous. Among the inhabitants of Rome, the risk was known to be run over by a chariot on a street. An educated Roman knew the danger and behaved accordingly, if a slave was run over, then this was usually only property damage.

One of the oldest regulations is the "Locomotive Act 1861"; Chapter 70 introduced by the following: "An Act for regulating the Use of Locomotives on Turnpike and other Roads, and the Tolls to be levied on such Locomotives and on the Waggons and Carriages drawn or propelled by the same".

It provided the following principles (subchapter of the regulation):

- Scale of Tolls to be taken after passing of this Act.
- Repeal of former Enactments as to Tolls to be taken for Locomotives.
- As to the Size and Weight of Locomotives.
- As to the Weight on each Pair of Wheels.
- Power to Secretary of State to prohibit the Use of Locomotives destructive to Highways or dangerous to the Public.
- Use of Locomotives restricted over Suspension and other Bridges.
- Damage caused by Locomotives to Bridges to be made good by Owners.
- Locomotives propelled by Steam to consume their own Smoke.

- As to the Number of Persons in charge of Locomotive and Waggons.
- Exemption from Tolls of Waggons, &c. now exempt under any General or Local Act.
- Limit of Speed of Locomotives on public Highways, &c.
- Provisions of General Acts relating to Turnpike Roads to apply to Locomotives.
- Right of Action in case of Nuisance.

Source excerpts from "Locomotive Act, 1861", download: https://www.legislation.gov.uk/ukpga/Vict/24-25/70/enacted.

It gave the British government the right for tax and provided already restrictions for the use of street and bridges, speed limits and some concerns about noise, steam etc. It did address "Locomotives and on the Waggons and Carriages drawn or propelled by the same". In a revision from 1865, a man with a red flag was required to walk in front of the locomotive. The person with the flag fulfilled at least two safety tasks:

- Slowing down the vehicle, as it was forced to drive at walking speed, and
- Warned approaching pedestrians and horse riders of their presence.

A priority for other road traffic participants seems to be not of highest priority, of course not elsewhere in the world.

For more than 60 years, the car dominated the traffic until the first pedestrian crossings set the power of the strongest. In Berlin, there was probably in March 1952 a first zebra crossing. On 8 July 1952, twelve crosswalks in Munich were officially painted on the streets. On 24 August 1953, the legislature introduced the first paragraph of the Paragraph 26 of the Highway Code (StVO), the pedestrian crossings in Germany. However, the number of accidents increased. Priority was given to pedestrians only for those who were already on the crosswalk. The drivers did not have to stop, especially not as a precaution. Their free travel for free citizens caused many accidents. The situation did not improve until 1 June 1964, when pedestrian precedence was introduced; the "Lex Zebra; taking special care and stopping if necessary" became mandatory. As a result, accidents increased first, but deaths decreases significantly. Especially pedestrian protection became even later a priority as a vehicle design feature, e.g. design of the vehicle front so that injuries in case of accidents possible lead to less severe consequences.

There are many other laws that deal with traffic rules and aspects of road traffic law. The legal works have not been decided at some point, but the most diverse works have been consolidated only transnational later. One of the most consistent development to date is the German Road Traffic Act (Straßenverkehrsgesetz, StVG).

Many of the basic principles of German law have been incorporated in international agreements and have thus been incorporated into many other country-specific traffic laws.

1.3.1 The German Road Traffic Act (StVG)

The German Road Traffic Act is a very complete set of legal requirements, which addresses the road traffic regulation for the territory of the Federal Republic of Germany.

The German Road Traffic Act (StVG) is a federal law; it forms the bases of the traffic right in Germany. It regulates extensively this legal area together with the Driving Licence Act (Fahrererlaubnis-Verordnung, FeV), the Vehicle-Licensing Act (Fahrzeugzulassungsverordnung, FZV), the Road Traffic Regulations (Straßenverkehrsordnung) and the Homologation Act (Straßenverkehrszulassungsverordnung, StVZO).

Due to harmonisation with EU legislations, the various acts are under a process of rearrangement, but the topics to be addressed remain the same.

It consists on the following parts:

- PART I. General Traffic Rules
- Part 11.
  - Regulation of Motor Vehicles
  - Licensing and Registration of Motor Vehicles
  - Licences of drivers
  - Provisions as to Driving and Offences
  - Road Traffic
  - Accidents
  - Provision as to light, noise, horn, reflectors
  - Seat belts and protective helmets
  - Weighting of motor vehicles
  - Driving rules
  - Regulations
  - The point system.
- Part III. Regulations of Public Passenger Vehicles and Road Licences
  - Road Licences for Public Passenger Vehicles
  - Conductors´ Licences
  - General Provision as to Licences
  - Drivers, Conductors and Passengers.
- Part IV. Regulations of Commercial Motor Cars
- Part V. International circulation of Motor Vehicles
- Part VI. Roads and Vehicles Generally
- Part VII. General
- Part VIII. Special Powers of Enforcements and Administration
  - Traffic Tickets
  - Traffic Wardens
  - Special Enforcements Provisions.

Some of the parts and examples about their contents:

Part 1 traffic regulations

Here automobiles and trailers (closer executed in the homologation act) include the basic traffic regulations on the licensing as well as for the licensing of people to the traffic as a base of the driving licence right (specifies executed in the driving licence act). An important regulation is §6 StVG which authorises the federal traffic ministry to remit other statutory acts for the regulation of the traffic.

Part 2 liability

Here the bases are regulated to the liability for physical injuries and property damages with a traffic accident.

Part 3 offense and penalty regulations

Here the offense and penalty regulations are described, for example, for the driving without driving licence, the indicator abuse and the driving under influence of psychoactive substances, alcohol, etc.

Part 4 driver register

The driver allowance register and their management are also defined here ("Flensburg-Punkte").

Part 5 vehicle register

Here the vehicle register, which contains data of the vehicles and vehicle owners of all automobiles admitted in the Federal Republic of Germany is described.

Part 6 driving licence register

Here the local ones and the central driving licence register are regulated which administers data of all driving licences spent in the Federal Republic of Germany and documents whether these are still valid or not.

Part 7 of other

Here common regulations and transitional rules are included.

The Germany Federal Ministry of Transportation and digital Infrastructure (BMVI) provides information and a translation of the German Road Traffic Act on following web-pages: https://www.bmvi.de/SharedDocs/EN/publications/german-road-traffic-regulations.html.

Part I. General Traffic Rules Section

§1 Basic Rules (1)

Use of the road requires constant care and mutual respect. (2) A person using the road shall act in such a way as not to harm or endanger or, more than is unavoidable in the circumstances, to hinder or inconvenience any other person.

Source https://www.bmvi.de/SharedDocs/EN/publications/german-road-traffic-regulations.pdf?__blob=publicationFile, download July 2019.

The basic rules require to drive with constant care and in such a way that others are not endangered or hinder more than unavoidable. In many court trials, compensations of damages or penalties for accidents had been decided based on that first paragraph.

1.3.2 *Origin of the German Road Traffic Act*

As in any country, road traffic regulations have a history, and their development based on the need to regulate or provide guidelines for the dedicated territory.

The German Road Traffic Act had been published in the first revision, from 3 March 1909 called "Gesetz über den Verkehr mit Kraftfahrzeugen" (law about the traffic by motor vehicles).

The first paragraph defined the applicability:

§ 1.

Motor vehicles to be put into service on public roads or squares must be approved for traffic by the competent authority.

For the purposes of this law, motor vehicles are cars or bicycles that are moved by machine power without being tied to railway tracks.

(translated by the author)

On January 23, 1953, the traffic law (StVG) of the Federal Republic of Germany came into force. It already contained some regulations, which are still valid until today. For example, the rules for liability for producer, owner and driver had been formulated.

1.3.3 *Adaptation of the Traffic Right for the Globalisation Trend*

Nearly any nation have their own road traffic law, they all have a different history and other needs to define priorities. The focus of all national regulations is very similar as in the German Road Traffic regulations.

The German Road Traffic Act (StVG) must submit for the challenges of the Europeanisation as well as the worldwide economic areas to continuous adaptation processes. Hence, programmes were put on to achieve here other harmonisations. These were, primarily, the growing also international traffic and the need for the recognition of regulations, acts, etc. worldwide.

Statement from European Union:

Opinion of the European Economic and Social Committee on 'A European highway code and vehicle register

(2005/c 157/04)

Chapters 1 and 2 provide the motivation and the reasoning to start the harmonisation, and §3 provides the foundation for future regulations.

3. Brief Summary of Legislation Set Out in Conventions and the Scope Thereof

3.1

International road traffic rules are enshrined in a number of conventions, the most important ones being the 1926 Paris Convention, the 1949 Geneva Convention and the 1968 Vienna Convention.

3.2

The International Convention relative to Motor Traffic was signed by 40 States in Paris on 24 April 1926. The Convention aimed to facilitate international tourism and is currently in force in over 50 countries.

3.2.1

The Main Objectives of This Convention Are:

a)

To Define Minimum Technical Specifications for Motor Vehicles, Including Registration, Lights and Vehicle Identification in Each State Visited;

b)

To Lay Down Rules on the Issuing and Validity of International Motor Vehicle Certificates, so that Drivers Can Legally Enter and Drive in the Territory Covered by the Convention;

c)

to recognise certain domestic driving permits and define the features of international driving permits valid in Contracting States that do not recognise other States' domestic permits, without any intention of replacing the latter;

d)

To Establish (six) Danger Signs Which Must Be Used by the Contracting States on Roads Within Their Territory;

e)

To Set up a System for Exchanging Information on Drivers Holding an International Certificate and Permit and Involved in a Serious Accident or Violating Domestic Road Traffic Laws.

3.2.2

While This Convention Simplified Customs Procedures, It Did not Exempt Drivers from Knowing and Observing Domestic Road Traffic Laws.

3.2.3

Moreover, the Convention did not enter into force until it had been ratified by each State and the respective ratified convention deposited. As a rule, the Convention is only in force in the main territory of the Contracting State; an explicit declaration is required for it to take effect in other territories under its administration.

3.3

The Convention on Road Traffic was signed by 17 States in Geneva on 19 September 1949. It is currently in force in over 120 countries and replaces the 1926 Convention in relations between the Contracting and Signatory States.

3.3.1

This Convention fleshed out the principles enshrined in the earlier convention, in line with developments in the car industry, and revealed a growing concern for road traffic safety.

3.3.2

While it did not stipulate the use of specific traffic signs, it did require that States adopt a homogeneous system of road signs and signals and only use them where strictly necessary.

3.3.3

This Convention did not lay down very many driving rules or introduce much which was new, other than precautions to take when passing oncoming traffic, rules on priority and the use of lights.

3.3.4

The Convention's entry into force was subject to the same conditions as previously, while moves to introduce harmonisation were tempered by the fact that States were not bound to adopt certain provisions and could also reject amendments to it.

3.4

The Convention on Road Traffic of 8 November 1968 was signed in Vienna by 37 States and is currently in force in around 100. Once ratified and deposited, this Convention replaced the 1926 and 1949 Conventions in relations between the Contracting States.

3.4.1

This Convention is the most comprehensive in terms of road traffic regulations, dedicating a whole 30-article chapter to the rules of the road and laying down rules on the main manoeuvres found in modern-day highway codes. The 1968 Convention went further than the previous—minimalist—conventions, which had just dealt with the passing of oncoming traffic and the associated signs and signals; it not only laid down the principles to be observed by drivers when performing the most dangerous manoeuvres (e.g. overtaking, change of direction, precautions to take vis-à-vis pedestrians, etc.) but also regulated standing and parking, passengers boarding and alighting from vehicles, driving in tunnels—in short, the whole gamut of typical situations that drivers face.

3.4.2

This Convention went further than previous texts requiring the Contracting and Signatory States to bring the substance of their domestic legislation into line with the driving rules laid down in it. The advantage of this for drivers was that they would be familiar with the main driving rules when driving in other Signatory States.

3.4.3

Nonetheless, States Were Still Given the Option of Rejecting Amendments to the Convention.

3.5

As this overview clearly shows, three international conventions are in force in the European Union (which now includes ten new members), although not every Member State has signed up to all three Conventions (12). The EU is therefore still far from having harmonised road traffic rules, particularly if twenty-five constantly evolving domestic bodies of law are added to the conventions mentioned above (13).

3.6

A number of barriers have been—or are in the process of being—overcome, such as the abolition of border controls, conditions for the type-approval of vehicles and their components, and the mutual recognition of—and harmonisation of rules on—driving licences. However, the fundamental aspects of road traffic remain unresolved: driving rules and road signs and signals.

3.7

As far as the rest of the world is concerned, while these conventions facilitate customs procedures and driving within the Community, third-country nationals visiting the EU come across as many different driving rules and regulations as they do States.

Source https://eur-lex.europa.eu/legal-content/EN/TXT/HTML/?uri=CELEX:52004IE1630&from=EN, Download July 2019.

2005/C 157/04 is still the motivation and basic background for European road traffic legislations. All the derived regulations and guidelines follow the basics of the basic statement.

1.3.4 Activities in Germany for Future Mobility Solutions

Nearly any nation in the world decided to support the development of automated driving. In Germany, the government already implemented some basic rules for automated driving.

The German Bundesanstalt für Straßenwesen (BASt) published the following report:

> Report on the research project F 1100.5409013.01 of the operational programme of the Federal Highway Research Institute:
>
> Legal consequences of an increase in vehicle automation
>
> Translation into English language of Part 1(BASt-Report F83)
>
> As a result of this various modifications of the German Road Traffic Act (StVG) had been initiated.
>
> On May 12, German Federal Parliament (Bundesrat) approved amendments to the Road Traffic Act (StVG) setting legal requirements for automated driving beyond "partial automation" in Germany. The revised legislation came into effect on June 21.
>
> The revised Act allows drivers to operate cars with Levels 3 and 4 functionalities, i.e. "conditional automation" and "high automation" according to the nomenclature in the BASt-Report F83.
>
> On 6 June 2017, the German Federal Parliament published the following amendment for automated driving (AD), "Eight Act Amendment of Road Traffic Act".
>
> *Source* https://www.bmvi.de/SharedDocs/EN/Documents/DG/eight-act-amending-the-road-traffic-act.pdf?__blob=publicationFile.

At the moment in time this change for Germany is a hard burden and no real progress for wide implementation of automated driving for mass production vehicles. For 2019, a review of the "Eight Act Amendment of Road Traffic Act" had been announced, but until now there is no real progress.

> "Section la Motor vehicles with a highly or fully automated driving function"
>
> The operation of motor vehicles by means of a highly or fully automated driving function shall be permissible if this function is used for its intended purpose.
>
> Motor vehicles with a highly or fully automated driving function within the meaning of this Act are vehicles equipped with technology that:
>
> 1. when activated, is able to control the motor vehicle—including longitudinal and lateral control—to perform the driving task (vehicle control);
>
> 2. is able, while the vehicle is being controlled in the highly or fully automated mode, to comply with the relevant traffic rules and regulations for operating a vehicle;
>
> 3. can be overridden or deactivated manually by the driver at any time;
>
> 4. is able to identify the need for the driver to retake manual control of the vehicle;
>
> 5. is able to indicate to the driver—by means of a visible, audible, tactile or otherwise perceptible signal—the need to retake manual control of the vehicle with a sufficient time buffer before it returns control of the vehicle to the driver; and
>
> 6. indicates that use is running counter to the system description.
>
> The manufacturer of such a vehicle shall state in the system description, in a binding manner, that the vehicle meets the requirements set out in the first sentence.

Source https://www.bmvi.de/SharedDocs/EN/Documents/DG/eight-act-amending-the-road-traffic-act.pdf?__blob=publicationFile.

At the moment in time this change for Germany is a hard burden and no real progress for wide implementation of automated driving for mass production vehicles. For 2019, a review of the "Eight Act Amendment of Road Traffic Act" had been announced, but until now there is no real progress. The intend behind the changes seems to be more to achieve releases for prototype testing.

The technical systems for mass production cars are not designed in a way that they comply with those regulations. The technical impact especially in conjunction with human capabilities, e.g. if a driver could control the vehicle technology in relevant driving situation and even more in case of failure, seems to be challenging.

Since this provides only a view on the German interpretation in other countries also if they or part of the European Union, a homologation for automated driving function might be possible also for mass production vehicles. The product liability seems to be a much higher burden, than the possible technical challenges. A secondary question is what is the target of automated driving? A lot of vehicle manufacturer did already announce that they would skip the levels for partly automated vehicles. It seems to be an immensely high technical impact to bring the driver back into the vehicle control loop, instead of designing the system in a way that also provide automated limp home functions in case of technical failure in the vehicle.

1.3.5 Genevan and Viennese Convention

All the national activities in the early days were recognised as insufficient for global road traffic regulations. Road traffic regulation worldwide took notice on the Geneva and Vienna Convention for road traffic.

The official name of the Vienna Convention (VC) is "Convention on Road Traffic". It came into force on 21 May 1977 and had been concluded in Vienna on the United Nations Economic and Social Council's Conference on Road Traffic (7 October—8 November 1968). It had been ratified by 78 countries, but even countries which did not ratify the convention follow today the sense of the so-called Vienna Convention. Today's road signs and signals have also their origin on a second convention from that conference.

UN ECE published the valid text on their home-page: https://www.unece.org/fileadmin/DAM/trans/conventn/crt1968e.pdf.

China and the United States for example belong to the non-signatory countries. Especially the validity of driving licences and the usage of cars from other countries are still today remarkable impacts for non-signatory countries.

The Vienna Conventions (VC) continuously evolved to meet the road safety needs of today and tomorrow.

A second agreement, the Convention on Road Traffic in Geneva, entered into force 26 March 1952; it had been published on 19 September 1949. It has been agreed by much more countries, especially in America, Africa and Asia.

A copy could be find also on the homepages from Korean government: https://dl.koroad.or.kr/koroadfiles/commonFiles/template/geneva.pdf.

The basic rules of international driving have their origin in this agreement.

The Vienna Convention defines the expected behaviour of mobile devices in road traffic. The basic version (source: https://treaties.un.org/doc/Treaties/1977/05/19770524%2000-13%20AM/Ch_XI_B_19.pdf) from 1968 requires in Article 8 the following:

Art. 8 (1) VC:

Every moving vehicle or combination of vehicles shall have a driver.

Art. 8 (5) VC:

Every driver shall at all times be able to control his vehicle or to guide his animals.

2016 saw a breakthrough for automated driving technologies with the entry into force of an amendment to the Convention on Road Traffic allowing the transferring of driving tasks to the vehicle, provided that these technologies are in conformity with the United Nations vehicle regulations or can be overridden or switched off by the driver. In September 2018 (https://assets.publishing.service.gov.uk/government/uploads/system/uploads/attachment_data/file/679994/MS1.2018_CM9570__Convention_on_Road_Traffic_WEB.pdf), a new resolution on the deployment of highly and fully automated vehicles in road traffic was adopted. It was decided on 23 March 2016 by the amendment "Report of the sixty-eighth session of the Working Party on Road Traffic Safety" (Source: https://www.unece.org/fileadmin/DAM/trans/doc/2014/wp1/ECE-TRANS-WP1-145e.pdf, 2014), and it define rules for automated driving technology:

Amendment of Article 8: A new paragraph (i.e., paragraph 5bis) is to be inserted into Article 8. The paragraph 5bis shall read as follows: 5bis. Vehicle systems which influence the way vehicles are driven shall be deemed to be in conformity with paragraph 5 of this Article and with paragraph 1 of Article 13, when they are in conformity with the conditions of construction, fitting and utilization according to international legal instruments concerning wheeled vehicles, equipment and parts which can be fitted and/or be used on wheeled vehicles Vehicle systems which influence the way vehicles are driven and are not in conformity with the aforementioned conditions of construction, fitting and utilization, shall be deemed to be in conformity with paragraph 5 of this Article and with paragraph 1 of Article 13, when such systems can be overridden or switched off by the driver.*

This amendment transfers the ability to control a vehicle or guide an animal to a system. It addresses of course also all the obligation of the driver to that system.

Art. 13 (1) VC:

Every driver of a vehicle shall in all circumstances have his vehicle under control so as to be able to exercise due and proper care and to be at all times in a position to perform all manoeuvres required of him […].

This requirement precisely says that the driver has to have under all circumstances the vehicle under control. Of course, it means that the "driver" also have to control the transition from animal driven, system driven or driver driven.

Art. 14 VC:

> 1. Any driver wishing to perform a manoeuvre such as pulling out of or into a line of parked vehicles, moving over to the right or to the left on the carriageway, or turning left or right into another road or into a property bordering on the road, shall first make sure that he can do so without risk of endangering other road-users travelling behind or ahead of him or about to pass him, having regard to their position, direction and speed.

This requirement requires to monitor all area of risk derived due to the drivers driving intend.

> :3. Before turning or before a manoeuvre which involves moving laterally, the driver shall give clear and sufficient warning of his intention by means of the direction indicator or direction-indicators on his vehicle, or, failing this, by giving if possible an appropriate signal with his arm. The warning given by the direction-indicator or direction-indicators shall continue to be given throughout the manoeuvre and shall cease as soon as the manoeuvre is completed."

Any lateral movement in various driving situations requires a transparent indication of the drivers intend.

Art. 17 VC:

> No Driver of a Vehicle Shall Brake Abruptly Unless It is Necessary to Do so for Safety Reasons.

The requirements require driving ahead and suddenly braking only in a safety–critical situation.

All these requirements address the basics of road traffic safety, and they are the basis for any road traffic act worldwide.

1.4 European Union (EU) and Legislation

To make easier a border-covering traffic, trade et cetera, many countries of Europe have agreed on the fact that certain rules, acts and directives are to be considered for all member countries. The directives of the European Union are considered as legislative acts.

The European Union (EU) defines the regulations as follows:

> The aims set out in the EU treaties are achieved by several types of legal act. Some are binding, others are not. Some apply to all EU countries, others to just a few.
>
> Regulations
>
> A "regulation" is a binding legislative act. It must be applied in its entirety across the EU. For example, when the EU wanted to make sure that there are common safeguards on goods imported from outside the EU, the Council adopted a regulation.

Directives

A "directive" is a legislative act that sets out a goal that all EU countries must achieve. However, it is up to the individual countries to devise their own laws on how to reach these goals. One example is the EU consumer rights directive, which strengthens rights for consumers across the EU, for example by eliminating hidden charges and costs on the internet, and extending the period under which consumers can withdraw from a sales contract.

Decisions

A "decision" is binding on those to whom it is addressed (e.g. an EU country or an individual company) and is directly applicable. For example, the Commission issued a decision on the EU participating in the work of various counter-terrorism organisations. The decision related to these organisations only.

Recommendations

A "recommendation" is not binding. When the Commission issued a recommendation that EU countries' law authorities improve their use of videoconferencing to help judicial services work better across borders, this did not have any legal consequences. A recommendation allows the institutions to make their views known and to suggest a line of action without imposing any legal obligation on those to whom it is addressed.

Opinions

An "opinion" is an instrument that allows the institutions to make a statement in a non-binding fashion, in other words without imposing any legal obligation on those to whom it is addressed. An opinion is not binding. It can be issued by the main EU institutions (Commission, Council, Parliament), the Committee of the Regions and the European Economic and Social Committee. While laws are being made, the committees give opinions from their specific regional or economic and social viewpoint. For example, the Committee of the Regions issued an opinion on the clean air policy package for Europe.

Source https://europa.eu/european-union/eu-law/legal-acts_en, download July 2019.

Directives will remit according to the subject of the directive on account of one of the procedures intended in the contracts. It is distinguished between legislation acts, carrying out directives of the commission and delegated directives. In contrast to EU regulations, EU directives are not immediately effective and obliging, but they must be implemented by national legislative acts to become effective.

The EU addresses many technical legislations to any industry. They compiled the relevant directives to "Technical Regulations". The Directive (EU) 2015/1535 provides an overview.

Directive (EU) 2015/1535 applies to all draft technical regulations.

Technical Regulations Include:

- technical specifications;
- other requirements;
- rules on services; and
- regulations prohibiting the manufacture, importation, marketing or use of a product or prohibiting the provision or use of a service, or establishment as a service provider.

In the Meaning of the Directive, 'product' Concerns Any Industrially Manufactured Product and Any Agricultural Product, Including Fish Products.

As regards services, the Directive applies only to Information Society services defined as any service provided for remuneration, at a distance, by electronic means and at the individual request of a recipient of services.

A 'technical specification' is a specification contained in a document which lays down the characteristics of a product such as dimension, labelling, packaging, level of quality, conformity assessment procedures etc. *This term also covers production methods and processes.*

'Other requirements' cover requirements imposed on a product for the purpose of protecting for instance consumers or the environment and which affects its life cycle once placed on the market, such as conditions of use, reuse or recycling. These conditions have however to influence significantly the composition or nature of the product or its marketing.

Source https://ec.europa.eu/growth/tools-databases/tris/en/about-the-20151535/what-is-a-technical-regulation/.

1.4.1 EU Directives Related to Road Traffic

The German or any other members of the EUs Road Traffic Act is officially just a one derivation of one or several EU directives.

The European Commission has released a road safety policy which provides guidance on the following topics:

- EU Road safety policy
- Driving licence
- Enforcement in the field of road safety
- Alcohol, drugs and medicines
- Road infrastructure
- Emergency calls
- Accident data collection
- Units of measurement
- Transport—training
- Transport—working conditions
- Transport—tachograph
- Transport—check of the working
- Transport—check of the working
- Third countries driver attestation
- Transport of dangerous goods
- Vehicles—type approval
- Vehicles—registration
- Vehicles—front protection of vulnerable users
- Vehicles—safety belts and other restraint systems
- Vehicles—tires
- Vehicles—daytime running lights
- Vehicles—blind spot mirrors

- Vehicles—conspicuity
- Vehicles—weights and dimensions

Source https://ec.europa.eu/transport/road_safety/specialist/policy_en, Download July 2019.

For the vehicle in 2007, a new framework had been released, which is the basis for the roadworthiness of the moving equipment, including the vehicle itself.

Directive 2007/46/EC of the European Parliament and of the Council of 5 September 2007 establishing a framework for the approval of motor vehicles and their trailers, and of systems, components and separate technical units intended for such vehicles (Framework Directive).

The Directive 2007/46/EC did replace many other directives and provides also the basis for the type-approval of vehicles.

In parallel, there are also further directives like the amending regulation (EC) No 715/2007 on type approval of motor vehicles with respect to emissions from light passenger and commercial vehicles (Euro 5 and Euro 6) and on access to vehicle repair and maintenance information.

Another regulation that becomes more in the focus is the following regulation:

REGULATION (EU) 2018/858 OF THE EUROPEAN PARLIAMENT AND OF THE COUNCIL of 30 May 2018

on the approval and market surveillance of motor vehicles and their trailers, and of systems, components and separate technical units intended for such vehicles, amending Regulations (EC) No 715/2007 and (EC) No 595/2009 and repealing Directive 2007/46/EC.

Regulation (EU) No 2018/858 on the approval and market surveillance of motor vehicles and their trailers, and of systems, components and separate technical units intended for such vehicles becomes valid from 1 September 2020; it replaces the Directive 2007/46/EC.

Subject Matter

1. This Regulation lays down the administrative provisions and technical requirements for the type-approval and placing on the market of all new vehicles, systems, components and separate technical units, referred to in Article 2(1), and for individual vehicle approvals.

   This Regulation also lays down provisions for the placing on the market and the entry into service of parts and equipment that may pose a serious risk to the correct functioning of the essential systems of the vehicles referred to in Article 2(1).
2. This Regulation lays down the requirements for the market surveillance of vehicles, systems, components and separate technical units that are subject to approval. This Regulation also lays down the requirements for the market surveillance of parts and equipment for such vehicles.

Source REGULATION (EU) 2018/858, Article 1.

The new regulation addresses particularly new principles required for the upcoming automation of automated driving functions.

Whilst this new Regulation does not contain any significant changes to the technical requirements applicable to vehicles, systems and components, it does contain major revisions to the applicable processes and procedures and introduces numerous additional checks to ensure the robustness of the type approval process. The following is a very brief summary of some of the main changes that have be introduced by this new Regulation:

- ***Designation of responsible Authorities***—Each Member State is required to designate an Authority in their country that is responsible for type approval and an Authority that is responsible for market surveillance. Where one Authority is responsible for both activities, there must be a strict separation of roles and responsibilities and the two activities must be managed autonomously as part of separate structures. Each Member State must also review and assess the functioning of their Type Approval Authority and their Market Surveillance Authority at least once every four years and provide the results of this assessment to the Commission.
- ***Market surveillance requirements***—New requirements on market surveillance are introduced to ensure that production vehicles and components being placed on the market comply with the type approval requirements, are safe and are not harmful to the environment. In each Member State, the designated Market Surveillance Authority must carry out compliance verification inspections and tests on vehicles and components placed on the market in their country. Each Market Surveillance Authority must carry out at least one compliance verification test for every 40,000 new motor vehicles registered in their Member State during the preceding year, with a minimum of five compliance verification tests per year. For Market Surveillance Authorities carrying out more than five compliance verification tests per year, at least 20% of their tests must be emissions tests. In addition to the testing undertaken by Market Surveillance Authorities, the European Commission will also undertake their own compliance verification inspections and tests on vehicles and components placed in the market anywhere within the European Union.
- ***Assessments on Type Approval Authorities by the Commission***—The European Commission will carry out assessments of the procedures put in place by each Type Approval Authority for the granting of type approvals, for carrying out conformity of production and for the designation and monitoring of Technical Services every five years and publish the results of these assessments.
- ***Forum for the Exchange of Information and Enforcement***—The European Commission will establish, chair and manage a Forum for the Exchange of Information on Enforcement which will meet on a regular basis. This Forum will be attended by representatives from the Type Approval Authorities and Market Surveillance Authorities of each of the Member States. The Forum will allow for the exchange of information on best practices, enforcement problems, interpretation issues, etc. *and allow for coordination of enforcement activities such as compliance verification testing.*
- ***Responsibilities of manufacturers, manufacturer's representatives, importers and distributors***—The new Regulation more clearly defines the specific responsibilities applicable to manufacturers, manufacturer's representatives, importers and distributors with regard to both type approval and market surveillance.
- ***Frequency of Conformity of Production checks***—The new Regulation specifies that Type Approval Authorities must carry out Conformity of Production checks and tests on manufacturers at least once every three years unless a specific frequency for such Conformity of Production checks and tests is specified in the Regulatory Act for which Conformity of Production is being assessed.
- ***Termination of validity of whole vehicle type approval certificates***—The Regulation introduces a new requirement on the termination of validity of whole vehicle type approvals. In the case of M1 and N1 category vehicles, if the type approval information package has not been updated for seven years, the issuing Type Approval Authority

must verify that the vehicle type still complies with all of the applicable Regulatory Acts and terminate the validity of the type approval if the vehicle type does not comply with all of those Acts. This same requirement is applicable to whole vehicle type approvals for other categories of vehicle, but only if the type approval information package has not been updated for ten years.

- ***Certificates of Conformity in electronic format***—The new Regulation still requires each vehicle to be issued with a paper format Certificate of Conformity. However, from 5th July 2026, the new Regulation requires manufacturers to make the Certificate of Conformity available as "structured data in electronic format". Precise details of the electronic format for the Certificate of Conformity will be specified within a separate Implementing Regulation to be adopted by the European Commission.
- ***Safeguard measures and recalls***—The new Regulation includes more detailed provisions on the implementation of safeguard measures (e.g. prohibition of sale) and on the initiation of recalls, and on the coordination of such activities between Member States. These new provisions specify the conditions under which Member States can implement these measures unilaterally and on the process for informing other Member States. The new provisions also allow for the European Commission to implement safeguard measures and initiate recalls on an EU wide basis.
- ***Access to Repair and Maintenance Information***—The current requirements on the provision of access to onboard diagnostic (OBD) system information and other repair and maintenance information for independent operators are specified within the EU emissions Regulations applicable to light and heavy duty vehicles; (EC) No. 715/2007 and (EC) No. 595/2009 respectively. However, to ensure consistency in their application, these requirements are deleted from the emissions Regulations by this new Regulation. To replace them, enhanced requirements on the provision of access to onboard diagnostic (OBD) system information and other repair and maintenance information for independent operators, are included in the new Regulation.
- ***Designation, monitoring and review of Technical Services***—To ensure the competency, impartiality and independence of Technical Services, the new Regulation specifies detailed requirements on the designation, monitoring and review of Technical Services. Prior to their designation, Technical Services must be assessed by an assessment team consisting of representatives from the Type Approval Authority from which designation is sought, two other Type Approval Authorities and the European Commission. Designation will only be valid for five years, after which time the assessment exercise must be repeated. During the designation period, the designating Type Approval Authority must continually monitor the performance of the Technical Service and carry out an on-site assessment of the Technical Service at least every 30 months.
- ***Penalties for non-compliance***—The new Regulation specifies that all Member States must implement rules on the financial penalties to be applied to manufacturers, importers, distributors and Technical Services for infringements of the new Regulation's requirements which are effective, proportionate and dissuasive. Furthermore, the draft Regulation specifies that the European Commission may impose administrative fines of up to €30,000 per vehicle, system or component for non-compliances with the type approval requirements.

Source https://www.interregs.com/articles/spotlight/new-eu-type-approval-framework-regulation-published-000195, download June 2020.

Another legal framework is the EC Regulation No. 765/2008, which regulates market surveillance. It is setting out the requirements for accreditation and market surveillance relating to the marketing of products (Market Surveillance Regulation). The basis for the market surveillance of the federal states is the EC Regulation No.

765/2008, which has been in force since 1 January 2010 and forms the general legal framework for monitoring the safety of products in the EU. Based on this regulation, the individual EU Member States and the Commission organise the system of market and product surveillance. This is based on the guiding principle that products in the EU should have free access to the market, but that these products should not pose a risk to health, environment and safety. The market surveillance of products according to the harmonised waste management legislation of the EU concerns the surveillance of vehicles, electrical and electronic equipment, batteries and accumulators, packaging and packaging waste. The subject of the monitoring is compliance with the conditions for placing products on the market, such as substance bans, limit values and labelling requirements.

1.4.2 *European Vehicle Class*

The European Community has provided in 1970 a definition of the vehicle classes by which groups can be arranged by vehicles EC-wide uniformly. The basis was the EC directive 70/156 / EEC since the 29 of April, 2009 by in 2007/46/the EC was substituted. The EC regulations over vehicles refer to these EC-vehicle classes. Thus, for example, the third stoplight is prescribed for the vehicle class M1 and is explained for other EC-vehicle classes for allowed. Also, exhaust emission regulations differ after these vehicle classes.

Automobiles and trailers are divided by the directives of the European Community into the following classes: (4) drivers of a vehicle is also that which activates a highly automated or fully automated driving function for the purposes of the sales 2 and uses to the vehicle control, even if he steers the vehicle, not within the scope of the designated use of this function personally (Table 1.1).

Table 1.1 EC-vehicle classes from EC directive 70/156

Class L	Two-wheeled or three-wheeled automobiles as well as light four-wheeled automobiles
Class M	To the transportation of human beings laid out and built automobiles with at least four wheels
Class N	Automobiles to the carriage of goods with at least four wheels
Class O	Trailer (including road semi-trailers)
Class R	Regional or forest trailers
Class P	pulled replaceable regional or forest machines
Class T	Classes T1 to T4.3: Traction engines for regional or forest purposes (act 167/in 2013/the EC)
Class C	regional or forest traction engines on tracks (by analogy with the classes T1 to T4 defined)

The current interpretation is as follow:

- M1, Used for the carriage of passengers, with no more than eight seats in addition to the driver seat, also known as passenger cars.
- M2, Used for the carriage of passengers, having a maximum mass not exceeding 5 tonnes.
- M3, Used for the carriage of passengers, having a maximum mass exceeding 5 tonnes.
- N1, Used for the carriage of goods, having a maximum mass not exceeding 3.5 tonnes, also known as Light Commercial Vehicle (LCV). Also includes Pick-Up trucks.
- N2, Used for the carriage of goods, having a maximum mass exceeding 3.5 tonnes but not exceeding 12 tonnes
- N3, Used for the carriage of goods, having a maximum mass exceeding 12 tonnes.
- L, Motor vehicles with less than four wheels and some lightweight four-wheelers.
- L6, A vehicle with four wheels whose loaded mass is not more than 350 kg, not including the mass of the batteries in case of BEV's, whose maximum design speed is not more than 45 km/h, and whose power does not exceed 4 kW.
- L7, Other than that classified for the category L6, whose loaded mass is not more than 450 kg (650 kg for those intended to carrying goods), not including the mass of batteries in the case of BEV's, whose power does not exceed 15 kW.

The list shows the classes L6 and L7 for light Battery Electric Vehicles (BEV).

In Asia and America, the classification is different again. Here one example from UN ECE (Table 1.2):

Even in the European Union, for tax, etc., other vehicle classes are relevant. A classification starting with A for small vehicle like Fiat 500 till F like a Mercedes S class; additional classes are defined for sports cars (S), SUV (J) and multi-purpose-cars (M).

Table 1.2 Vehicle category according to UN ECE

Categories	Japan	Europe	United States of America
Passenger Car	Passengers of 10 or less	Passengers of 9 or less (M1)	Passengers of 10 or less (Passenger Car) (MPV: truck chassis or off-road use)
Bus	Passengers of 11 or more	Passengers of 10 or more M2:GVM $\leq$ 5t M3:GVM > 5t	1. Passengers of 11 or more
Truck	Quantitative Definition Floor area (Passenger < Cargo) Weight (Passenger < Payload) Loading/Unloading openings (dimension/area)	Qualitative Definition ("designed and constructed for the carriage of goods") N1:GVM $\leq$ 3.5t N2:GVM $\leq$ 3.5t – 12t N3:GVM > 12t *Each country has different criteria	Qualitative Definition 2. ("carrying load or commercial goods")

Source https://www.unece.org/trans/main/wp29/wp29wgs/wp29gen/wp29classification.html

1.4.3 EU Directive for Future Mobility

There are many other EU regulations that already influence and much more that will influence future mobility solutions. Since protection against explosion of batteries, fuel cells are somehow addressed in automotive regulations, EU regulations will affect also the vehicle design, when it comes to interactions with the infrastructure. In case of people transportation or trucks, whenever protection of people comes in the focus further EU legislation could not be neglected.

Examples are:

- Directive 2014/34/EU of the European Parliament and of the Council of 26 February 2014 on the harmonisation of the laws of the Member States relating to equipment and protective systems intended for use in potentially explosive atmospheres (recast).
- Directive 2012/18/EU of the European Parliament and of the Council of 4 July 2012 on the control of major-accident hazards involving dangerous substances, amending and subsequently repealing Council Directive 96/82/EC.

In addition, trade is a well regulated in the European Union in order to understand the regulations the "Technical Regulation Information System (TRIS)" had been established.

The targets of TRIS are:

> In the Single Market, quantitative restrictions on the movement of goods and measures having an equivalent effect are not allowed. As the Single Market is an area without internal borders, the free movement of goods, persons, services and capital should be guaranteed.
>
> Technical regulations can lead to trade barriers. Therefore, Member States should ensure as much transparency as possible over national technical regulations they are planning to adopt.
>
> In 1983 a Transparency Procedure Was Set up and the Technical Regulation Information System (TRIS) Established. TRIS Enables the Member States and the Commission to:
>
> - inform and be informed about new draft technical regulations
> - examine these drafts
> - detect potential barriers to trade before they have any negative effects
> - pinpoint protectionist measures
> - comment on the draft regulations
> - have an effective dialogue when assessing the notified drafts and
> - identify the need for harmonisation at the EU level.
>
> In This Way They Can Ensure that: These Texts Are Compatible with EU Law and the Single Market Principles.
>
> At the same time, Member States can: draw on the ideas of their partners to solve common problems over technical regulations, using TRIS as a benchmarking and better regulation tool.
>
> TRIS is also Accessible to the Public so that Anyone Can Access All Notified Draft Technical Regulations and Send Contributions with Their Views on Any Notification.

Source https://ec.europa.eu/internal_market/scoreboard/performance_by_governance_tool/tris/index_en.htm.

Similar to explosion protection, usually all in-vehicle standards are provided by the relevant road traffic act; another example is the electromagnetic compatibility or the machinery directive. Since there is for a homologated vehicle adequate regulations in place, the EU regulations are not applicable for those systems or parts of the vehicle. If there are not vehicle specific regulations in place or they are not applied, EU regulations have to be respected.

Another Example is the Radio Equipment Directive (RED).

The radio equipment directive 2014/53/EU (RED) establishes a regulatory framework for placing radio equipment on the market. It ensures a single market for radio equipment by setting essential requirements for safety and health, electromagnetic compatibility and the efficient use of the radio spectrum. It also provides the basis for further regulation governing some additional aspects. These include technical features for the protection of privacy, personal data and against fraud. Furthermore, additional aspects cover interoperability, access to emergency services and compliance regarding the combination of radio equipment and software.

Even radar assembled at a vehicle need to consider the directive because no adequate automotive regulation is in place.

1.5 Licensing Act

Licensing acts like the German Straßenverkehrszulassungsverordnung (StVZO) are usually part of the national road traffic acts. The process how to achieve the licence for the vehicle and what are the operational constraints for the specific territory are defined in the national licencing acts. All the protection equipment and basic systems to make the car road-worthy are harmonised by the UN ECE (United Nations Economic Commission for Europe). Countries like China, Japan, Australia, Russia et cetera also apply these European licensing regulations.

UN ECE defines their mission as follow:

Mission

The United Nations Economic Commission for Europe (UNECE) Was Set up in 1947 by ECOSOC. It is One of Five Regional Commissions of the United Nations. The Others Are the:

- Economic Commission for Africa (ECA),
- Economic and Social Commission for Asia and the Pacific (ESCAP),
- Economic Commission for Latin America and the Caribbean (ECLAC),
- Economic and Social Commission for Western Asia (ESCWA).

UNECE's major aim is to promote pan-European economic integration. UNECE includes 56 member States in Europe, North America and Asia. However, all interested United Nations member States may participate in the work of UNECE. Over 70 international professional organisations and other non-governmental organisations take part in UNECE activities.

Source https://www.unece.org/mission.html, *download July 2019.*

UN/ECE Regulations demand an arrangement about the acceptance of uniform technical regulations for bicycle vehicles, pieces of equipment and parts which can be inserted in bicycle, vehicles (n) and/or be used and the conditions on the mutual recognition of the approvals which were given according to these regulations. These on account of the arrangement settled and the arrangement to connected regulations contain technical regulations for bicycle vehicles, pieces of equipment and parts which can be inserted in bicycle vehicles (n) and/or be used, test methods as well as conditions on the granting by type approvals for these bicycle vehicles, pieces of equipment and parts as well as approval signs and conditions on the warranty of the conformity of the production.

Most prominent UN ECEs are the one for brake systems (UN ECE R13H, Hydraulic Brake Systems) and for steering systems (UN ECE R79). The latest version and current work of UN ECE for vehicles could be found here: https://www.unece.org/trans/main/welcwp29.html.

In which connection or how these directives are applied by the vehicle licensing could be in any country different. In Germany, usually the TÜVs provide certificates for the German "Kraftfahrbundesamt, KBA", sometimes even the different TÜVs provide or ask for different interpretations.

It is always necessary to evaluate the entire context which vehicle for which market in what country a homologation should be achieved. What requirements derive are always a result of a necessary detail analysis.

1.6 US Road Traffic Regulations

In America, the Vienna Convention follows another interpretation. Canada and USA are using very similar principles since Middle and South America are very heterogeneous, some follow more UN ECE, some more the North American approach. The US Department of Transportation provides on their homepage information about the history and the ideas and targets behind.

Who We Are

The National Highway Traffic Safety Administration (NHTSA) was established by the Highway Safety Act of 1970, as the successor to the National Highway Safety Bureau, to carry out safety programs under the National Traffic and Motor Vehicle Safety Act of 1966 and the Highway Safety Act of 1966.

NHTSA works every day to help Americans drive, ride and walk safely. We do this by promoting vehicle safety innovations, rooting out vehicle defects, setting safety standards for cars and trucks, and educating Americans to help them make safer choices when driving, riding, or walking.

History

Fifty years ago, on September 9, 1966, President Lyndon B. Johnson signed the National Traffic and Motor Vehicle Safety Act and the Highway Safety Act into law. These laws created the National Highway Safety Bureau, the precursor to the National Highway Traffic Safety Administration (NHTSA).

Four years later, NHTSA was established by the Highway Safety Act of 1970 to carry out safety programs under the National Traffic and Motor Vehicle Safety Act of 1966 and the Highway Safety Act of 1966. The Vehicle Safety Act has subsequently been re-codified under Title 49 of the U. S. Code in Chapter 301, Motor Vehicle Safety. NHTSA also carries out consumer programs established by the Motor Vehicle Information and Cost Savings Act of 1972, which has been re-codified in various Chapters under Title 49.

On December 4, 2015, the Fixing America's Surface Transportation (FAST) Act (Pub. L. No. 114–94) was enacted into law, providing long-term funding certainty for highway and motor vehicle safety.

These laws have led to one of the most effective public health and safety efforts of the past half century. NHTSA's efforts have led to hundreds of thousands of lives saved by making vehicles safer and helping people choose to drive more safely. The following is a list of some of NHTSA's formative events and accomplishments:

NHTSA's Office of Emergency Medical Services (EMS) offers the first national guidelines for training emergency medical technicians (1971). NHTSA continues to advance a national vision for EMS through projects and research; fosters collaboration among Federal agencies involved in EMS planning; measures the health of the Nation's EMS systems; and delivers the data EMS leaders need to help advance their systems.

The Presidential Commission on Drunk Driving was established in 1982, the same year as the first dedicated national grant program to combat drunk driving.

The first child passenger safety law was enacted in Tennessee in 1978. All States enacted laws by 1985.

New York State enacted the first seat belt laws in 1984. All States (except New Hampshire) enacted laws by 1995. The passive motor vehicle restraints rule, which requires all vehicles to be equipped with seat belts took effect with model year 1987. NHTSA launched the Vince and Larry, "You Could Learn A Lot from a Dummy," media campaign in 1985 to boost seat belt use.

In 2002, NHTSA launched its first nationwide Click It or Ticket seat belt campaign. By 2015 seat belt use across the United States reached an all-time high rate of 87 percent.

The 21-year-old minimum drinking age laws were achieved nationwide in 1988 and nationwide zero-tolerance drunk driving laws for drivers under 21 were achieved in 1998. In 2003, the first national media campaign to fight drunk driving was launched. In 2005, the 0.08 Blood Alcohol Content (BAC) drunk driving laws were enacted nationwide. In the following decade, as a result of the laws, high-visibility enforcement, and media efforts, drunk driving fell 23 percent.

The New Car Assessment Program (NCAP) was established to provide crashworthiness ratings for new vehicles. Five-Star Safety Ratings were offered in 1993 to help consumers compare ratings and safety features of new cars and trucks. Research and Rulemaking efforts are underway to update the NCAP program to accommodate emerging vehicle safety technologies.

Driver and passenger side air bags were required in cars and light trucks by model year 1998. In 2013, a NHTSA study stimated that 43,000 lives were saved by air bags.

In the past decade, NHTSA has expanded our efforts across traffic safety including an Electronic Stability Control (ESC) rule to prevent vehicle rollover; preventing distracted, drowsy, and drugged driving; new fuel economy rules; the "Where's Baby?" media campaign to prevent heatstroke in cars; and an increased emphasis on pedestrian and bicyclist safety.

Source https://www.transportation.gov/transition/understanding-national-highway-traffic-safety-administration-nhtsa, Download July 2019.

USA and Canada follow similar principles in the road traffic act, while USA provides the Federal Motor Vehicle Safety Standards (FMVSS), and the Canadian Authority provides Canadian Motor Vehicle Safety Standards (CMVSS).

Since the CMVSS are very similar in structure and contents to FMVSS, the Canadian Government provides a different framework for the road traffic act. The CMVSS are embedded in the Motor Vehicle Safety Regulations (C.R.C., c. 1038). In Canada, we have most regulations on country level since in the USA each state is responsible to implement the road traffic act in their own understanding and needs.

The FMVSS and the CMVSS are comparable with the vehicle homologation regulations provided by the UN ECE. The application of FMVSS and CMVSS for new vehicles is very different in comparison to UN ECE regulations. Since UN ECE or the authorities in the applying countries require certificates by accredited bodies, Canada and the USA apply self-certification. In addition to any FMVSS regulation, US Department of Transportation provides a "Laboratory Test Report", which describes how to perform a compliance test.

The valid list of those laboratory test reports could be find with that link: https://www.carsandracingstuff.com/library/n/nhtsa_fmvsscompliance.php.

1.7 UNECE

The UN ECE is commission of the United Nations. They formulate their mission as follow:

Mission

The United Nations Economic Commission for Europe (UNECE) was set up in 1947 by ECOSOC. It is one of five regional commissions of the United Nations. The others are the:

- Economic Commission for Africa (ECA),
- Economic and Social Commission for Asia and the Pacific (ESCAP),
- Economic Commission for Latin America and the Caribbean (ECLAC),
- Economic and Social Commission for Western Asia (ESCWA).

UNECE's major aim is to promote pan-European economic integration. UNECE includes 56 member States in Europe, North America and Asia. However, all interested United Nations member States may participate in the work of UNECE. Over 70 international professional organizations and other non-governmental organizations take part in UNECE activities.

Source https://www.unece.org/mission.html, download February 2020.

American and UN ECE are working since many years aligning the relevant standards. For most of the UN ECE, there is a 1:1-mapping of the regulations in North America possible. From the contents point of view, they differ from due to a different understanding of right systems and historic developments of current legislation.

Worldwide Harmonization

Since the end of the 1950s the UN ECE in Geneva has been working on cross-border harmonization of technical regulations. The UN ECE agreement dated 20 March 1958 concerning the adoption of uniform technical prescriptions for wheeled vehicles, equipment and parts that can be fitted and/or be used on wheeled vehicle and the conditions for reciprocal recognition of approvals granted on the basis of these prescriptions, laid a milestone on the path towards uniform technical registration regulations. The agreement now comprises 131 technical regulations. These concern not only systems and components for active and passive safety but also environmentally relevant regulations. Meanwhile the agreement has been signed by 51 countries.

Within the EU, there is complete harmonization of registration regulations and mutual recognition. Outside the EU, not all signatory states apply all regulations. Many countries such as Russia have integrated the regulations into their national registration law. The situation is similar in Japan. However, even if only a few of the ECE regulations are applied, this already represents a first step towards harmonization.

The 1959 agreement is based on the type approval process. An independent test institution is producing a report with the objective of confirming the validation of ECE requirements. A public authority issues the type approval on this basis. As a result, the USA is fundamentally barred from joining the 1958 agreement. This is because the process of self-certification used in the United States is incompatible with the type approval. For this reason, a further agreement has been launched under the auspices of the UN: these are referred to as global technical regulations under the UN ECE agreement of 1998 ("Global technical regulations for wheeled vehicles, equipment and parts that can be fitted and/or be used on wheeled vehicles"). This agreement has now been signed by 33 countries—including some EU member states, the People's Republic of China, South Korea and the United States of America.

Since 1998, 13 global technical regulations (GTRs) have been worked out in the expert committees of the UN ECE. These refer to, for example, door locks and door retention components—GTR1, head restraints—GTR 7, electronic stability control systems—GTR 8, proposal to develop a global technical regulation concerning heavy-duty vehicle exhaust-emissions—Appendix to GTR 4, pedestrian safety—GTR 9 as well as regulations concerning hydrogen and fuel cell vehicles—GTR 13.

Most GTRs have been implemented in the set of rules contained in the 1958 agreement, as a result of which they are accepted by the signatory states to that agreement and are implemented via ECE regulations. Furthermore, a few GTRs have been adopted into the national law of the states that signed the 1998 agreement. However, the 1998 agreement has a "congenital defect." According to its Article 7, the contracting state that votes in favor of establishing a specific GTR shall be obligated to implement its provisions into the national laws of that contracting state. Furthermore, regular reports should be submitted to the UN about the further course of the process. There is no obligation for the agreement to be implemented. Many signatory states including the USA only got halfway towards implementing the GTRs in their national law. It remains to be seen whether the negotiations that started at the end of 2013 regarding a bilateral free-trade agreement between the European Union and the United States (TTIP, Transatlantic Trade and Investment Partnership) will lead to mutual recognition of technical regulations and certifications. Examinations are under way as to whether and to what extent a functional equivalence can be validated between the regulations of the EU and the USA, at least with regard to vehicle safety. This would be a basic precondition for imposing the same requirements on certification in the EU and the USA.

The committees of the ECE at the United Nations in Geneva are working in parallel towards the objective of developing the 1958 agreement further and increasing the number of states using this agreement. The agreement should be made more attractive for other states to join. There should also be dialog with developing countries that are not yet in a position to apply the ECE regulations with full stringency. As a result, a process has been launched for expanding the agreement through to an international whole vehicle type approval (IWVTA).

This is being worked out in a working group of the UN WP.29, the world forum for harmonization of technical vehicle regulations. The 1958 agreement is being expanded so as to create not only harmonized regulations for systems or components, but for the whole vehicle. For this purpose, an ECE regulation 0 is being formulated, which describes a whole vehicle approval.

In order to make it easier for developing countries to use the IWVTA, and thus make joining the agreement more attractive, the draft that is currently under discussion allows scope for the contracting parties to take a more flexible approach to the strictness and scope of requirements within the individual countries. Although this will not allow complete mutual recognition of approvals initially, it does represent a first step in that direction. As a result, the first level would involve an approval with limited recognition in the event that it only certifies a lower scope of technical features. The result of this might be that it is not possible to provide validation of compliance with individual regulations. In contrast to this, however, there might also be universal approvals certifying compliance with all requirements (on the highest level). The great advantage compared to the EU type approval already practiced today is that the ECE regulations would apply in a significantly wider area than the European Union.

Contracting parties that intend to apply type approvals with restricted recognition must notify the UN Secretariat which deviations from a universal IWVTA they are prepared to accept. As a first step, the IWVTA will not cover all subsystems of a vehicle, meaning that it is only possible to refer to a partial type approval on an initial basis. Nevertheless, the objective is for the approval to be expanded successively. The goal is and remains whole vehicle approval. On the long path leading to "approved once—accepted everywhere," at least the first IWVTA approval should be able to be granted by the end of the current decade.

Source https://www.vda.de/en/topics/safety-and-standards/harmonization-of-admission-requirements/worldwide-harmonizationm download July 2019.

This information from the German VDA provides many information about the targets. Motor cycle brakes and some pedestrian safety measures where successfully converted into GTR, and they are used for homologation. UN ECE provides the further candidates for GTRs under this link: https://www.unece.org/trans/main/wp29/wp29wgs/wp29gen/wp29glob_candidate.html.

The most successful example seems to be UN GTR No. 13—Global Technical Regulation concerning the hydrogen and fuel cell vehicles

Applying GTRs in USA or Canada is still an issue since no official laboratory test reports are provided from the American authorities. A proper tailoring of the FMVSS laboratory test reports for FMVSS 301–304 could be applied, but crash impact and treatment of hydrogen requires different measures as to be applied for LPG or CNG.

UN ECE published the following document: WORLD FORUM FOR HARMONIZATION OF VEHICLE REGULATIONS (WP.29) HOW IT WORKS HOW TO JOIN IT

Fig. 1.1 WP.29 PARTICIPANTS. *Source* https://www.unece.org/fileadmin/DAM/trans/doc/2012/wp29/WP29-157-03e.ppt

https://www.unece.org/fileadmin/DAM/trans/main/wp29/wp29wgs/wp29gen/wp29pub/wp29pub2002e.pdf.

It provides detailed motivation and targets for the harmonisation.

Some very interesting information had been provided by Dr. B.V.Kisulenko, WP.29 Chairman, The Russian Federation APEC Kazan car Dialogue, in 2012.

Source https://www.unece.org/fileadmin/DAM/trans/doc/2012/wp29/WP29-157-03e.ppt (Fig. 1.1).

The figure shows the worldwide recognition of the activity and the members in 2012 (Fig. 1.2).

The table provides some ideas what kind of safety requirements derive from the legal standards. It shows that the figures for busses are quite low; even they are transporting most people.

For trucks, much less crash regulations apply, with alternative fuels such as hydrogen, this must be questioned.

1.8 Legal View on Future Mobility

The key targets for future mobility are zero emission and zero accidents, by maximum flexibility and minimal travel time.

ESTABLISHMENT OF SAFETY REQUIREMENTS

The number of objects of requirements applied to the different vehicle categories

Safety Kinds	Vehicle Category		
	Cars M1	Buses M2,M3	Trucks N1-N3
Active Safety	13	10	13
Passive Safety	16	10	16
Post-crash Safety	2	3	2
Ecological Safety	11	7	11
Total	42	30	42

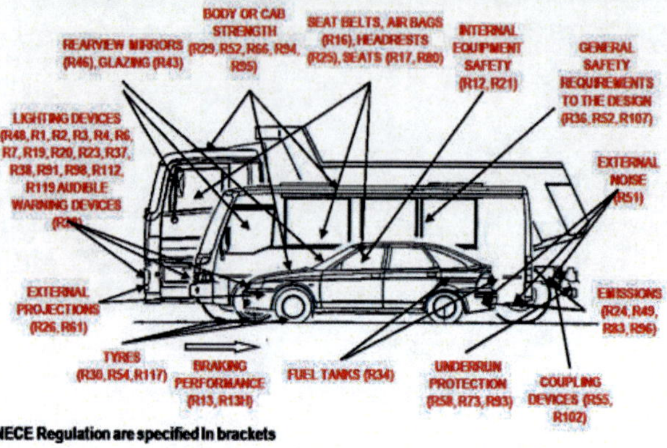

The numbers of the UNECE Regulation are specified in brackets

7

Fig. 1.2 Kinds of safety for the different vehicle categories. *Source* https://www.unece.org/fileadmin/DAM/trans/doc/2012/wp29/WP29-157-03e.ppt

At the moment for zero emission the E-mobility is in focus, maybe with fuel cell as range extender.

The second target is zero accidents and a big discussion how automated driving could support that target. No doubt people make mistakes or are wrong. Human driver make a lot of failure, but if they cause damages, they have to compensate them. Mass transportation systems like busses or railway cause less accidents than individual drivers. However, if they cause accidents are to be feared with catastrophic consequences and several deaths.

Maximum flexibility means new cooperation concepts between individual traffic and mass transportation systems.

Minimum travel time could require higher speed, but also less traffic jam and faster change from one transportation system to another. Alternatively some have the idea to bore tunnels and pneumatic tube seems to be the fastest locomotion.

Influence of the traffic law on the future mobilisation

Origin of all these law initiatives was the safety of the other road users. In particular, the splits appeared for the state power between support of the desired or also necessary mobility, also towards the singles, and the necessary protection of the respective weaker road user.

From which also Section 1 of the road traffic regulations developed which is as follows:

§1 Basic Rules

(1) *Participation in traffic requires constant care and mutual consideration.*
(2) *Who takes part in the traffic has to behave in such a way that no one else endangers injured, or more than is hindered after the circumstances inevitably or is bothered.*

This section is often denoted also elastic section and mess around an eager sporty driver acquaintance with this section has already made if he has realised his ambitions in the public sphere of traffic a little bit too intensely.

Generally, we had to do in Germany gladly his one universal law basis to have which regulates the mobility in the traffic. In the introductory examples on top, the basic dependence on the different arrangements and laws becomes worldwide of course clear. Vehicles, driving licences, rules about the behaviour in the traffic make locally also no sense because it can come only for a very limited solution of the range of application. This cannot be of course the target of an industry which would like to bring new mobilisation draughts in the market. One needs certain planning safety around also evolutionary new mobility draughts can develop.

Unfortunately, it is a fallacy that one can plan laws also for new technologies in advance. From millenniums of long experience, one has also initiated at the same time the product liability laws in many countries.

The future mobility and the legal limitations are particular workout in WP.29.

WP.29 the Following Outcomes of Their Activities:

- Adopted UN Regulations, UN Rules and UN GTRs
- UN Regulations adopted by AC.1 (1958 Agreement)
- UN Rules adopted by AC.4 (1997 Agreement)
- UN Global Technical Regulations adopted by AC.3 (1998 Agreement)

On their home pages, you will find the current status of relevant agreements. The documents called "status of the Agreement" provides useful information about the agreements and the regulations/rules adopted in the framework of these Agreements.

Here Are the Links to the Three Relevant Documents:

- Status of the 1958 Agreement
- Status of the 1997 Agreement
- Status of the 1998 Agreement

In These Documents, You Will Find the:

- Date of entry into force of an Agreement/Regulation/Amendment etc.,
- List of Countries (Contracting Parties) applying an Agreement/Regulation/Amendment
- Reference of the documents constituting an Agreement/Regulation/Amendment
- Countries applying an Agreement/Regulation/Amendment
- List of Technical Services for a given UN Regulation in a given Country
- Contact data of Type Approval Authorities and Technical services.

Resolutions

- Resolutions adopted by WP.29:
- The Consolidated Resolution R.E.3 (related to the 1958 Agreement)
- The Special Resolutions (related to the 1998 Agreement)
- The Mutual Resolutions (related to the 1958 and 1998 Agreements)
- The Resolution R.E.6 (related to PTI and the 1997 Agreement)

Other Resolutions Are Under Development, Such as a Resolution Containing Information on Standardized Light Sources, One Related to the 1997 Agreement Etc.
Others

- Dummy specifications
- Deceleration Calculator (Regulation No. 117)
- Electric Vehicle Regulatory Reference Guide
- Classification and vehicle categories
- Guidelines for Transitional Provisions
- Guidelines for document preparation
- Accronyms, abbreviations and symbols

Source https://www.unece.org/trans/main/wp29/presentation_regulations.html, download March 2020.

WP.29 Established Six Permanent Working Parties (GRs), I.E. Subsidiary Bodies that Consider Specialized Tasks, Consisting of People with a Specific Expertise:

- Noise (GRB)
- Lighting and Light-Signalling (GRE)
- Pollution and Energy (GRPE)
- Brakes and Running Gear (GRRF)
- General Safety Provisions (GRSG)
- Passive Safety (GRSP).

The World Forum WP.29 also established (and continues to establish) a number of Informal Working Groups (IWGs) with a time-limited mandate to deal with certain technical issues. Depending on the case and the terms of reference under which they are conceived, such groups report their work to one of the GRs or directly to WP.29. Informal Working Groups can be structured in subgroups, performing detailed tasks within the primary mandate.

Source https://www.unece.org/trans/main/wp29/introduction.html, download March 2020.

Informal Working Groups (IWGs) under GRSG

Since June 2012, all documents concerning the meetings of IWGs under WP.29 and its Working Parties are uploaded in this working environment.

The Documentation Concerning Specific IWG Meetings Can Be Accessed by Browsing the Menu Located on the Left Hand Side of This Webpage or Below:

- Service Doors, Windows and Emergency Exits (SDWEE)
- SDWEE Previous Sessions

- SDWEE 8th session
- GTR on Motorcycle Controls, Tell-tales and Indicators (MCSYM)
- MCSYM Previous Sessions
- Introduction of Plastic Glazing (IPG)
- Camera Monitor Systems (CMS-II)
- Task Force on Liquefied Natural Gas vehicles (TF-LNG), under GRPE
- Automatic Emergency Call Systems (AECS)
- Panoramic Sunroof Glazing (PSG)
- Preview Meeting of NHTSA questionnaire status
- Awareness of Vulnerable Road Users proximity in low speed manoeuvres (VRU-Proxi)
- Behaviour of M2 & M3 General Construction in Case of Fire Rvent (BMFE).

Source https://wiki.unece.org/pages/viewpage.action?pageId=2523225, download March 2020.

For future automated driving function, the working group GRVA is working on some significant topics.

Working Party on Automated/Autonomous and Connected Vehicles (GRVA) and their Informal Working Groups (IWGs) provide the following schedule of Informal Working Group (IWGs) and Task Force (TF):

List of IWGs and TF Meetings:

- Functional Requirements for Automated and Autonomous Vehicles (FRAV)
- DSSAD/EDR
- GRVA—WebEx (2019 Q1)
- Validation Method for Automated Driving (VMAD)
- Task Force on Cyber Security and (OTA) software updates (CS/OTA)
- Automatically Commanded Steering Function (ACSF)
- Modular Vehicle Combinations (MVC)
- Automatic Emergency Braking and Lane Departure Warning Systems (AEBS/LDWS)

Source https://wiki.unece.org/pages/viewpage.action?pageId=63310525, download March 2020.

GRVA's priorities include:
Safety and security of vehicle automation and connectivity:
(See the Framework Document on Automated Vehicles)

- Functional requirements ("FRAV")
- Validation method for automated driving ("VMAD")
- Cyber security (and software updates)
- EDR/data storage system for automated driving (currently)

Adas:

- Remote control manoeuvring
- Automatically commanded steering systems
- Dynamics (steering, braking, etc.):
- Advance emergency braking systems
- Anti-lock braking system for motorcycles
- Electronic stability control.

The targets of Functional Requirements on Automated Vehicles (FRAV) are as follows:

- Develop functional (performance) requirements for automated/autonomous vehicles, in particular, the combination of the different functions for driving: longitudinal control (acceleration, braking and road speed), lateral control (lane discipline), environment monitoring (headway, side, rear), minimum risk manoeuvre, transition demand, HMI (internal and external) and driver monitoring. This work item should also cover the requirements for Functional Safety;
- Do this in line with the following principles/elements a. System safety, b. Failsafe Response, c. HMI/Operator information d. OEDR (Functional Requirements) described in document ECE/TRAN/WP29/2019/34.

Source ECE/TRANS/WP.29/GRVA/2019/15/Add.1.

The targets of Validation Methods for Automated Driving (VMAD) are as follows:

2. As Noted in Document ECE/TRANS/WP29/2019/34, the Informal Group Shall:
   - develop assessment methods, including scenario's, to validate the safety of automated systems, based on a multi pillar approach including auditing, simulation, virtual testing, test track testing, real world testing.
   - in line with the following principles/elements d) assessment method/test for Object Event Detection and Response and f) Validation for System Safety.
3. The Informal Group Shall Take Full Account of Developments and Work in Full Cooperation with Other Subsidiary Working Parties of WP.29 in Particular ACSF and FRVA.

Source Proposal for an Informal Working Group on Validation Methods for Automated Driving (VMAD), GRVA-03–19, 3rd GRVA, 3–4 June 2019.

1.8.1 New Licensing Approach for Automated Vehicles

On initiative of the OICA (organisation international of the Constructeurs d'Automobiles, lobby organisation with more than 30 national member's associations, it was founded on initiative of the German manufacturers in 1919 in Paris) a draft UN ECE was formulated which was intended to be ready for voting in 2019.

The draft UN ECE was published in November 2018, called "ECE/TRANS/WP.29/GRVA/2019/13", it could be found in the UN ECE repository: https://www.unece.org/trans/main/wp29/wp29wgs/wp29grva/grva2019.html.

The regulation has not yet been adopted in this form.

The known SAE levels could be subdivided in the following main classes:

- The classes 1 and 2 are described as the up to now known functions of the automated driving, which have arisen extensively from enlargements of the assistance functions. Efforts for a so-called L2 + class are not considered. In L2 classes, it is considered that the driver is still responsible to control the vehicle, the traffic environment and the condition and status of the vehicle and associated systems.
- From class 3 are derived conditioned automated driving function. The systems should be capable to manage the hand-over from manual and system driving and system driving including traffic observation, vehicle and systems monitoring and diagnostics etc..
- Classes 4 and higher are summarised vehicles into those the driver partly or completely are just passengers in the vehicle.

Beginning with class 3, the following main pillars are to be considered for homologation or the licensing of subsystems:

- special audits and assessments at manufacturers side (not only at the vehicle manufacturers side, since for the AD-system a supplier could be responsible for the product.)
- determined physical licensing tests and
- real-world test driving catalogues defines (Fig. 1.3)

The three pillars are detailed and described in a pyramide (Fig. 1.4).

The first column is based on audits and assessments with the following measures:

- Audits of the processes of development (methods and standards);
- Assessments of the safety draughts (functional safety and safety-in-use) and the derived measures;
- Review of the integration of the safety requirements and attention of the relevant traffic rules;
- Knowledge from simulations (loan from intensive validation driving tests (driven kilometre) to handle ability with critical situations which are not checkable on test areas or on public streets);
- Assessments of the developing data and field test results, as well as manufacturer self-commitments.

The column of the physical licensing tests sees the following targets:

As well, audit and assessment results fit to the behaviour in the real world

Assessment and validation tests of the functions on test tracks or public streets are not checkable

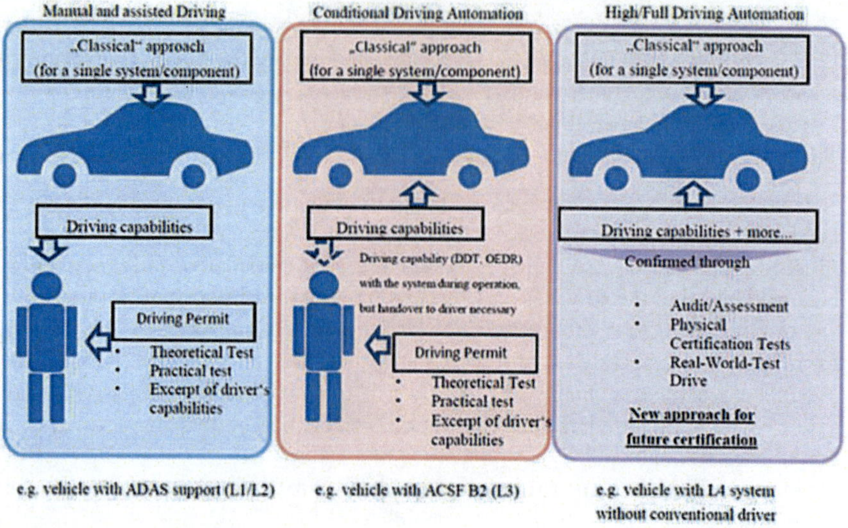

Fig. 1.3 New certification (homologation) approach on the basis of a proposal OICA. *Source* ECE/TRANS/WP.29/GRVA/2019/13″, 19.November.2018

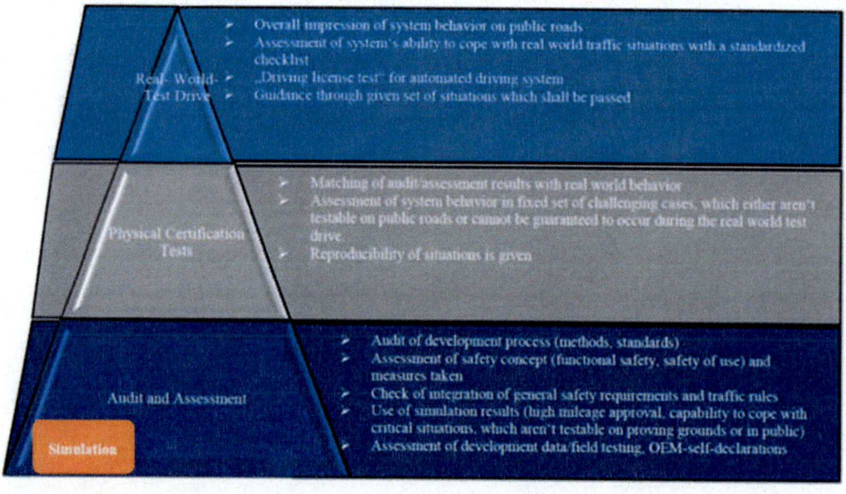

Fig. 1.4 Column to OICA for the licensing of vehicles AD. *Source* ECE/TRANS/WP.29/GRVA/2019/13″, 19.November.2018

Arguments for the reproducibility of the measures in the situations (how well are returned the looked situations and her criticality).

The "real-world check" looks at the following aspects:

- Overall impression of the behaviour on public streets;
- To control evaluation of the ability the real traffic situations and use of standardised test lists;
- Driving licence check for the automation system;
- Directives and default for situations to be looked;

To three columns, an example is also given (Figs. 1.5 and 1.6).

The example sees before this a person the street crossed.

The situation in the zebra crossing by an oncoming AD vehicle is looked as a usual situation.

For the physical check, it is planned that the person goes out into the streets between parking vehicles.

For the audit and assessment, the ability of the system and the analyses and simulations is looked, with oncoming traffic correctly between make way with journey in the oncoming traffic and the accident with the person to weigh.

The new ASCF (ALKS, Automatic Lane Keeping Systems) and VMAD are the first results of that new approach.

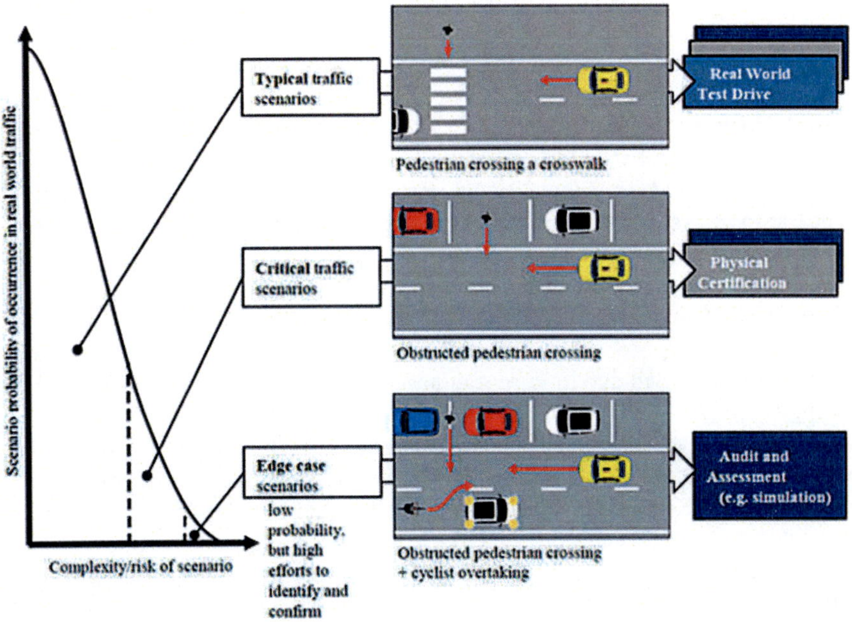

Fig. 1.5 Example for new approach

Fig. 1.6 Examples of different test measures for automated driving functions
Source ECE/TRANS/WP.29/GRVA/2019/13″, 19.November.2018.

1.8.2 ITS Law

ITS stands for "Intelligent Transportation System". The Intelligent Transport Systems Act was published in the Federal Law Gazette 2013 Part I, No 29, p. 1553 on 20 June 2013 and entered into force on 21 June 2013. Directive 2010/40/EU of the European Parliament and of the Council of 7 July 2010 on the framework for the deployment of intelligent transport systems in the field of road transport and for interfaces with other modes of transport (OJ L 207 of 6 August 2010, p. 1) was transposed into German law by the Intelligent Transport Systems Act. "Intelligent transport systems" mean that traffic-related data are collected, transmitted, processed and utilised by using information and communications technologies with the objective of organisation, information and traffic management.

The First Act amending the Intelligent Transport Systems Act entered into force on 25 July 2017.

The Act is amended due to European provisions (delegated regulations supplementing Directive 2010/40/EU). Accordingly, a National Body will be established, which is to review the road, traffic and trip data made available by data providers as to their conformity with the requirements of the delegated regulations. In Germany, the Federal Highway Research Institute (BASt, Bundesanstalt für Straßenwesen) will be responsible for and will carry out the tasks incumbent on the National Body.

Directive 2010/40/EU stipulates that the Member States of the European Union shall apply the specifications issued by the Commission when implementing priority actions as defined by Article 2. Priority areas within the meaning of the directive are the optimal use of road, traffic and travel data, continuity of traffic and freight management ITS services, ITS road safety and security applications as well as linking the vehicle with the transport infrastructure.

In 2020 a standardisation activity starts in ISO cooperation with CENELEC: "Mobility Safety-Technologically advanced infrastructures"; CEN/TC 278 "intelligent transport systems", coordinating with CENELEC, ETSI and ISO to ensure a coherent approach to standardization.

eCall, which must be introduced by all Member States, represents an exception. In the form of a delegated regulation, these are directly applicable EU law; no transposition into national law is required to make them binding. Therefore, the Member States are obliged to enforce these acts.

To develop a common perspective of all players for the Europe-wide introduction of cooperative intelligent transport systems (C-ITS), the European Commission (DG MOVE) has established the "C-ITS Platform". The task of this advisory body is to support the European Commission in developing a "C-ITS Masterplan". After the conclusion of the first stage in January 2016, during which the C-ITS platform exclusively concentrated on vehicle-to-vehicle or vehicle-to-infrastructure communication, in the second stage it will focus on issues related to automated and connected driving.

Cooperative ITS—safe and intelligent Noticing traffic jams before you see them. Detecting risks before they become a threat. Arriving at your destination safe and sound. This vision of safe and intelligent mobility can be realised by wireless connecting vehicles and infrastructure. In technical terms, it is achieved with so called cooperative Intelligent Transport Systems (C-ITS—also known as V2X communication for vehicle-to-vehicle and vehicle-to-infrastructure communication). Cooperative systems enable direct communication between vehicles, roadside infrastructure and traffic control centres.

The benefits of V2X communication are numerous. It enables anticipatory and safe driving, as drivers are informed about the current traffic situation and danger zones in time. In addition, traffic centres receive precise and comprehensive information on the traffic situation from vehicles. In this way, it is possible to control the traffic flow more differentially, efficiently and quickly, resulting in an improved flow of traffic. The effect: more safety, less accidents, improved use of the road network, less congestion and a decrease in CO2 emissions.

Intelligent mobility—beyond national borders The basis for the pan-European deployment of cooperative ITS is already in place. The cooperative ITS technology has been developed within research and development projects (R&D) and are evaluated in field operational tests (FOTs). The majority of the enabling technology is already standardised. The non-technical aspects (e.g. organisational structures, safety concept, legal aspects) are currently addressed in public private partnerships in preparation for the market launch. On this basis, now the road operators in Germany, the Netherlands and Austria start joint deployment of cooperative ITS in Europe with partners from industry.

Cooperative ITS require the commitment of many partners from different branches of industry and politics. The Amsterdam Group, a strategic alliance of road operators and industry on a European level, is coordinating the efforts towards deploying cooperative ITS. Involved are CEDR as an organisation of public road operators, ASECAP as an umbrella association of the toll road operators, POLIS as an umbrella association of cities and the Car2CarCommunication Consortium representing the automotive manufacturers and associated industries.

Source Cooperative ITS Corridor Joint deployment.

Communication standards have been developed to use sensors and communication systems in the infrastructure.

ITS programmes are established world-wide; in the USA, a "Joint Program Office" has been founded for ITS.

They are answering the question: "What Are Connected Vehicles and Why Do We Need Them?" as follow:

The U.S. Department of Transportation's (USDOT's) Connected Vehicle program is working with state and local transportation agencies, vehicle and device makers, and the public to test and evaluate technology that will enable cars, buses, trucks, trains, roads and other infrastructure, and our smartphones and other devices to "talk" to one another. Cars on the highway, for example, would use short-range radio signals to communicate with each other so every vehicle on the road would be aware of where other nearby vehicles are. Drivers would receive notifications and alerts of dangerous situations, such as someone about to run a red light as they're nearing an intersection or an oncoming car, out of sight beyond a curve, swerving into their lane to avoid an object on the road.

Source https://www.its.dot.gov/cv_basics/cv_basics_what.htm.

ITS in the US

In the USA, the US Department of Transportation sponsors the programme ITS4US. The focus of that programme is to provide affordable mobility for all US citizens.

They formulate their targets as follow:

The lack of transportation options for all travelers, including travelers from underserved communities is a persistent challenge for access to jobs, education, healthcare, and other activities. The Complete Trip—ITS4US Deployment Program aims to solve mobility challenges of all travelers, regardless of location, income, or disability, in accessing jobs, education, healthcare and other activities.

As a response to these challenges, the USDOT is increasing its investments in innovations that enhance access and mobility for all travelers. USDOT is launching a new department-wide initiative to expand access to transportation for people with disabilities, older adults, and individuals of low income. The Complete Trip portfolio will identify ways to provide more efficient, affordable, and accessible transportation services for people with disabilities, older adults and other underserved communities that often face greater challenges in accessing essential services.

As part of the Complete Trip Portfolio,—the Complete Trip—ITS4US Deployment Program is a multimodal effort let by ITSJPO with support from OST, FTA, and FHWA. This programme will make up to $40 million available to enable communities to showcase innovative business partnerships, technologies, and practices that promote independent mobility for all (Figs. 1.7 and 1.8).

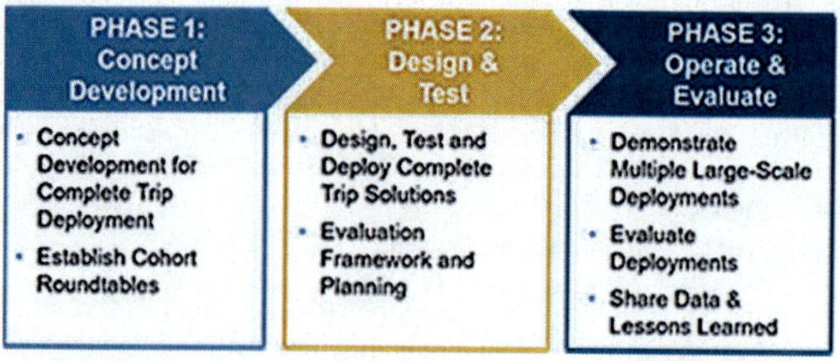

Fig. 1.7 Program ITS4US. *Source* https://www.its.dot.gov/its4us/index.htm

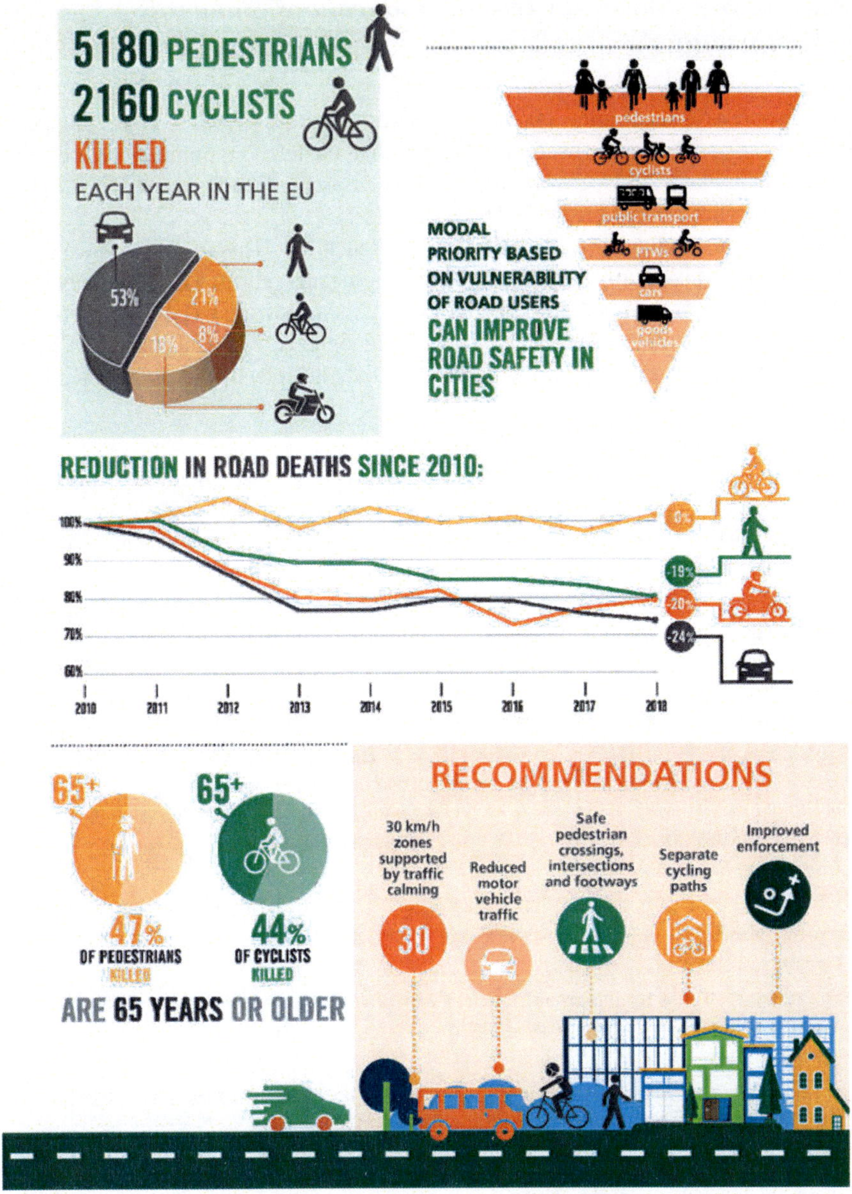

Fig. 1.8 Poster from ETSC to make aware of cyclists and pedestrian care

1.8.3 *ETSI—European Telecommunications Standards Institute*

The European Telecommunications Standards Institute (ETSI) is an independent, not-for-profit, standardisation organization in the telecommunications industry (equipment makers and network operators) in Europe, headquartered in Sophia Antipolis, France, with worldwide projection.

ETSI is registered as an association under French law. This means that we are a not-for-profit organisation. Any surplus income is usually returned to our members as credit towards their membership contributions for the following year. It is a European Standards Organization (ESO). They are the recognized regional standards body dealing with telecommunications, broadcasting and other electronic communications networks and services.

They See Themselves as Follow:

We have a special role in Europe. This includes supporting European regulations and legislation through the creation of Harmonised European Standards. Only standards developed by the three ESOs (CEN, CENELEC and ETSI) are recognized as European Standards (ENs).

Our Global Impact

We Were Initially Founded to Serve European Needs, but We Have a Global Perspective. Our Standards Are Now Used the World Over.

We collaborate and work in partnership with different types of organizations around the world. This makes us well placed to support our members who operate in an increasingly international and competitive environment.

In addition, we are a partner in the international Third Generation Partnership Project (3GPPTM). Through this project, we are helping to develop 4G and 5G mobile communications. We also work with partners around the globe in the oneM2M partnership project to develop standards for machine-to-machine communications.

Our Vision

ETSI is a Leading Standardization Organization for Information and Communication Technology (ICT) Standards Fulfilling European and Global Market Needs.

Our Mission

To Provide Platforms for Interested Parties to Work Together to Produce Standards for ICT Systems and Services that Are Used Globally.

Our Long-Term Strategy

ur Long-Term Strategy is Developed by Our Members. It Defines Our Vision, Our Mission and the Nine Basic Principles that Constitute Our Core:

- ETSI's fundamental values about the work it undertakes are timeliness, quality and responsiveness to market needs
- ETSI is inclusive. Our global membership represents a wide range of stakeholders. We have a global network of partnerships and a consensus approach in our working methods
- ETSI is fully compliant with WTO/TBT provisions and EU Regulation (EU) 1025/2012
- ETSI works across all sectors of industry and society that make, use or rely on ICT technology
- ETSI works at the forefront of developing and emerging technologies

- ETSI produces technical standards aimed at being adopted by the most competitive markets
- ETSI promotes the adoption of its technical standards worldwide for the benefit of its members' competitiveness
- ETSI, in its role as an ESO, provides technical standards supporting EU regulatory requirements and policy
- ETSI produces technical standards that use spectrum effectively, minimize interference and avoid undesirable effects. We produce our standards in accordance with appropriate regulatory frameworks

Our Five Strategic Objectives Are:

- Being at the heart of digital
- Being an enabler of standards
- Being global
- Being versatile
- Being inclusive

Each of These Objectives is Further Detailed in Our Long-Term Strategy.

ETSI Plays a Key Role in Supporting Regulation and Legislation with Technical Standards and Specifications. To Do This We Co-Operate with Other Organizations Including:

- the European Commission (EC)
- the European Free Trade Association (EFTA)
- the Electronic Communications Committee (ECC) of the European Conference of Postal and Telecommunications Administrations (CEPT)

Harmonised Standards

Standardization has been a key factor in the creation of the single European market, the biggest unified market place in the world. The single market has benefited all involved—manufacturers, network operators and users.

Harmonised Standards Enable Manufacturers and Suppliers to Access European Markets

The EC harmonizes the requirements for Information and Communication Technologies (ICT) products and services via*European Directives, Regulations and Decisions, which are enforced by legislation.*

The EC/EFTA issues standardization requests to ETSI and the other two European Standards Organizations (ESOs), CEN and CENELEC, with proposals to develop Harmonised Standards (European Standards (ENs) with a special status). The ESOs agree together whether and how they want to respond to a specific standardization request, for example, which of the ESOs will carry out or lead the work.

These Harmonised Standards provide the technical detail necessary to achieve the 'essential requirements' of a Directive. By conforming to these standards, manufacturers and suppliers can demonstrate that they comply with relevant legislation. They may then access the whole of the European market. A product may be certified as being compliant with an ETSI standard, but ETSI does not provide certification services nor endorse any product based on an ETSI standard.

We Produce Harmonised Standards for:

- the Radio Equipment Directive (RED)—2014/53/EU
- the EMC Directive—2014/30/EU

By adhering to these harmonised standards, manufacturers and service providers can demonstrate that they have followed the essential requirements of the directive and are able to claim 'presumption of conformity'. This allows them to put their products and services on the market in Europe.

Harmonised Standards Take Effect When They Are Cited in the Official Journal of the European Union (EU).

Lists of ETSI Harmonised Standards

- Harmonised standards covering the essential requirements of the Radio Equipment Directive (RED)—2014/53/EU—cited in the Official Journal of the EU
- Harmonised standards covering the essential requirements of the EMC Directive—2014/30/EU—cited in the Official Journal of the EU
- Harmonised standards covering the essential requirements of the RED & EMC Directives—not yet cited in the Official Journal of the EU

Use of Harmonised Standards

National governments also use our standards to enforce regulations. Administrations act against non-compliant equipment, so that, for example, legitimate users can use spectrum without interference.

ETSI Supports Various European Directives Through the Production of Harmonised Standards Including Those Concerning:

- radio equipment
- lectromagnetic compatibility (EMC)

We Provide All the Harmonised Standards Needed to Implement the European Directives for the Use of Radio Spectrum and Access to Emergency Services.

Community Specifications

We also produce European Community Specifications (CSs) under the Single European Sky Interoperability Regulation (i.e. in civil aviation). CSs have a similar status to Harmonised Standards.

System Reference Documents

ETSI works closely with the EC and CEPT/ECC to co-ordinate European requirements for radio spectrum and to ensure that the necessary spectrum is available for ETSI radio standards.

We Have a Memorandum of Understanding with the CEPT/ECC

- the Telecommunication Conformity Assessment and Market Surveillance Committee (TCAM)
- he Radio Spectrum Committee (RSC)
- the Radio Spectrum Policy Group (RSPG)

We Produce a Specific Type of Technical Report Called a 'System Reference Document' (SRdoc) Which:

- provides technical, legal and economic background to new radio systems, services or applications.
- advises on the need for an allocation of spectrum, either when a change in the current frequency designation or its usage, or a change in the regulatory framework for the proposed band(s) is needed to accommodate a new radio system or service.

SRdocs are typically produced alongside Harmonised Standards to ensure radio equipment is constructed in such a way as to avoid harmful interference. We usually develop SRdocs in parallel with the standards we are developing. In this way, ETSI is the route for industry to drive the regulatory process so that it meets market needs.

Source https://www.etsi.org/about.

This self-statement from ETSI shows the range we need to consider safety in the field of connectivity.

1.8.4 OECD, Organisation for Economic Co-operation and Development

Organisation for Economic Co-operation and Development (OECD) introduced themselves as follow:

The Organisation for Economic Co-operation and Development (OECD) is an international organisation that works to build better policies for better lives. Our goal is to shape policies that foster prosperity, equality, opportunity and well-being for all. We draw on almost 60 years of experience and insights to better prepare the world of tomorrow.

Together with governments, policy makers and citizens, we work on establishing evidence-based international standards and finding solutions to a range of social, economic and environmental challenges. From improving economic performance and creating jobs to fostering strong education and fighting international tax evasion, we provide a unique forum and knowledge hub for data and analysis, exchange of experiences, best-practice sharing, and advice on public policies and international standard-setting.

Source https://www.oecd.org/about/.

They also hosting the ITF (International Transportation Forum), they introduce themselves as follow:

Who We Are

The International Transport Forum at the OECD is an intergovernmental organisation with 60 member countries. It acts as a think tank for transport policy and organises the Annual Summit of transport ministers. ITF is the only global body that covers all transport modes. The ITF is administratively integrated with the OECD, yet politically autonomous.

What We Do

The ITF works for transport policies that improve peoples' lives. Our mission is to foster a deeper understanding of the role of transport in economic growth, environmental sustainability and social inclusion and to raise the public profile of transport policy.

Source https://www.itf-oecd.org/about-itf.

They published an important paper about automated driving and the relevant regulations.

The summery of the document is as follow:

Many cars sold today are already capable of some level of automated operation, and prototype cars capable of driving autonomously have been—and continue to be—tested on public roads in Europe, Japan and the United States. These technologies have arrived rapidly on the market and their future deployment is expected to accelerate. Autonomous driving promises many benefits: improved safety, reduced congestion and lower stress for car occupants, among others. Authorities will have to adapt existing rules and create new ones in order to ensure the full compatibility of these vehicles with the public's expectations regarding safety, legal responsibility and privacy. This report explores the strategic issues that will have to be considered by authorities as more fully automated and ultimately autonomous vehicles arrive on our streets and roads. It was drafted on the basis of expert input and discussions amongst project partners in addition to a review of relevant published research and position papers.

Source Automated and Autonomous Driving Regulation under uncertainty, OICA International Transportation Forum, 2015.

They support the industry according to their position paper with the following:

A number of issues will have to be addressed in order to support the deployment of high-automation scenarios. These include:

- Vehicle-to-X connectivity (V2X): Connectivity is an important element of the automated vehicles especially secure V2X communication requiring low latency. V2X technologies encompass the use of wireless technologies to achieve real-time two-way communication among vehicles (V2V) and between vehicles and infrastructure (V2I). The convergence of sensor-based solutions (current advanced driver assistance—ADAS) and V2X connectivity will promote automated driving.
- Decision and control algorithms: These include decision, planning and control algorithms for a cooperative, safe, human compatible traffic automation.
- Digital infrastructure: Digital infrastructure (for road automation) includes static and dynamic digital representations of the physical world with which the automated vehicle will interact to operate. Issues to address include: sourcing, processing, quality control and information transmission.
- Human factors: Human factors in automation relate to understanding the interaction(s) of humans with all aspects of an automated road transport system (ART), both from within a vehicle, when taking the role of a driver/operator and also as a road user, when interacting with automated vehicles. Knowledge and theories from social-psychological and behavioural sciences are useful to understand how humans interact with such systems.
- Evaluating road automation: Automation of road vehicles has the potential to impact on lifestyles and society. Economic impacts too will be important and it will be necessary to gauge these impacts in a common cost–benefit framework with other transport investments when assessing public expenditure on supporting infrastructure or services.
- Roadworthiness testing: Roadworthiness testing, understood as the necessary tests to evaluate if a vehicle is legally allowed to drive on public roads is of capital importance for the deployment of new automated driving functionalities.

Source Automated and Autonomous Driving Regulation under uncertainty, OICA International Transportation Forum, 2015.

In the paper, they are addressing the connectivity between vehicles and to the infrastructure, but also human factors, in order to make the automation concept applicable for the benefit of humans.

1.8.5 ETSC European Transport Safety Council

European Transport Safety Council (ETSC) is a Brussels-based independent non-profit making organisation dedicated to reducing the numbers of deaths and injuries in transport in Europe. Founded in 1993, ETSC provides an impartial source of expert advice on transport safety matters to the European Commission, the European Parliament, and Member States. It maintains its independence through funding from a variety of sources including membership subscriptions, the European Commission, and public and private sector support for various activities.

They work in the following areas:

- Speed
- Drink-driving
- Vulnerable road users
- Enforcement
- Work-related road safety
- Road safety performance indicators
- Vehicle safety
- Cycling
- Post-collision care, etc.

They formulate, provide evidences or analyse and promote statements like:

- Speed is the number one killer on Europe's roads. Read our latest report on how EU countries are tackling the problem.
- Drink-driving is responsible for the deaths of thousands of Europeans every year.
- Hype around automated cars is dangerously confusing drivers says Euro NCAP.
- Urgent action needed to tackle deaths of pedestrians and cyclists

The following example shows a typical field of their activities:
Commenting on the Report, Graziella Jost, Projects Director at ETSC Said:

> The EU is facing a multitude of challenges: the climate emergency; road deaths and serious injuries; air pollution and obesity. Policies that improve the safety of cycling and walking can also make a major contribution to tackling all these challenges. Some EU countries, the Netherlands and Denmark in particular, are showing the way forward. If they can do it, so can the rest of the EU.

- How safe is walking and cycling in Europe? etsc.eu/pinflash38
- 21% of road deaths are pedestrians, 8% are cyclists, reflecting the much higher numbers of people walking

- When a person walking falls down and dies, it is not considered as a road death. A cyclist that crashes into an obstacle and dies may also not be counted as a road death, particularly if the police are not called to the scene.

They are driving projects like:

- REVIVE project is looking at best practice in emergency response across Europe.
- Intelligent Speed Assistance (ISA) is a life-saving technology that helps drivers stick to the speed limit.

Organisations like ETSC do not make the laws or regulations, but their expertise and research, are key input for the regulations for not only the European Union and their Member States.

1.8.6 IT Security Act

Major attacks on corporate IT systems have laid bare the vulnerability of critical infrastructure (see info box "Critical infrastructure"). As early as 2015—even before the European Directive on Security of Network and Information Systems (NIS Directive)1 entered into force—German legislators created corresponding regulations, which they inserted into the German Act on the Federal Office for Information Security (Gesetz über das Bundesamt für Sicherheit in der Informationstechnik) by way of the German IT Security Act (IT-Sicherheitsgesetz). In the Regulation Amending the German Regulation on the Identification of Critical Infrastructure (Änderungsverordnung zur BSI-Kritis-Verordnung)2 of 21 June 2017, legislators have specified in greater detail the criteria according to which financial sector companies qualify as operators of critical infrastructure.

Critical infrastructure is defined as follow:

Critical infrastructure within the meaning of the Act on the Federal Office for Information Security comprises facilities, systems or parts thereof belonging to the energy, information technology and telecommunications, transport, health, water, food, finance or insurance sectors and are essential for the functioning of society, because their failure or disruption would lead to considerable supply shortages or risks to public safety and security in Germany.

The German Security Act based on the EU directive 2016/1148, with the following title:

> DIRECTIVE (EU) 2016/1148 OF THE EUROPEAN PARLIAMENT AND OF THE COUNCIL of 6 July 2016, concerning measures for a high common level of security of network and information systems across the Union

In the translation from the German legislation, the "German IT-Sicherheitsgesetz" addresses safety, security and other critical aspects, not only security.

1.9 Product Liability

In the area of today's Federal Republic of Germany, there were already legal systems in the middle ages required for different activities and products to follow adequate duties for care and marketing. On the basis of the Civil Code come into force on the 1st of January, 1900, the general care duty was defined as one of the first interpretations by the supreme court of the German Empire in Leipzig in 1903. It is a duty to the action or omitted, for avoiding, preventing or decreasing of turn away risks for users or third and forms the basis for the later legal regulations.

Product liability is a part of the German offence right as paragraph §1 to 19 of the product liability law (ProdHaftG). It should be distinguished from Civil Code §823 which addresses "manufacturer's responsibility". §823 is the implementation of EU Directive 85/374/EC on product liability (Product Liability Directive) in Germany.

The product liability assumes neither a contract between the manufacturer, dealer, distributor and the end-user. Compensation of damages could only claimed in case of credible remarkable deviation from the possible expectations of the user. For supplier, a remarkable aspect is "culpa in contrahendo" ("cic", Fault before contract conclusion) which addresses occurring damages before a contract had been agreed.

The EU provides the following information for product safety directive:

General Product Safety Directive

EU rules on product safety are defined in the general product safety directive. Under the directive a product is safe if it meets all statutory safety requirements under European or national law.

If there are no regulations or EU standards, the product's compliance is determined according to other reference documents such as national standards, Commission recommendations, codes of practice.

Safety of Products Sold Online

- More and more consumers shop online. The proportion of sales online has been growing compared to the total of sales. Online shopping is convenient for consumers but it also poses certain challenges for product safety. For this reason, on 1st August 2017 the Commission issued a Notice on the market surveillance of products sold online to help public authorities with their work.
- On 25th June 2018, several online marketplaces have voluntarily committed to improving the safety of non-food consumer products sold on their online marketplaces by third party sellers with the signature of a Product Safety Pledge. The initiative sets out specific voluntary actions that go beyond what is already established in the EU legislation. The ultimate goal is to improve the detection of unsafe products marketed in the EU before they are sold to consumers or as soon thereafter as possible, and to improve consumer protection.

Product Safety Pledge

Exceptions

The Directive Does not Cover the Following Products, Which Are Regulated Separately

- pharmaceuticals
- medical devices
- food

Responsibilities of businesses and authorities

It is the responsibility of businesses and national authorities to ensure that only safe products are for sale.

Businesses must

- only place products which are safe on the market
- inform consumers of any risks associated with the products they supply
- make sure any dangerous products present on the market can be traced so they can be removed to avoid any risks to consumers

Appointed national authorities are responsible for market surveillance. They

- check whether products available on the market are safe
- ensure product safety legislation and rules are applied by manufacturers and business chains
- apply sanctions when necessary

National Authorities also Send Information About Dangerous Products Found on the Market to the Safety Gate, the Rapid Alert System for Dangerous Non-Food Products.

Source https://ec.europa.eu/info/business-economy-euro/product-safety-and-requirements/consumer-product-safety/product-safety-rules_en.

On that web-page also a link to the derived regulations for UK is available.

In Germany, the Product Liability Act provided by Directive 85/374/EC on product liability (Product Liability Directive) implements into national law and provides for a strict liability regime. In addition, the Civil Code (sections 823 et seq) provides for a fault-based liability system under tort law.

Another important aspect is product safety. In Germany, the Product Safety Act (Produktsicherheitsgesetz, "ProdSG") defines products as goods, substances or preparations created by means of a production process. The ProdSG derived from *Directive 2001/95/EC of the European Parliament and of the Council of 3 December 2001 on general product safety.*

Key element of ProdSG is the "CE marking"; it defines products or product groups, which have to comply with the regulations, and how they should be marked etc.

According to the ProdSG, manufacturers, importers and traders may only place such non-food products on the market that fulfil statutory requirements in respect of safety and the protection of health of persons. It applies where products (with the exception of antiques) are made available, presented or used for the first time on the market on a commercial basis. If a product is already subject to specific sector-based stipulations (e.g. the machinery directive or homologation standards etc.), then such stipulations take priority.

According to the ProdSG, a product may only be made available on the market if it presents no danger to safety or health of persons when used as intended or as may be expected. As assessment criteria, the product properties, the way a product affects and/or interacts with other products, product related statements and the consumers and/or user groups that are at special risk are taken into account.

In US Consumer Product Safety Act (CPSA) covers the product safety acts. CPSA is the umbrella statute enacted in 1972. This law established the agency, defines CPSC's basic authority and authorizes the agency to develop standards and bans. It also gives CPSC the authority to pursue recalls and to ban products under certain circumstances.

The main aspects of all product safety acts are:

- Product field monitoring
- Recall obligation or
- Ban of products

Typically, governmental organisations are assigned to control and/or register the field events and initiate field measures or assures that measures are initiated.

Muss noch was getan werden!!!!!

expectations from the end-user. accountable damage, a fault is still necessary for the liability of the manufacturer. Rather the user has to go before certain risks going out from a defective product, regardless of a fault of the manufacturer are protected, even if these have appeared only after marketing of the product. It concerns a pure danger liability. In the absence of contract or contract between manufacturer and the user who has acquired the product as a rule from intertrader claims retire from warranty, positive breach of contract (pVV) and culpa in contrahendo ("cic", fault before contract completion). Also, a contract in favour of the third is not considered regularly because the user to the manufacturer and the intertraders is not known yet and is not included, hence, in the contract closed between them.

Generally, product liability is only relevant if it really to damage comes. This is an essential difference to the respective traffic laws, where to mark out the socially accepted framework the legislator tried preventively. Every driver must take note of this if he goes faster than permissibly and is got. Whether it is avenged then only one breach of the law, juridically more massively or is pursued even by different juridical authorities, often depends also on the degree of the danger which would have caused by the too high speed can become or be caused.

Then in the detailing of the product liability, one will often find no default what one has to follow everything with the product development concerning the contents but only after which graduations one is measured in the case of damage and is valued. If one develops a product which has proved itself in the use or in dealing with the product already in the market, one is well-advised to find out before about the "state of the art" and to consider this with the product development.

If one develops products which are new in dealing with themselves or in her use, one can decrease not on to say, there is no state of the art, but experienced product-like recommend here the state to consider of science and technology.

Whether it shows then enterprise risk or desperation shall not be discussed in the case of damage any more, substantially only the question, because of the placing on the market, the manufacturers, the traders are careless or even roughly negligently has acted. Should it concern coarse carelessness, one is also confronted as a developer personally with the laws of the criminal law.

Table 1.3 Directive and well-chosen examples to their implementation in the ProdSV

EU directive	ProdSV
Low-voltage directive in 2006/95/the EC	Act about electric firm means (1. ProdSV)
Toys directive in 2009/48/the EC	Act about the safety of toys (2. ProdSV)
Directive about easy receivers in 2014/29/the EU	Act about easy receivers (6. ProdSV)
Directive about gas consumption equipment 90/396/EEC	Gas consumption equipment act (7. ProdSV)
Directive about personal protective equipment 89/686/EEC	Act about the marketing of personal protective equipment at the market (8. ProdSV)
Machine directive in 2006/42/the EC	Machine act (9. ProdSV)
Directive over sports boats 94/25/EEC	Act over sports boats and water motorcycles (10. ProdSV) [1]
ATEX Product directive in 2014/34/the EU	Explosion protection act (11. ProdSV)
Lift directive in 2014/33/the EU	Lift act (12. ProdSV)
Directive about aerosol packets 75/324/EEC	Aerosol packet act (13. ProdSV)
Pressure device directive 97/23/the EC	Pressure device act (14. ProdSV)

Because the legislator knows that there can be no preventive entire protection of products and devices there are beside the product liability laws also the Product safety laws.

This is the Product Safety Law in Germany (ProdSG), it describes regulations to the safety requirements of technical working means and consumer products before. It substitutes for the devices and Product Safety Law since the 1st of December, 2011 (GPSG).

Among the rest, the following acts were remitted after the GPSG and were adapted from the 1st of December, 2011, formally to the ProdSG (Table 1.3):

Besides the product safety, the ProdSV with the special acts also regulated the marketing of different goods, which must fulfil special security properties (machines, toys, sports boats, electric appendices in the atmosphere capable of an explosion among other things). With it, a basis was created to promote the movement of goods about harmonised security requirements in the EU.

From the devices and product safety law, basic determinations also follow to the supervision-destitute appendices whose establishment and firmware regulated now in the essentials in the operational safety act.

A further aspect of the block of topics is the field and market observation obligation, which has been known for years from product liability laws. This is often perceived as recalls.

Other regulations are narrow with the product safety laws linked like they:

- Directive in 2014/30 / the EU (EMC)
- Directive in 2014/53 / the EU about the provision of radio communication systems at the market

For the EMC directive, there are for automobiles suitable counterparts, the radio directive has like the ECE R10 (EMC), as we see, nevertheless, to this and also to the low-voltage directive (for automobiles the ECE R100) essential overlapping's.

The following understands by a radio communication system the directive in 2014/53/the EU about the provision of radio communication systems:

"Radio communication system" for the purposes of the directive is an electric or electronic product which emits in accordance with the requirements radio waves for the purpose of the radio communication and/or the radiolocation and/or conceives, or an electric or electronic product, the accessories, about one aerial needs, so that it can emit in accordance with the requirements radio waves for the purpose of the radio communication and/or the radiolocation and/or conceive (article. 2.1).

In the respective Article 7 and 8, the directive indicates the overlapping with the directives to low-voltage and EMC:

(7) In the directive in 2014/35/the EU (low-voltage) to stipulated targets for security requirements are for radio communication systems enough; hence, in the present directive should be expelled to them and her use be planned. So that no unnecessary duplication of the regulations, with which it concern not around those which concern the basic requirements, originate, the directive should not be valid in 2014/35/the EU, nevertheless, for radio communication systems.

(8) In the directive in 2014/30/the EU (EMC) to stipulated basic requirements in the area of the electromagnetic compatibility are for radio communication systems enough; hence, in the present directive should be expelled to them and her use be planned. So that no unnecessary duplication of the regulations, with which it concern not around those which concern the basic requirements, originate, the directive should not be valid in 2014/30 / the EU, nevertheless, for radio communication systems.

Not only the vehicle-to-infrastructure communication (Car2X) and the vehicle-to-vehicle communication (Car2Car) become here relevant, also for sensors as for example the radar is not sufficient the EMC consideration, also potential interferences of the radio traffic by devices must be looked.

Besides the radio directive the subject Ocular security with a Lidar will have an essential influence on the development, also moreover there is for stationary systems already extensive directive and standards, but for the fixing in vehicles, there will be in the future still a lot of new knowledge.

Now the curve can be stretched arbitrarily further, whenever it will come multiple or often to dangers or damages of people, the respective legislator will react.

Without suitable analysis to the state of science and technology, there will be no security that the products can be brought in particular within the scope of mass productions or serial productions in the market.

1.10 Common Law and Civil Law

Legal systems in countries around the world generally fall into one of two main categories:

- common law dominated systems and
- civil law dominated systems.

Legal systems of about 150 countries could be described as primarily civil law systems, whereas there are about 80 common law countries.

Typical common law countries are Anglo-Saxon countries like, the USA, UK, India, or Canada.

Germany, Japan, France, Spain, or China are primary Civil Law Countries.

The main difference between the two systems is that in common law countries, case law in the form of published judicial opinions is of primary importance. In civil law systems, codified statutes predominate. In fact, many countries use a mix of features from common and civil law systems. Many common law countries have an English history or are still member of the commonwealth.

Civil law in many European nations could trace back to the code of laws compiled by the Roman Emperor Justinian. Countries with a strong Roman history and impact by the Catholic Church are typical for their civil law preference.

Civil-dominated countries provide a dedicated legal framework with less room for interpretations. Common law systems offer a wider range of interpretation of legal requirements, but in case of occurring damages compensations in product liability cases lead to intensive court trials.

1.11 Legal Regulation in China

The division standard (division "Guójiā Biāozhǔn", in German "national standard") forms the basis for product tests, to the products liable to CCC within the scope of the CCC certification. The CCC certification is comparable with the Ghost certification for Russia, this must be applied on defined products which are introduced in China. This is also valid for cars and their essential supply components. The division standard will give change from "Standardization Administration of the People's Republic of China", although this is a standard institute comparable with the German Institute of standardisation (DIN), they are made obliging, for example, by the commercial laws juridical (Table 1.4).

The original list is based on supply components in China VR were to be delivered. As for themselves in China own automobile industry developed ones these requirements became applied also on the new vehicles, which were manufactured in China.

The GBs and GB/Ts are harmonised with UN ECEs regulations. The GB/Ts for electric vehicles resemble in essential details of an amendment of the UN

Table 1.4 Table of China regulations related to safety

Product name	CNCA regulation	GB Standard
Whole vehicle homologation	CNCA-C11-01	GB/T 3730.1 GB/T 17,350 GB 7258 GB/T 15,089
Fire trucks	CNCA-C11-01/A1	
Motorcycle	CNCA-C11-02	GB 7258 GB/T 5359.1 GB/T 15,089
Motorcycle engine	CNCA-C11-03	
Seat belt	CNCA-C11-04	*GB 14,166*
Headlamp	CNCA-C11-07	GB 4599 GB 4660 GB 11,554 GB 5920 GB 15,235 GB 17,509 GB 18,409 GB 18,099 GB 18,408 GB 25,991
Front position lamp/rear position lamp, parking lamp, outline marker lamp, braking lamp		
Turning-signal lamp		
Reversing lamp		
Front fog lamp, rear fog lamp		
Rear license-plate light		
Side-marker lamp		
Rear-view mirror	CNCA-C11-08	*GB 15,084*
Interior trimming material: floor covering, seat shield, decorating scale boards (inside door shield/panel, front wall inner shield, side wall inner shield, rear wall inner shield, roof liner)	CNCA-C11-09	*GB 8410*
Door lock and door hinge	CNCA-C11-10	*GB 15,086*
Seat and head restraints	CNCA-C11-12	GB 15,083 GB 11,550 GB 13,057 GB 8410
Motorcycle helmets	CNCA-C11-15	GB 811 > /td >
Tire	CNCA-C12-01	GB 9746 GB/T-2978
Safety glass products	CNCA-C13-01	*GB 9656*
Restraining Devices for child occupants	CNCA-C22-03	*GB 27,887*

Source https://www.chinacertification.com/gbstandards/

ECEs. Nevertheless, there are also examples like the ISO 6469–4:2015 "Electrically propelled road vehicles—Safety specifications, with its parts:

Part 1: On-board rechargeable energy store system (RESS)
Part 2: Operational safety for vehicles
Part 3: Protection of people against electric blow (Electric security in 2018)
Part 4: Post-crash electrical safety

Now there are various GB/Ts (in the status as legal recommendations), which become reworked as mandatory GBs. Here the entire regulations around battery systems change to mandatory regulations, since fire becomes a significant issue due to growing electrification of the powertrain.

In the field of fuel-cell and hydrogen technology, China in cooperation with Korea and Japan working on various regulations and standards.

The Chinese regulations consider similar to the aviation industry a process that qualifies for applying for the rules and regulations according FAA or EASA; here one looks separately whether one can develop products or produce products for airplanes. For both, a supplier could be qualified separately. There are similar principles and criterion behind like automotive process according to ISO in 9001 and IATF, which defines criterion for the ability to provide qualified products in terms of quantity and quality of the relevant products.

There is a Rule 27 in the Chinese regulations, which defines a subset of GB and GB/Ts that are relevant for the qualification as manufacturer for electric vehicles. This qualification goes out in China the manufacturer's registration in Germany. The rule 27 defines besides driving tests, functional tests of the dedicated functions, system feature also fault-injection test. It is stipulated that the vehicles must fulfil certain reliability criteria what must be proved by prescribed driving cycles. For an entire vehicle homologation, a rule 44 defines the entire scope of rules and regulations, for electric vehicles.

The part certification should be implemented after the CNCA implementing regulations, because, otherwise, the products can be held on by the import to China with the customs duty.

In the coarse a licensing of a vehicle functions as also in Germany. Primarily, one must qualify as first as a manufacturer. Also in Germany, one cannot become just like that a manufacturer of a vehicle. As a German manufacturer one must be registered with the strength driving Federal Office with which, for example, a register is opened for the manufacturer in the vehicle register. Then, therefore, the strength driving Federal Office to whom also knows in Germany it to itself in the case of damage has to turn.

UN ECE regulations built also for China the basis for homologation. The basic idea is the same in all countries, there are requirements they must be fulfilled so that vehicles are suitable for the traffic. That means homologation addresses particular the "road-worthiness" of the transport media.

CCC for Complete Vehicles

All products mentioned in the official catalogue published by CNCA and AQSIQ require certification. In addition to the catalogue, the CNCA and the AQSIQ regularly announce new products that are subject to certification.

Due to the differences in testing criteria, there are four different types of CCC marks. The designated mark for safety is "S", electromagnetic compatibility is "EMC", electromagnetic compatibility and safety is "S&E" and fire protection is "F" (Fig. 1.9).

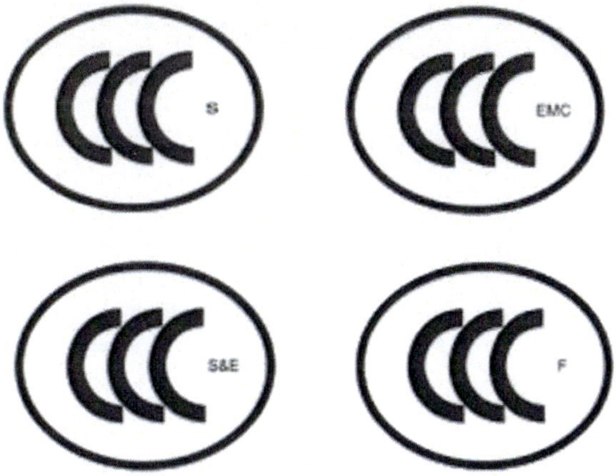

Fig. 1.9 Marking for the certification process according to CCC. *Source* The booklet, China Certification Corporation, www.china-certification.com/en, May in 2014

The CCC-Regulation also intend a certification of a complete vehicle. In particular, for electric vehicles, there is very extensive interpretation for a successful CCC homologation which must be applied. The CCC-marking is in the focus of trade law. However, the regulations behind are the same GBs and GB/s.

1.12 Insurance

Unless causes of accidents could not traced to cross-negligence, insurances are a valid solution for risk sharing. Of course, insurances are not a real risk-reduction measure for the society but for an individual company or enterprise an appropriate method to balance risk.

Comparison Aviation and a Regulation Attempt for Robot-Taxies

The Forbes Magazine published an interesting comparison from an insurance point of view. Here one has the governments, licensing methods and the interest-looked to the insurances in the traffic system in risk cover (Fig. 1.10).

A public authority can only define and release roads for automated transportation systems, like air corridors. Also due to the discussion with new flight corridors for automated air taxies, such combined transportation systems are more often discussed.

A regulation of the suitable traffic routes like in the aviation will go only by a state place be authorised or via a place those by the national authorities. Of course will feel in Europe in addition appoint the EU what will try to develop one also with the member states structured. The attempt on top from the OICA will be for the vehicle licensing also a conceivable basis. How it will go with the harmonisation of the traffic right, also has to go in Europe in other activities of the relevant EU directives to tired, otherwise automated driving will always remain an island solution. In this case, an operator model must be found in terms of product liability, since there will be no attractive business models for a pure manufacturer. Public transport goes to all areas

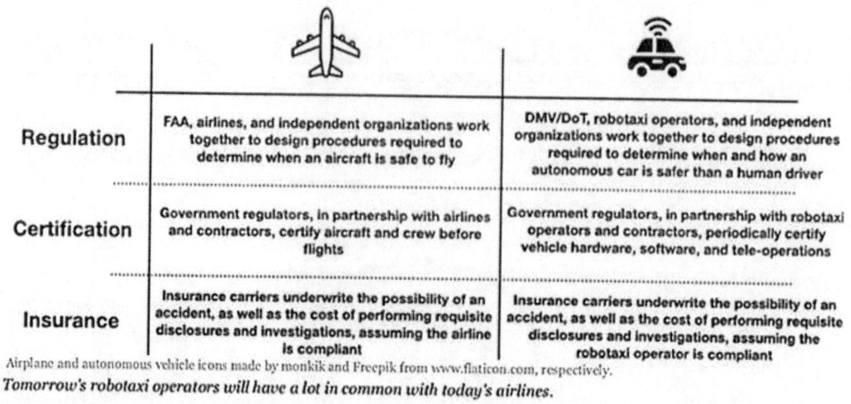

Fig. 1.10 Risk management comparison aviation and future Robot-Taxies *Source* The Next Billion-Dollar Businesses in Autonomous Cars_ Safety and Compliance (Forbes Magazine, 21.5.2018)

of the mobility whether navigations, Buss, roads or air traffic only about operator's models, individual traffic by autonomous vehicles goes juridical up to now in no industry.

These rules of the game, for the infrastructure, the vehicles and the operators must be specified and are monitored. So that these rules of the game are enough defined and are monitored, there must be an organisation, which admits the vehicles and the operators for the different spheres of traffic. Into English one speaks of certification (certification); however, it is in our usual language skills about the licensing for vehicles and operators in the respective sphere of traffic. Uber (American companies) also had to learn in Germany that to be able to operate a licensing needed around taxi companies one in Germany.

In the third aspect, it is about risks, which foresee as those not directly, but happen in every traffic system repeatedly. Because the risk can come for the singles fast to too much high financial burden, the risk is helped carry in all industries also about insurances. We know this in Germany for many decades, without liability insurance; no vehicle is admitted for the public traffic. One knows too many that insurances can demand back the damage sum or simply not with coarse carelessness pay need. Hence, the insurances will also have an essential influence on the security requirements, which are relevant for the automated driving. Whether the insurances have an effect in sum about the legislators in the licensing of operator and vehicles or find own individual regulations, will appear. However, without certain insurances, there will be able to be no operators or vehicle licensing.

Chapter 2
Risk Management

People associate risk always with negative effects. The Greek origin of the word Risk is also derived into many languages for the word Risks. Product risk describes the cross-product of possible damages and the probability of occurrence as any other dimension of risk. There are different opinions to the terms and definition of the risk in the economic literature. Definitions differ from definitions like "risks, by product faults and errors" how the mathematical definitions which are derived from "likelihood × severity" or vary from "danger of a variance of error" to the mathematical definition "risk = probability x severity". The general definition is as follows: the probability of damage or loss as consequence of a distinct behaviour or events; this refers to hazardous/dangerous situations in which unfavourable consequences may occur but do not necessarily have to. On the one hand, risk can be traced back etymologically to "riza" (Greek = root, basis); see also "risc" (Arabic = destiny). On the other hand, risk can be referred to "ris(i)co" (Italian); "The cliff, which has to be circumnavigated". "Safety" derives from Latin and could be translated as "free from worry" (se cura = without worry). Today, the topic of safety applies in various different contexts, for example, in regards to economic safety, environmental safety, admittance and access security but also in terms of work safety, plant or machinery safety and vehicle safety. The term safety varies significantly from just only functional safety.

Technical systems or products could be only considered to be sufficient free from intolerable risk. It is always a question, as with the scales of Justicia, of the balance between tolerable risk on the one hand and benefits and profitability on the other (Fig. 2.1).

Absolute freedom of risk could not be defined; especially people see also chances to win if they take higher risk than others do. Entire stock markets based on risk estimation with a certain probability between winning and losing.

The Oxford Business Dictionary provides the following definition: "Decisions to accept exposure or to reduce vulnerabilities by either mitigating the risks or replying cost-effective controls".

H.-L. Ross, *Safety for Future Transport and Mobility*,
https://doi.org/10.1007/978-3-030-54883-4_2

Fig. 2.1 Definition of safety

If risk lead to damages of other authority protect the weak parties. They provide basic legislation, so that no one could make profit by exploiting others.

Consequently, risk management became in any industry an essential discipline of management systems.

Dr. P.K. Gupta gave the following definition of risk: "Risk management is an integrated process of delineating specific areas or risk, developing a comprehensive plan, integrating the plan, and conducting the ongoing evaluation."

Insurers dealing commercially with that risk and they define rules to control and manage the risk of projects, products or any subject matter of insurance.

They ask questions like:

- What are the possible sources of loss?
- What is the probable impact of a loss should it at all occur?
- What should be done when a loss takes place?

Are damages even allowed or something should be done to avoid, minimise or mitigate them? The question of protection in the best possible way and the question of evaluating the future possibility of such events need an answer.

The probable expenditure or the economy of loss prevention, (it should be remembered that any extra expenditure for loss prevention would be economically justified so long the expenditure made is smaller than or at best equal to the savings made by way of loss reduction.

As already mentioned, in insurance, the risk is isolated from the whole business venture and the pure risk portion of it is assumed entirely by a different group of people of organisation (insurer) in a most technical, expert and economic way.

Basic risk management based on the steps that derived the definition:

- Risk evaluation,
- Emission and exposure control,
- Risk monitoring.

A systematic approach used to identify, evaluate and reduce or eliminate the possibility of an unfavourable deviation from the expected outcome and thus prevent the damages because of negligence and the loss of financial assets resulting from such damages.

Keeping all factors in view would come up the question of properly rating a risk, as this would be the basis of charging of quantifying risk. Mathematical valuation of risk becomes a major factor of risk management.

It is important to distinguish between products that meet standards, that are safe, and those that merely feel safe by the society.

For products quite often, the society recognizes products as safety since they are declared as compliant to a standard. The product or design meets applicable standards and practices for design and construction or manufacture, regardless of the product's actual safety history.

If the real world and the history have proven for the society that the product is safe. The compliance to a specify standard is not in the focus for the safe recognition.

Perceived or subjective safety refers to the users' level of comfort and perception of risk, without consideration of standards or safety history. Traffic signs or speed limits provide a confidence that other accepted the rules and traffic participants feel safe due to their experiences.

Social security or public safety, security addresses the risk of harm due to intentional criminal acts such as assault, burglary or vandalism. Society perceives security very much through public discussion. If we report on risks in the press quite often, then people also perceive a higher risk. However, it also leads to unease when rumours arise that certain risks are concealed.

Safety is often interpreted as implying a real and significant impact on risk of death, injury or damage to property. In response to perceived risks, many interventions may be proposed with engineering responses and regulation being two of the most common.

In the context of decision theory, the acceptance of risks is often referred to as the "degree of a positive or negative assessment of perceived risks". Risks and hazards are perceived, assessed, evaluated and accepted in very different ways. The risk acceptance decides by a reasoned induction to what extent the existing risk can be accepted or whether the cause responsible for triggering the risk can be considered unacceptable because the existing risk is too great. The acceptance of risks is significantly influenced by the perception of the risks. Risk perception is often understood as an everyday process by which people can assess and evaluate risks without the possibility of resorting to long data series and exact calculation models. Risk perception is often the intuitive or purely experience-based, unstructured perception of the possibilities for success and failure and of possible connections between actions and consequences. However, the subject area of risk acceptance and risk perception is complex. Therefore, it is not unusual to find various, sometimes even contradictory, explanatory approaches being argued.

Most common individual response to perceived safety issues is insurance, which compensates for or provides restitution in the case of damage or loss. Since the risk for people is not easy to be quantified, the insurance companies found many relations how to calculate the premiums for their insurance licences.

2.1 Risk Management Cycle

Risk management is always a question in what position the organisation, the project or even the product in the delivery chain.

In new mobility concepts, business models in all classical contexts should be taken into account. Especially if you deal with the infrastructure, authorities have to be considered as potential customers, or at least authorities are part of the decision-making process. Since motor vehicle manufacturer consider selling their cars via internet portals, service like ride-services or additional or temporary feature via cloud services, business to end-user (direct customer business) is getting more and more a beneficial business model for the future.

Consequently, typical risk management cycles for future mobility products require further analysis and decision-making steps.

Seven steps of risk management and their aspects of new mobility could be considered:

- Development of the context,
- Identification of risk,
- Assessment of the risk in the target context,
- Development of a strategy and treatment of potential risk,
- Development and defining of activities and their objectives,
- Development of an implementation and realisation strategy,
- Review and evaluation of the plan.

2.1.1 Development of the Context

Stakeholder analysis is always the first step in order to develop the context for business, an organisation, a project or a product. New services to mobility requires to consider not just any context, because most of the stakeholder see the product (product is that for that someone pays for) under very specific and very different views and perspectives (Fig. 2.2).

Establishing the context includes planning the remainder of the process and mapping out the scope of the exercise, the identity and objectives of stakeholders, the basis upon which risks will be evaluated and defining a framework for the process and agenda for identification and analysis.

2.1.2 Identification of Risk

After establishing the context, the next step in the process of managing risk is to identify potential risks. Risks are about events that, when triggered, will cause problems or reduce the expected profit.

Hence, risk identification can start with the source of problems, unknown impacts or with the problem itself.

Risk identification requires knowledge of the organisation, the market in which it operates, the legal, social, economic, political and climatic environment in which

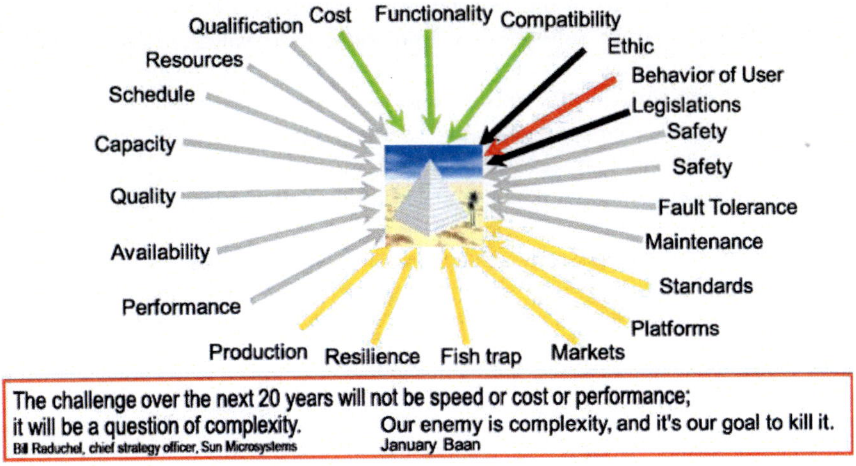

Fig. 2.2 Stakeholder of products. *Source* Driving forces of technical architectures (Source: IBM)

it does its business, its financial strengths and weaknesses, its vulnerability to unplanned losses, the manufacturing processes and the management systems and business mechanism by which it operates. This overall context based on a single-use product, mobility products or service are strongly depending on its operation context, which could be a specific operating system for software or specific operating condition for a system-function.

Any failure or deviation from the expected design of one of the mayor stakeholder at this stage to identify risk may cause a major loss for the organisation that intend to deal with the product or service.

Risk identification provides the foundation of risk management.

The identification methods are formed by systematic analysis of the stakeholder requirements and their expectations and the right to claim their right on, e.g. legal basis.

2.1.3 *Assessment of the Risk in the Target Context*

Once risks have been identified, they must then be assessed as to their potential severity of loss and to the probability of occurrence. Furthermore, the political, legal and moral impact in case of losses or damages need to be evaluated. Especially if we consider safety-related damages, product liability legislation could lead to significant compensations and even penalties. Risk–benefit calculations in case of damages to humans could end in bankruptcy and criminal lawsuit.

For any new product or service in a new context, the fundamental difficulty in risk assessment is determining the rate of occurrence since statistical information is not available on all kinds of past incidents.

Evaluating the severity of the consequences or the impact is often quite difficult for immaterial assets like software-based products. Asset valuation for immaterial products and their average of occurring damages are difficult to evaluate since cause and effect are not always sufficient transparent to lawyer, assessors or surveyor as well as the product developers.

Thus, best-educated opinions and available statistics are the primary sources of information, but always a significant remaining risk of failing with the valuation of the risk.

Risk assessment should produce such information for the management of the organisation that the primary risks are easy to understand and that the risk management decisions may be prioritised.

Quantifying of risk defines a lot of formulas based on cost–benefit or expected compensations, a typical theory based on the annualized loss expectancy (ALE) which is the product of the annual rate of occurrence (ARO) and the single loss expectancy (SLE) or

$$\mathrm{ALE} = \mathrm{ARO} \times \mathrm{SLE}$$

Most of them fail if a product has to be taken from the market because of product liability issues.

2.1.4 Development of a Strategy and Treatment of Potential Risk

Once the result of the risk assessment leads to continue with the product ideas, the question how to control the risk needs an answer.

Risk management strategies based on the following categories or combinations:

Risk Avoidance: Avoid the risk or the circumstances, which may lead to losses in another way, includes not performing an activity that could carry risk. Avoidance may seem the answer to all risks, but avoiding risks also means missing the potential gain that accepting or retaining the risk may consider. Not entering a business to avoid the risk of loss also avoids the possibility of earning the profits.

Risk Control: Risk can be controlled either by avoidance of causes of risk, by the propagation of the causes or by controlling losses or their impact. Control of risk always means considering an impact to the performance of the product or service or benefit reduction to the customer or user.

Risk Transfer: Risk transfer means that the expected party transfers whole or part of the losses consequential of risk exposure to another party for a cost. The insurance contracts fundamentally involve risk transfers. Apart from the insurance device, there are certain other techniques by which the risk may be transferred.

Risk Retention: Risk retention implies that the losses arising due to a risk exposure shall be retained or assumed by the party or the organisation. Risk retention is generally a deliberate decision for business organizations inherited with the following characteristics. Self-insurance and captive insurance are the two methods of retention.

In between risk transfer and risk retention, there are many options of development partners to share the risk or distribute the risk or the occurring compensations in case of their occurrence. Especially if the product or service based on a supply chain, adequate arrangements for risk management need to be negotiated.

2.1.5 Development and Defining of Activities and Their Objectives

Risk strategy often based on combination of measures to deal with the risk and the occurring damages. Decide on the combination of methods to be used for each risk. Each risk management decision should be recorded and approved by the appropriate level of management. Beside financial losses and possible criminal lawsuits for the organisation and individuals, the damage to one's image is not negligible. The risk management plan is a lifecycle document and need to be recorded until the product has been withdrawn from the market.

Provision or reserve for guaranties is necessary measures for any company, and these are positions in any balance sheet of stock market companies.

Caution with risk retention measures, their documentation could be a mayor argument in a product liability trail.

Law often requires field observation or continuous monitoring of the risk management measures for products.

2.1.6 Development of an Implementation and Realisation Strategy

Do as planed should be the next step. Implementation is not just doing, negotiations with development partners with the decision-making instance of the stakeholder, customers and user experiences will lead to mayor challenges as well as any design decision of the product will be criteria for the validity of the risk management plan.

Any new insight needs a process cycle through all steps of the risk management plan because new risk or new aspects and impacts of risk could question the validity of all valuations of the product strategy.

Purchase insurance policies for the risks that have been decided to be transferred to an insurer, avoid all risks that can be avoided without sacrificing the entity's goals, reduce others and retain the rest. There are even insurances versus development faults and occurring damages in case of recalls. However, any recall means higher insurance rates as it is with any insurance.

2.1.7 *Review and Evaluation of the Plan*

Risk management is a lifecycle approach until the end and decommissioning of the product. Even then occurring damages that are traceable to services or products could lead to compensation far beyond the product lifecycle.

Practice, experience and actual loss results will necessitate changes in the plan and contribute information to allow possible different decisions to be made in dealing with the risks being faced.

Risk analysis results and management plans should be updated periodically because.

- The control measures are still applicable and sufficient effective, and
- Possible risk level changes in the business environment or the context of the product or service.

The risk management plan is a mayor artefact of the product lifecycle history.

In the publication from the EU about "EU general risk assessment methodology", the steps are very similarly described, but they focus more on the dedicated product development (Fig. 2.3).

The paper seems to be typically developed for a stand-alone product, but not for functional units, which are part of complex system and strongly context depended like a road vehicle.

2.2 Technical Risk

The definitions of risks and the kind of risks are very different in many industries. Therefore, there are also in the different industries very different approaches like those that risks are to be avoided and which risks are tolerated by the society.

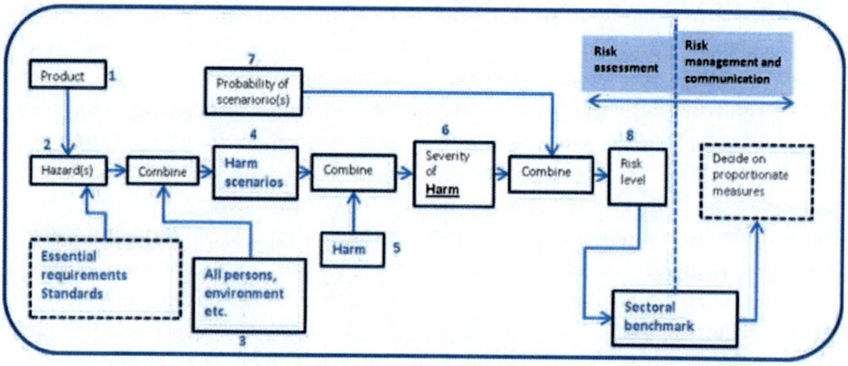

Fig. 2.3 Overview of the risk methodology for a product. *Source* COMMISSION 2013

EU published a basic document abut risk of technical products:

The Following Terms Are Key for the Purpose of This Methodology:

- Risk: combination of the probability of occurrence of a hazard generating harm in a given scenario and the severity of that harm.
- Harm: injury or damage to the health of people or damage to property, economic damage to consumers, damage to environment, security and other aspects defined in the scope of New Approach Directives.
- Hazard: potential source of harm. The hazard, or danger, is intrinsic to the product. Probability of occurrence of that harm: the likelihood of the harm occurring.
- Risk level: Degree of risk, which may be 'serious', 'high', 'medium' or 'low'. When different levels of risks in different scenarios have been identified, "the risk" of the product is given by the highest risk.

Source COMMISSION 2013.

In railway industry, a risk matrix consists of the two classes:

- Hazard frequency classification
- Consequence and severity classification.

It is just the confirmation for the risk definition in any other industry. For example, railway uses a very similar approach (Fig. 2.4).

The figure shows the system definition in the application condition as the entry point of the lifecycle assumptions.

The safety integrity for the railway, the mechanical engineering, power station technology as well as now also for the automotive industry is derived since end of the last millennium of the IEC 61508. Nevertheless, in these standards only the risks, which derive from malfunctions of electrics and electronics, are looked.

The railway standard EN 50126 "specifications and evidence of the reliability, availability, maintainability and safety (RAMS)" defines the discussed terms as follows:

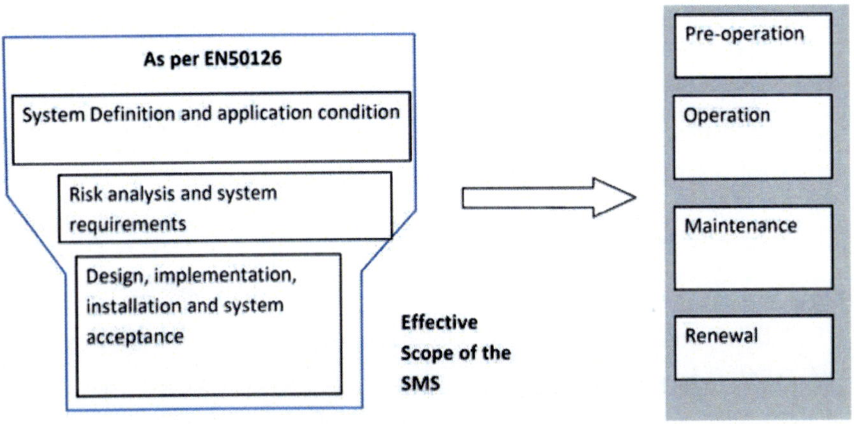

Fig. 2.4 Railway risk approach similar to EN50126. *Source* EN50126 2016

RAMS is an acronym for reliability, availability, maintainability and safety.

- Reliability: Probability that an item can perform a required function under given conditions for a given time interval.
- Availability: Ability of a product to be in such a state to perform a required function under given conditions at a given time interval.
- Maintainability: Probability that a given active maintenance action, for an item under given conditions of use can be carried out within a stated time interval.
- Safety: Freedom from unacceptable risk of harm.

In the aviation industry, there are two certification bodies, European Aviation Safety Association (EASA) and Federal Aviation Association (FAA) are accredited to make the necessary rules for air traffic. The FAA expels to CFRs (the US Civil Flight Regularisations) which prescribe the risk classification, measures et cetera for the different flight systems. EASA as well as FAA publish the so-called Certification Standard (CS) or Advisory Circular (AC), which give methods and rules for the safety assurance case of the respective systems and the relevant airplane classes.

CS (Certification Standards, AESA definition) and AC (Advisory Circular, FAA definition) are using ARP 4761 "GUIDELINES AND METHODS FOR CONDUCTING THE SAFETY ASSESSMENT PROCESS ON CIVIL AIRBORNE SYSTEMS AND EQUIPMENT", as certification standard. They describe the assessment process and the methods to be applied for civil flight systems. The activities, comparable with the software and hardware part of safety standards from other industries, are for example DO 254 (electronics hardware) or DO 178C (software). Typical design principles for the different parts or systems in the airplane or the requirements for the integration environment mostly base on the ATA-Regulations (ATA = Air Traffic Association). Technical communication systems, however, also the EE-architecture of flight control units are published by a company so-called ARINC standards (Aeronautical Radio Incorporated from Anna's polity, Maryland, the USA).

Some further notifiable standards are:

- SAE ARP4754A Guidelines for Development of Civil Aircraft and Systems, 2010,
- RTCA DO-278 Guidelines for Communication, Navigation, Surveillance and Air Traffic Management (CNS/ATM) Systems, Software Integrity Assurance, 2002,
- RTCA DO-278A Guidelines for Communication, Navigation, Surveillance and Air Traffic Management (CNS/ATM) Systems, Software Integrity Assurance, 2011,
- RTCA DO-330 Software Tool Qualification Considerations, 2011,
- RTCA DO-331 Model-Based Development and Verification Supplement to DO-178C and DO-278A, 2011,
- RTCA DO-332 Object-Oriented Technology and Related Techniques Supplement to DO-178C and DO-278A, 2011,
- RTCA DO-333 Formal Methods Supplement to DO-178C and DO-278A, 2011.

ISO 13824:2009, bases for design of structures—general principles on risk assessment of systems involving structures specifies general principles on risk assessment of systems involving structures. The focus is on strategic and operational decision-making related to design, assessment, maintenance and decommissioning of structures (Fig. 2.5).

Very interesting standard is ISO/IEC 31010—Risk management—Risk assessment techniques. It defines the following steps:

- Identifying the risk and the reason for its occurrence.
- Identifying the consequences if the risk occurs.
- Identifying the probability of the risk occurring once more.
- Identifying factors that reduce the consequences or probability of the risk.

For risk management, it addresses to the steps in ISO 31000. Very interesting are the assessment methods, which are listed in the attachment (Fig. 2.6).

The examples from the standards and the proposal from the European Commission (EC) show very similar principles. How to distinguish commercial risks, society risks and technical risks can only be shown in the target context by a dedicated analysis.

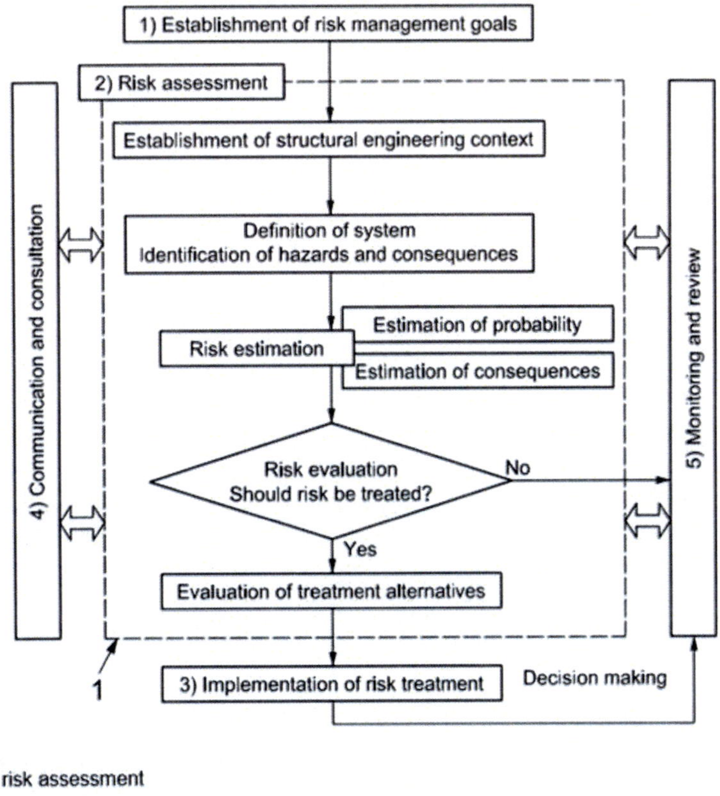

Fig. 2.5 Risk assessment process and scope of risk assessment of system involving structure. *Source* ISO 13824:2009, Bases for design of structures—General principles on risk assessment of systems involving structures

1. Brainstorming	16. Cause and consequence analysis
2. Structured or semi-structured interviews	17. Cause-and-effect analysis
3. Delphi method	18. Layer protection analysis (LOPA)
4. Checklist	19. Decision tree
5. Preliminary hazard analysis (PHA)	20. Human reliability analysis (HRA)
6. Hazard and operability study (HAZOP)	21. Bow tie analysis
7. Hazard analysis and critical control points (HACCP)	22. Reliability centered maintenance
8. Toxicity assessment	23. Sneak circuit analysis
9. Structured What If Technique (SWIFT)	24. Markov analysis
10. Scenario analysis	25. Monte Carlo simulation
11. Business impact analysis	26. Bayesian statistics and Bayes nets
12. Root cause analysis	27. FN curve
13. Failure mode and effects analysis (FMEA)	28. Risk index
14. Fault tree analysis	29. Consequence/probability matrix
15. Event tree analysis	30. Cost/benefit analysis
	31. Multi-criteria decision analysis (MCDA)

Fig. 2.6 List of 31 risk assessment methods. *Source* ISO/IEC 31010, Annex B, 2009

The context for a product development is composed as follows:

- What is the purpose of the product?
- How should the product be used?
- In which areas should the product be used?
- Which promises of benefits do I offer the user?

The intended purpose for the product, the way in which the product is intended to be used and what is the intended scope and environment for the product are major input for a technical risk analysis.

2.3 Risk in Road Traffic

One of the mayor concerns for manual and automated driving is that people do not behave, as expected in the various heterogeneous road traffic situations. Humans could be in various roles in the different road traffic scenarios. If we now develop new automation function and involved people do not understand the context for the new functionality, it leads already to risk. The secondary concern is that external impacts or combinations of impacts lead to dangerous situations occur, where humans react erroneous or wrong.

The first risk of driving is the vehicle itself. Moving a mass of several thousand kilograms in public traffic environment means risk for other road traffic participants and the occupancy in the vehicle.

In the first glance, risk seems to be very heterogeneous and complex.

In an enlightened society with responsible forms of government, the legislation is the framework for the security of society. This should be also the case for the road traffic environment.

In any nation, the dedicated road traffic act is the basis for safety in road traffic environment. Any violation of this legal framework leads to risk for people or the environment.

Consequently, the road traffic act and derived regulations provides the level of tolerable risk for the road traffic. In order to avoid risk due to unknown scenarios the product liability acts provide regulations and penalties to compensate damages due to unforeseeable product risk. Any cross-negligence usually additionally addresses the dedicated criminal law.

The German Road Traffic Act (StVG) consists of the following parts:

1. Traffic regulations
2. Liability
3. Provisions on penalties and fines
4. Driver aptitude register
5. Vehicle register
6. Driver license register
   6a. Data processing
7. Common rules and transitional provisions.

The traffic regulations provide the rules for any road traffic participant how to assure a common safe behaviour in the road traffic environment.

The liability chapter addresses the responsibility of any party involved in road traffic like the driver, owner and the producer of the vehicles.

Provision on penalties and fines addresses all road traffic participants, also pedestrians, cyclists, etc.

The driver aptitude register addresses not only the driver's abilities, but it also rules how the register is managed in Germany.

The vehicle register records all German vehicles and specific data for example the owner of the vehicle.

The drivers' licences register records all holders of Germans driving licences and rules how to achieve and how the permission to drive could terminate.

Common rules and transitional provisions addresses all preliminary or particular cases and the application of regulations for foreign drivers and vehicles.

Any player and any party, which could be involved, are somehow addressed in the Road Traffic Act unless a relevant case could be considered.

For any credible accident scenario or case of occurring significant damages, you could find a law in the German Road Traffic Act. If that is not the case the paragraph, one could be applied:

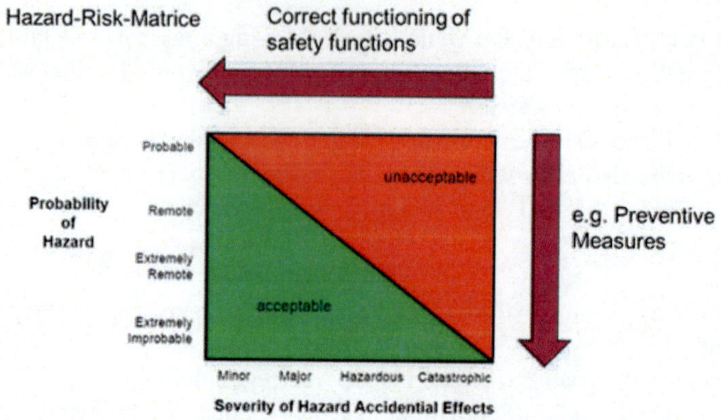

Fig. 2.7 Area of risk

> §I. General Traffic Rules
>
> Section 2.1 Basic Rules
>
> (1) Use of the Road Requires Constant Care and Mutual Respect.
>
> (2) a Person Using the Road Shall Act in Such a Way as not to Harm or Endanger or, More Than is Unavoidable in the Circumstances, to Hinder or Inconvenience Any Other Person.
>
> *Source* German Road Traffic Act 2017.

Assumption no. 1: A society is in the area of tolerable risk, if anybody in road traffic follows conservatively the given road traffic act.

Assumption No. 2: Humans follow the given regulations, if not they get penalties and pay for the occurring damages (Fig. 2.7).

The basic level of tolerable or acceptable risk from a society point of view, all human actors behave as expected by law and all involved systems operate or behave as specified, or correct and show no dysfunctional behaviour and are free of fault, errors or failure.

In the first iteration, a risk analysis, correct functioning of any involved system element should be considered in order to avoid too much complexity. In case of governmental approval or authorisation, the correct design and production could be part of the approval, but the safety assessment based usually on a correct functioning system.

2.3.1 Causes of Technical Risk

In mobility industry, product risks dominate the risk management. Even with immaterial goods like software, the technical risk is in the focus of vehicles, infrastructure and their environment. Since most of the products are goods with technical properties, features or functionality which are sold to various customers. Product liability is just one aspect of product risk that products simply are not meeting the stakeholder requirements becoming more and more driver of product development, distribution and production. What is the benefit of a driver assistance system, where the driver is impacted with even more risk than without such a system? If people get a move in a comfortable environment at the desired time to the right place, individual driving becomes a hard completion. All disruptive technologies inherent a very significant risk, and all proven technology and safety principles are questioned as well. Even if a product based on the same design and maybe even in the same technical context, how people use it and the effects of unintended use are not foreseeable.

There are very different causes for technical risk, depending on in which context they appear and in what relation people are in different situations, conditions or state of operation, etc., could lead to many kinds of different risk scenarios.

Based on the physical effects, cause could derive from following sources:

- Chemical reaction could lead to fire, explosions, cauterize, poisoning, etc., and could have impact to health and environmental damages.
- Toxic materials could lead to direct and indirect health risks, how poisoning (also carbon monoxide), injuries (also as a result of degassing of batteries, interference of the driver or workshop personnel), other damages, etc.
- High voltage could lead to thermal impacts and injuries in case of touching (touch protection).
- Radiation does not just cause impacts to semiconductor due to alpha particle, and high-frequency radian is also well discussed under health aspects.
- Thermionic could cause damages by overheating, burn, fire, smoke, etc.
- Kinetics is not just a topic of occupancy protection; the fact that moveable objects are kinetic energy that need to be controlled means a challenge for any mobility solution (Fig. 2.8).

A common way to classify hazards is by category:

- Biological—bacteria, viruses, insects, plants, birds, animals and humans, etc.,
- Chemical—depends on the physical, chemical and toxic properties of the chemical,

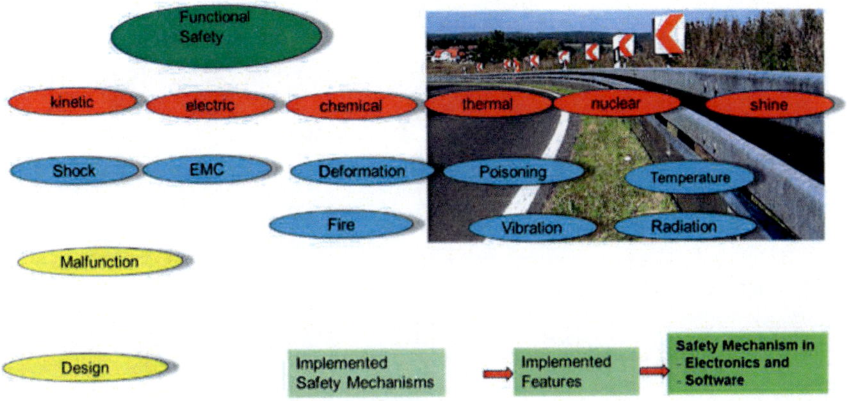

Fig. 2.8 Physical effects as sources for technical risk

- Ergonomic—repetitive movements, improper set up of workstation, etc.,
- Physical—radiation, magnetic fields, pressure extremes (high pressure or vacuum), noise, etc.,
- Psychosocial—stress, violence, etc.,
- Safety—slipping/tripping hazards, inappropriate machine guarding, equipment malfunctions or breakdowns.

The definition of "safety" is closely related to functional safety, but the other parameter are not covered with functional safety at all.

2.3.2 Control of Technical Risk

There are two basic principles how to control risk. The basic idea of functional safety is to implement further functions in order to control the possible sources of risk. This could be solved by means of a fence, so that faults could not propagate or leave their dedicated environment (Fig. 2.9).

The risk-causing object is behind a barrier, and unless people do not override the barrier (opening the fence or goes inside) or the barrier fails (dog jumps over the fence), the safety mechanism could be considered as sufficient safe. A fault in the barrier does not ultimately lead to a dangerous situation. The safety mechanism like a fence or a dog leash could be made of mechanical material or in electronics. An electric fence or an electronic ankle bracelet can support the mechanical function, monitor or independently represent a protective effect.

The principle alternative is reducing the possible degree of freedom or the distance, the energy, space of movement, etc. In this case, a critical threshold needs to be evaluated since what physical parameter (e.g. the length of the leach) risk could occur. Requirements on reliability of the safety mechanism (the leach and the fixation)

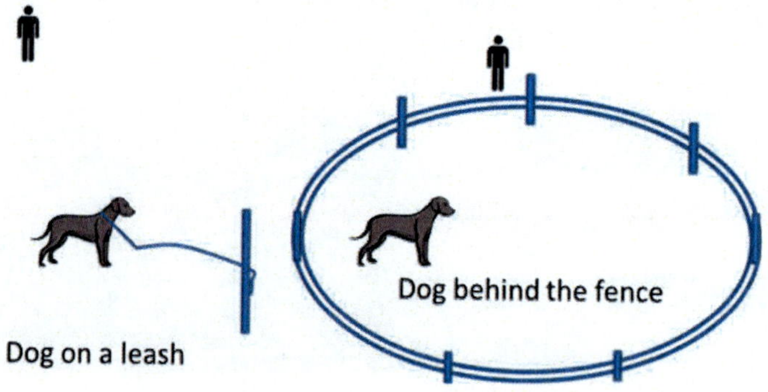

Fig. 2.9 Principles of safety mechanism

Industry	Automotive	Railway	Machinery	Medicine	Process	Energy	Aviation	Military
Society Safety	UN, EU-Directives, Civil Law, Road Traffic regulations	EU-Directives	Machinery Directive (2006/42/ec)	UN, US HHS, EU-Directives, FDA	UN, EU-Directives, Seveso-Directive	EU-Directives	UN, ICOA, IATA	UN, national law
Risk	ISO 26262	EN 50126 (RAMS)	ISO 12100, HSE	2001/95/EC, ISO 14971, FDA	RAPEX Guidelines, various EU-Directive (ATEX), ISO 13702, IEC 61511,	RAPEX Guidelines, 29 CFR 1910.269, ISO 31000	14 CFR 25.1309	MIL-STD-882E
Hazard Ident.	ISO 26262		ISO 12100		IEC 61511			
Worthiness	Homologation, FMVSS, UN ECE	Rolling Stock, ATOC, RIS-1530-PLT	mobile machines etc. defined in machinery directive				EASA or FAA regulations, (EU) No. 748/2012	EMACC Handbook
Control of Risk	ISO 26262. ISO 21448, ISO 21434, IATF 16949	EN 50126/28/29	harmonised standards, ISO 13849		IEC 61511	IEC 61513	ARP 4754A, ARP 4761	MIL-STD-338B
Safety Integrity	ISO 26262		ISO 13849, ISO 25119	IEC 61508, IEC 60601, IEC 62304,	IEC 61511, EN 50495,	IEC 61513	EUROCAE ED-12B, DO 160, DO 178C, DO 254	Defence Standard 00-56
Design Principles	EGAS, NHTSA-, VDA-Publications, ISO 6469		DIN EN 61784			IEC 62351	Certification standards, advisory circular	MIL-STD-217F

Fig. 2.10 Source of safety requirements in other industries

depends on the criticality of the dangerous cause (dog is always aggressive or only aggressive in case provoking). If special event (provoking) initiates the critical situation, the safety mechanism shall be effective in case of the specific event (demand mode). If the dangerous situation immediately occurs if the safety mechanism failed, higher reliability over time is requested because it must be continuously effective (continuous mode).

An alternative for such safety mechanisms is to develop a safe design (train the dog not to attack people); this is the principle of the safe passenger compartment of today's vehicle. During development, risk is the design-driving factor (risk-based design). Any neglected risk leads to design changes or implementation of additional risk reducing measures (Fig. 2.10).

There is no general principle how risk to be identified or where the society expectation on tolerable risk are defined. Any industry has their own rules and legislations or authorities and organisations for rule-making.

In the various industries, the degree of freedom, which is given, based on historical impact. If voluntary commitments or existing organisations preventively and sufficiently prevent causes of risk, legislation does not regulate risk and safety measures unless their effectiveness is not questioned.

A further driver of preventive risk management is the "United Nations, UN"; they provide a complete a basic framework to assure their understanding of "safety" (Fig. 2.11).

The five columns have their specific range, but how UNO, WHO, FAO, UNESCO or ILO affecting the various industries could not made transparent. Since the World Bank provides credits for many projects of the other organisations, the so-called self-governing bodies or implementing bodies are more directly linked to dedicated industries.

- ICAO—is a mayor driver of rule-making in the avionics industry
- WTO/OECD—harmonises trade mechanism, which is often linked to safety assurance measures of the importing bodies in many countries (e.g. CCC for China or GHOST for Russia)
- IMO—The maritime body provides several code of practice for maritime safety measures (IMO GAS Code, IGC for gas carrier)

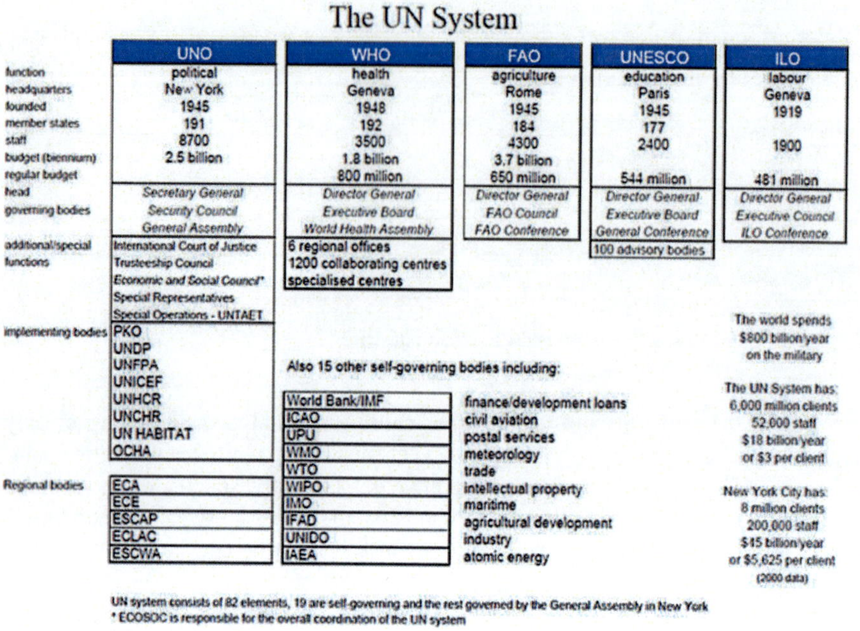

Fig. 2.11 UN, the public safety risk management framework. *Source* UN publication, WHO_strategy_safety_risk_management

- UNIDO—supports many projects as well in the relation to the Seveso-Directive
- OCHA—provides guideline for labour protection as well of protection of people in public road traffic
- IAEA—provides basic guidance for atomic or nuclear energy

Until today, the principles of individual traffic remains in the framework of UN ECE or the principles of FMVSS (US) or CMVSS (Canada), with their basis in the "Vienna Convention".

As more public road traffics get connected and even a passenger car becomes more, a mobile people transporter basic rules and practices from other transportation systems are not negligible.

This example from the legal view on technical risk management follows and a break down to product classes, industries and markets.

From standardisation point of view, the interface between risk and safety is ambiguous. Many standards derive from IEC 61508, and new standards for safety and risk are established in the various industries.

In medicine industry, risk is more associated to the hazards caused by non-technical impact like biological, physiological and chemical, etc.; however, IEC 61508 is the basis for integrity of EE-systems (Fig. 2.12).

It is recognized that there is a great variety of E/E/PES applications in a variety of application sectors and covering a wide range of complexity, hazard and risk potentials. In any particular

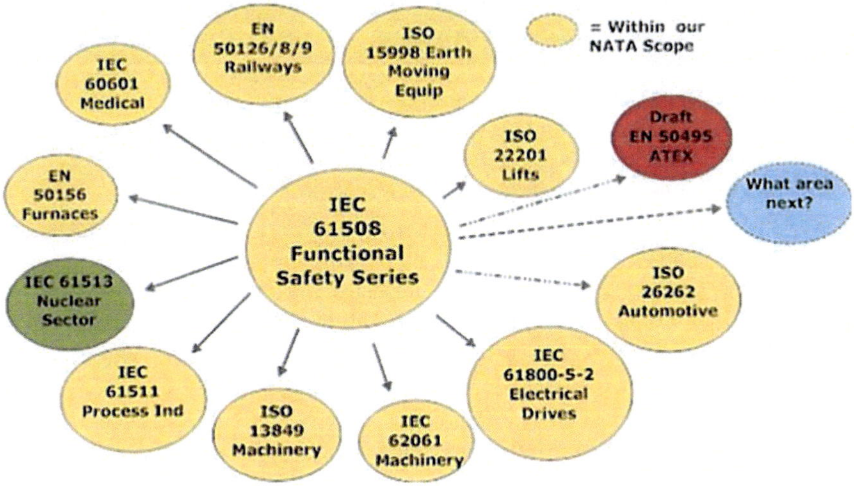

Fig. 2.12 Standards derived from IEC 61508

application, the required safety measures will be dependent on many factors specific to the application. This International Standard, by being generic, will enable such measures to be formulated in future application sector international standards.

Source IEC 61508, 1998.

IEC 61508 had been originally developed to derive other-sector specific standards, there for basic principles could be recognised in many other standards.

2.4 Standardisation of Safety for Road Vehicles

With the publication of ISO 26262, functional safety becomes the leading focus of safety in automotive industry. Other industries were with standardisation partially 20 years or how the aviation and astronautics roughly half a century ahead.

Society and legal frameworks as well as automotive customers, manufacturers, workshops and dealers demanded more and more functionality based on EE-functions for motor vehicles. Vehicle architecture, control units and all the control loops within the vehicle become more and more complex. The dependencies and the causes for failure could not anymore identified. The repair workshops could only change entire control units and flashing software updates, but the existing electronics could not anymore repaired. Mechanical engineering did dominate in the last millennium the vehicle design and electronic were just a supplementary technology.

The different systems in the vehicle should become substantially more dynamic and provide more performance by shifting the functionality closer to the design limits. Without electronic protective measures like Anti-Blocking System (ABS), further electronic occupancy protection systems like airbags and the growing traffic density, safety standards became unambiguous.

With the growing number of the vehicles and the number of the active safety systems in the respective vehicles also grew the number of the perceived electronics errors. Up to now, the de-energised state could be accepted as a "Safe State" for electronic systems because the driver was a fallback solution by law in the car.

With the new millennium, discussions start about steer- or brake-by-wire systems. The first and initial target for "by-wire-systems" was to replace the hydraulic, optimization of the vehicle compartment and simplifying the assembly of the vehicle. Steering systems replaced the hydraulic by an electronic power-pack, but the steering column could not be replaced. Most of the "Active Front Steering" Systems and "Rear Axle Steering" System had been significantly degraded by performance due to safety concerns. One of the concerns was that the driver as fallback could not react in case of failure because the feedback from the street was missing. In many driving situations, the driver could not perceive in time any malfunction from those steering systems, before the vehicle left significantly the drive corridor.

Another issue was that with the hydraulic many inherent protection mechanisms had been established, which were not so easy designing with the same performance by electronics. ABS was in the beginning a protection mechanism for the brake system. Hydraulic brake systems did control the brake pressure by orifices in the hydraulic lines; with higher brake pressure and the demand of higher pressure gradients, orifices fail in case of various road conditions. Blocking wheels lead to loss of steering and especially on the rear axle to sudden loss of stability. Depending on drive trajectory, brake wear-out, tire conditions, road surface, etc., such pressure gradients could not controlled by orifices or by the design of chassis. ABS provided a solution to prevent from locking wheels due to high brake pressure. Nevertheless, since today such protection mechanism, like ABS just prevent design inadequacies from hydraulic brakes but they are less effective for other cause of vehicle destabilisation. Today, it is still "state of the art" to use hydraulic brake systems instead of pure electronic brake systems. Some appearances with electromechanically brakes (EMB) become never robust enough for mass production and availability in case of power failure, precision and adaption on road conditions as well as synchronisation of all four wheel-brakes was always a challenge. Alternatively, to friction brakes eddy-current brakes are well discussed in the ages of E-mobility, but to safeguard the standstill of the vehicle, no real alternatives to friction brakes are in consideration.

The most overdue disruption in vehicles is the change in vehicle communication. In 2020, state-of-the-art communication is still CAN-BUS (Controller Area Network), which is anything else than a network solution. In the typical Golf-class, there are still about 40 control units communication via CAN.

If any car manufacturer and all supplier just developing on just for the big vehicle platforms, we could not change technology and not finance such a change. What solution would be compatible to others? What solution provides sufficient confidence in safety? There is no way out, standardisation is the answer since the beginning of the industrial era.

ISO 26262 was just a start with some basic principles of system safety engineering; many standards need to follow, before we find our solutions for new mobility. ISO

26262 targets only the risk due to malfunctions of E/E-Systems or their elements. Active safety: any kind of protection mechanism or even a basic control function like steering was not in the focus of the standard. The effectivity, effectiveness or sufficient performance of those systems developed evolutionary based on industry capability and had been supported by the legal framework.

Fulfilling the legal requirements is not enough, unless any motor vehicle follows their own strategy to protect any road traffic participant from accidents. Common approaches are necessary to synchronize in a safe manner all road traffic participants.

Current discussions on "SOTIF—Safety-of-the-Intended-Functionality" and the necessary operational safety in various standardisation committee need to be harmonised.

Iso/pas 21448:2019

Road Vehicles—Safety of the Intended Functionality

Abstract

The absence of unreasonable risk due to hazards resulting from functional insufficiencies of the intended functionality or by reasonably foreseeable misuse by persons is referred to as the "Safety Of The Intended Functionality (SOTIF)". This document provides guidance on the applicable design, verification and validation measures needed to achieve the SOTIF. This document does not apply to faults covered by the ISO 26262 series or to hazards directly caused by the system technology (e.g. eye damage from a laser sensor).

This document is intended to be applied to intended functionality where proper situational awareness is critical to safety, and where that situational awareness is derived from complex sensors and processing algorithms; especially emergency intervention systems (e.g. emergency braking systems) and Advanced Driver Assistance Systems (ADAS) with levels 1 and 2 on the OICA/SAE standard J3016 automation scales. This edition of the document can be considered for higher levels of automation, however additional measures might be necessary. This document is not intended for functions of existing systems for which well-established and well-trusted design, verification and validation (V&V) measures exist at the time of publication (e.g. Dynamic Stability Control (DSC) systems, airbag, etc.). Some measures described in this document are applicable to innovative functions of such systems, if situational awareness derived from complex sensors and processing algorithms is part of the innovation.

Intended Use and Reasonably Foreseeable Misuse Are Considered in Combination with Potentially Hazardous System Behaviour When Identifying Hazardous Events.

Reasonably foreseeable misuse, which could lead directly to potentially hazardous system behaviour, is also considered as a possible event that could directly trigger a SOTIF-related hazardous event.

Intentional Alteration to the System Operation is Considered Feature Abuse. Feature Abuse is not in Scope of This Document.

Source 22/SC32, 2019.

ISO/PAS 21448 does not address actuators like DCS (BMW abbreviation for ESC), but for any higher automation level than 2, where the EE-System does not act just like an electronic stability control system (ESC) or an electric power steering system (EPS) also the actuator need to provide new functionality. Most likely, the actuator will be at least partly controlled by wire and not anymore by a human driver. Current updates of the standard are in progress.

2.4.1 *Risk and Integrity Definition from the IEC 61508*

IEC 61508 was also the basis for ISO 26262, but not all aspects had been derived from the same point of view. The focus of ISO 26262 where software and electronics in road vehicles. First discussions even questioning the need of a functional safety standard, because the established quality management and related methods like technical risk analysis by FMEA (Failure Mode and Effect Analysis) or FTA (Fault-Tree-Analysis) or the assignment of "Special Characteristics" covers also risk derived due malfunctions of electrical and electronic systems.

In the first revision of IEC 61508, the standard defines its understanding of risk in part 5. The first viewpoint was to identify the role of electronic, electric and programmable electronic system (E/E/PE safety-related system) which are considered in their scope.

> A.3 Role of E/E/PE Safety-Related Systems
>
> E/E/PE Safety-Related Systems Contribute Towards Meeting the Necessary Risk Reduction in Order to Meet the Tolerable Risk. A Safety-Related System Both
>
> - implements the required safety functions necessary to achieve a safe state for the equipment under control or to maintain a safe state for the equipment under control, and
> - is intended to achieve, on its own or with other E/E/PE safety-related systems, other technology safety-related systems or external risk reduction facilities, the necessary safety integrity for the required safety functions (3.4.1 of IEC 61508-4).
>
> NOTE 1—The first part of the definition specifies that the safety-related system must perform the safety functions which would be specified in the safety functions requirements specification. For example, the safety functions requirements specification may state that when the temperature reaches x, valve y shall open to allow water to enter the vessel.
>
> NOTE 2—The second part of the definition specifies that the safety functions must be performed by the safety-related systems with the degree of confidence appropriate to the application, in order that the tolerable risk will be achieved.
>
> *Source* IEC 61508-5:1998, Chapter A3.

These two examples show how the considered solutions how the standard considers, that an E/E/PE safety-related system provide risk reduction. In the first example, the safety function is an inherent part of the considered operational intended process, which could for example leading to reduction of use of other constraints for example. The second part is an additional implementation of additional safety-related functionality which could limit, or degrade etc. the intended use or performance of the process. For the second example, the standard proposes a safety-related system.

There might be and there are other solutions, but those are the 2 cases the standard did considered for the application of the IEC 61508.

Beside the electronic system, also people are considered which could play a role within the operation of the dedicated system in the intended environment.

> A person could be an integral part of an E/E/PE safety-related system. For example, a person could receive information, on the state of the EUC, from a display screen and perform a safety action based on this information.

E/E/PE Safety-Related Systems Can Operate in a Low Demand Mode of Operation or High Demand or Continuous Mode of Operation (See 3.5.12 of IEC 61508-4).

Source IEC 61508-5:1998, Chapter A3.

The person considered in IEC 61508 are just an example that this person monitors and react based on this information. A driver of a vehicle could be also considered as a operator who receives information from systems and the environment and reacts on those information or by his own intend.

The standard also considers continuous operating systems and systems operating on demand. A steering system would be usually continuously operated by a driver, and a brake system usually would be activates based on the drivers demand.

In order to achieve necessary risk reduction by E/E/PE-systems the standard defines the term "Safety Integrity".

A.4 Safety Integrity

Safety integrity is defined as the probability of a safety-related system satisfactorily performing the required safety functions under all the stated conditions within a stated period of time (3.5.2 of IEC 61508-4). Safety integrity relates to the performance of the safety-related systems in carrying out the safety functions (the safety functions to be performed will be specified in the safety functions requirements specification).

Source IEC 61508-5:1998, Chapter A4.

IEC 61508 defines safety integrity as a probability to perform a required safety function. This provides a basis for a quantitative approach for risk reduction by safety-related functions. Unfortunately, ISO 26262 has not provided any adequate definition for EE-systems in road vehicles.

Systematic safety integrity; that part of safety integrity relating to systematic failures in a dangerous mode of failure (see 3.5.4 of IEC 61508-4). Although the mean failure rate due to systematic failures may be capable of estimation, the failure data obtained from design faults and common cause failures means that the distribution of failures can be hard to predict. This has the effect of increasing the uncertainty in the failure probability calculations for a specific situation (for example the probability of failure of a safety-related protection system). A judgement therefore has to be made on the selection of the best techniques to minimise this uncertainty. Note that it is not necessarily the case that measures to reduce the probability of random hardware failure will have a corresponding effect on the probability of systematic failure. Techniques such as redundant channels of identical hardware, which are very effective at controlling random hardware failures, are of little use in reducing systematic failures.

Source IEC 61508-5:1998, Chapter A4.

The standard explains, that systematic failures and common cause failures are difficult to predict by implemented design measures. Protection measures, their effectiveness and relation to systematic failure and situations provides uncertainties not only for the potential or for the degree of risk reduction. An occupancy protection mechanism like an airbag is a typical example. Even today, due to accidents, the technology require continuous optimisations since airbags had been introduced for passenger cars. The standards stated also, that homogeneous hardware provides little potential to reduce systematic failure, why it is not the same for homogeneous software, would remain a secret of the authors of the standard. Using dissimilar software on homogeneous redundant hardware could provide an advantage to control random hardware faults and systematic failure in the hard- and software.

> The Required Safety Integrity of the E/E/PE Safety-Related Systems, Other Technology Safety-Related Systems and External Risk Reduction Facilities, Must Be of Such a Level so as to Ensure that
>
> - the failure frequency of the safety-related systems is sufficiently low to prevent the hazardous event frequency exceeding that required to meet the tolerable risk, and/or
> - the safety-related systems modify the consequences of failure to the extent required to meet the tolerable risk.
>
> *Source* IEC 61508-5:1998, Chapter A4.

The standard considers certain levels of safety integrity to reduce the necessary risk, by E/EPE, external measures and measures of other technology. It addresses the frequency of the safety-related system as to be sufficient low and the ability of the safety-related system to influence the consequences of failure.

> Figure A.1 illustrates the general concepts of risk reduction. The general model assumes that.
>
> - there is an EUC and an EUC control system;
> - there are associated human factor issues;
> - the safety protective features comprise
> - external risk reduction facilities,
> - E/E/PE safety-related systems,
> - other technology safety-related systems.
>
> NOTE—Figure A.1 is a generalised risk model to illustrate the general principles. The risk model for a specific application will need to be developed taking into account the specific manner in which the necessary risk reduction is actually being achieved by the E/E/PE safety-related systems and/or other technology safety-related systems and/or external risk reduction facilities. The resulting risk model may therefore differ from that shown in figure A.1.
>
> *Source* IEC 61508-5:1998, Annex A1 (Fig. 2.13).

The figure provides a concept of risk reduction related to a EUC (Equipment under Control). The EUC could be considered as a typical machine, which requires protection and safety mechanism to assure a safe operation. A motor vehicle could be also considered as a EUC, but ISO 26262 did not address the EUC approach. A motor vehicle is undoubtedly a machine, which is intended to be developed to operate in the road traffic environment.

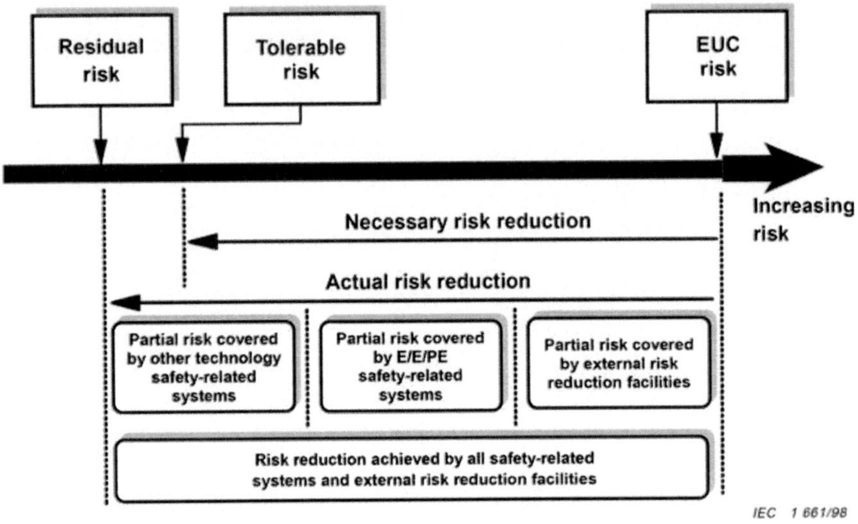

Fig. 2.13 Risk reduction: general concept. *Source* IEC 61508-5:1998, Figure A.1

Further citation from IEC 61508:

The Various Risks Indicated in Figure A.1 Are as Follows:

- EUC risk: the risk existing for the specified hazardous events for the EUC, the EUC control system and associated human factor issues—no designated safety protective features are considered in the determination of this risk (see 3.2.4 of IEC 61508-4);
- Tolerable risk; the risk which is accepted in a given context based on the current values of society (see 3.1.6 of IEC 61508-4);
- Residual risk: in the context of this standard, the residual risk is that remaining for the specified hazardous events for the EUC, the EUC control system, human factor issues but with the addition of external risk reduction facilities, E/E/PE safety-related systems and other technology safety-related systems (see also 3.1.7 of IEC 61508-4).

The EUC risk is a function of the risk associated with the EUC itself but taking into account the risk reduction brought about by the EUC control system. To prevent unreasonable claims for the safety integrity of the EUC control system, this standard places constraints on the claims that can be made (see 7.5.2.5 of IEC 61508-1).

The necessary risk reduction is achieved by a combination of all the safety protective features. The necessary risk reduction to achieve the specified tolerable risk, from a starting point of the EUC risk, is shown in figure A.1.

Source IEC 61508-5:1998, Annex A1.

The standard refers to human factors influencing the risk and the potential of measures of other technology. The authors of the standards seem to have the fence around the machine in mind; human impact and other factors of a human–machine interface (HMI) could have far more dimensions. In order to map the approach to automotive industry, we need to consider the entire vehicle design as measures of other technology, the safe compartment, the chassis design, etc., follow quite a lot of safety principles which provide protection mechanism for driver, occupancies inside the vehicle and pedestrians, cyclists and other road traffic participants around the motor vehicle. Many protection mechanisms in a vehicle are today an inherent part of the overall vehicle design. On the other hand, the entire road design becomes state of the art due to many insights from accident research. The road design and the road traffic regulations give protection and safety mechanisms, which prevent people from entering motorways or other critical road segments, etc. However, there are also regulations about the width of roads, cycle lanes, and the distances between road traffic participants each others.

Further citation from IEC 61508:

7.5.2.4 Where failures of the EUC control system place a demand on one or more E/E/PE or other technology safety-related systems and/or external risk reduction facilities, and where the intention is not to designate the EUC control system as a safety-related system, the following requirements shall apply:

(a) the dangerous failure rate claimed for the EUC control system shall be supported by data acquired through one of the following:

- actual operating experience of the EUC control system in a similar application;
- a reliability analysis carried out to a recognised procedure;
- an industry database of reliability of generic equipment;

(b) the dangerous failure rate that can be claimed for the EUC control system shall be not lower than 10–5 dangerous failures per hour;

NOTE 1—The rationale of this requirement is that if the EUC control system is not designated as a safety-related system, then the failure rate that can be claimed for the EUC control system shall not be lower than the higher target failure measure for safety integrity level 1 (which is 10–5 dangerous failures per hour; see table 3).

(c) All reasonably foreseeable dangerous failure modes of the EUC control system shall be determined and taken into account in developing the specification for the overall safety requirements;

(d) The EUC control system shall be separate and independent from the E/E/PE safety-related systems, other technology safety-related systems and external risk reduction facilities.

NOTE 2—Providing the safety-related systems have been designed to provide adequate safety integrity, taking into account the normal demand rate from the EUC control system, it will not be necessary to designate the EUC control system as a safety-related system (and, therefore, its functions will not be designated as safety functions within the context of this standard). In some applications, particularly where very high safety integrity is required, it may be appropriate to reduce the demand rate by designing the EUC control system to have a lower than normal failure rate. In such cases, if the failure rate is less than the higher limit target safety integrity for safety integrity level 1 (see table 3), then the control system will become safety-related and the requirements in this standard will apply.

Source IEC 61508-1:1998.

IEC 61508 in general considers already a controlled machine. The EUC control system could be also considered as an electric system, but also of hydraulic or pneumatic control system. The standard allows even to consider a certain risk reduction by the EUC control system and provides a limit of 10E-5 dangerous failures per hour. The EUC control system should be separated and independent from the E/E/PE safety-related system. This had been not considered in ISO 26262, the automotive derivate combines the basic control function and the safety-related function to provide the necessary risk reduction according the safety integrity level. ISO 26262 did not explain the idea of Safety Integrity, even that the ASIL (Automotive Safety Integrity Level) is the basic attribute for the risk reduction provided by the EE-function developed according to ISO 26262.

> 7.5.2.5 If the requirements of 7.5.2.4 (a) to (d) inclusive cannot be met, then the EUC control system shall be designated as a safety-related system. The safety integrity level allocated to the EUC control system shall be based on the failure rate that is claimed for the EUC control system in accordance with the target failure measures specified in tables 2 and 3. In such cases, the requirements in this standard, relevant to the allocated safety integrity level, shall apply to the EUC control system.
>
> NOTE 1—For example, if a failure rate of between 10–6 and 10–5 failures per hour is claimed for the EUC control system, then the requirements appropriate to safety integrity level 1 would need to be met.
>
> NOTE 2—See also 7.6.2.10.
>
> *Source* IEC 61508-1:1998.

ISO 26262 seems to be addressing the requirement 7.5.2.5. The EUC control system would be already a safety-related function. Since ISO 26262 does not explain the approach of EUC, the assignment of quantitative risk reductions has no reference. The absolute figures for the probabilistic metric for random hardware failure (PMHF) as defined in ISO 26262 has no reference or basis in the standard. Considering risk reduction without considering a context is questionable. ISO 26262 tried to define the context with the ITEM Definition and defined several requirements to specify the intended function and the constraints derived from the integration environment in the vehicle. Since the overall functionality of the vehicle does not change, its interface of the item under consideration does not lead to mayor change in the impact. However, due to the new technology for E-mobility, automation and connectivity, the interfaces change significantly. IEC 61508 defines clear understanding in order to distinguish between risk and safety integrity. IEC 61508 provides with the EUC approach a wider view, and the item according ISO 26262 considers only an EE-system. If the EE-system is used for the provision of the intended function, as a protection mechanism or any other kind of safety or security mechanism is only an architecture or design constrain for an item according ISO 26262.

Citation from IEC 61508, Part 5, A5:

> A.5 Risk and Safety Integrity
>
> It is important that the distinction between risk and safety integrity be fully appreciated. Risk is a measure of the probability and consequence of a specified hazardous event occurring. This can be evaluated for different situations (EUC risk, risk required to meet the tolerable risk, actual risk (see figure A.1)). The tolerable risk is determined on a societal basis and

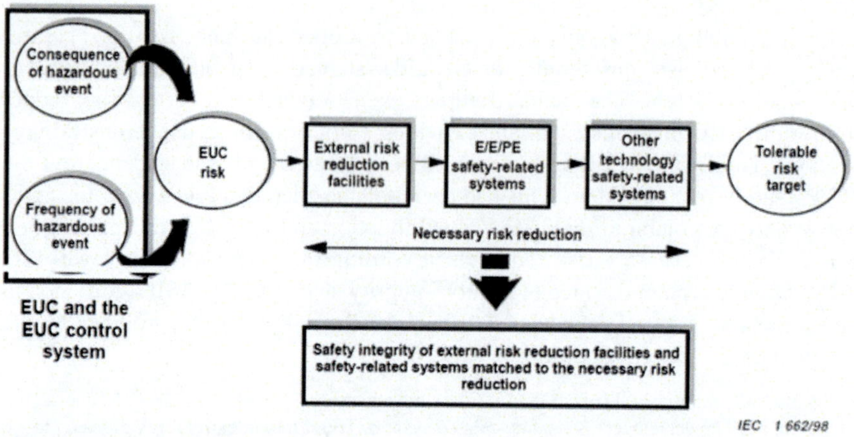

Fig. 2.14 Risk and safety integrity according IEC 61508. *Source* IEC 61508-5:1998, Figure A.2

involves consideration of societal and political factors. Safety integrity applies solely to the E/E/PE safety-related systems, other technology safety related-systems and external risk reduction facilities and is a measure of the likelihood of those systems/facilities satisfactorily achieving the necessary risk reduction in respect of the specified safety functions. Once the tolerable risk has been set, and the necessary risk reduction estimated, the safety integrity requirements for the safety-related systems can be allocated (see 7.4, 7.5 and 7.6 of IEC 61508-1).

NOTE—The Allocation is Necessarily Iterative in Order to Optimize the Design to Meet the Various Requirements.

The Role that Safety-Related Systems Play in Achieving the Necessary Risk Reduction is Illustrated in Figures A.1 and A.2.

Source IEC 61508-5:1998, Annex A5 (Fig. 2.14).

The standard defines methods to reduce the EUC risk to a level of tolerable risk. What the level of tolerable risk accepted by the society is another issue. In any nation, the relevant legislation defines the minimum tolerable risk of the society in the respective nation. The tolerable risk is usually if you apply state-of-the-art protection, safety and security measures, etc. For these, the EUC-model provides an accepted framework (Fig. 2.15).

The example in Annex C.1 shows how a dedicated protection mechanism provides the necessary risk reduction to reach the targeted tolerable risk level. IEC 61508 uses sometimes the terms.

- EUC control system,
- E/E/PE safety-related system,
- Safety-related system and
- Safety-related protection system.

The EUC control system provides the intended function of the EUC and could provide a limited level of risk reduction. Higher risk reduction by electric, electronic

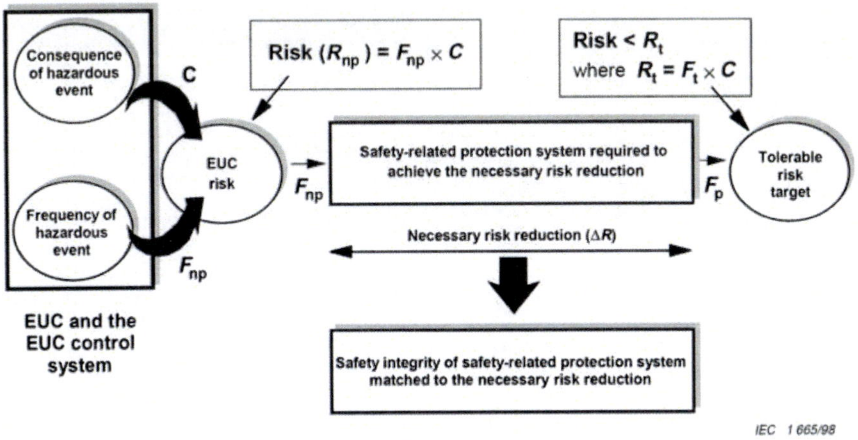

Fig. 2.15 Example from IEC 61508 for a safety-related protection system. *Source* IEC 61508-5:1998, Figure C.1

and/or programmable electronic systems would require following the requirements of IEC 61508 to reach the risk reduction as assigned by the SIL-attribute.

Safety-related systems are more general so that the risk reduction could also provide by means of other technology or by external measures (Fig. 2.16).

Protection is often the process of keeping something or someone free from risk or damages. Protection is more a preventive measure. Safe means more the condition or feeling of being safe; security; certainty. Machine protection, occupancy protection provide a barrier against possible impacts of risk. Rain protection cloths provide certain protection versus rain, but could not averse people from becoming wet at all. An airbag provides a certain level of protection in case of accidents, but it could not provide any protection in a burning vehicle.

The process and methodology behind IEC 61508 assures for a specific identified risk scenario a defined level of risk reduction by a safety-related function implemented on a specific system, developed according the standard and assigned with a specific safety integrity level (SIL).

The standard distinguished between a low and a continuous mode of operation (Fig. 2.17).

The SIL in a low demand mode of operation considers a critical event of a failure once a year, so that the average probability of failure to perform its design function on demand is very low. In other words, the safety function could fail very often, but it performs only not the dedicated function if the failure of the safety function occurs at the same time as the specific failure or the risk case in the EUC occurs. Since the risk case in the EUC less than once a year, the basic reliability for the E/E/PE safety-related system could be lower as for risk cases which occurs more frequent or often (Fig. 2.18).

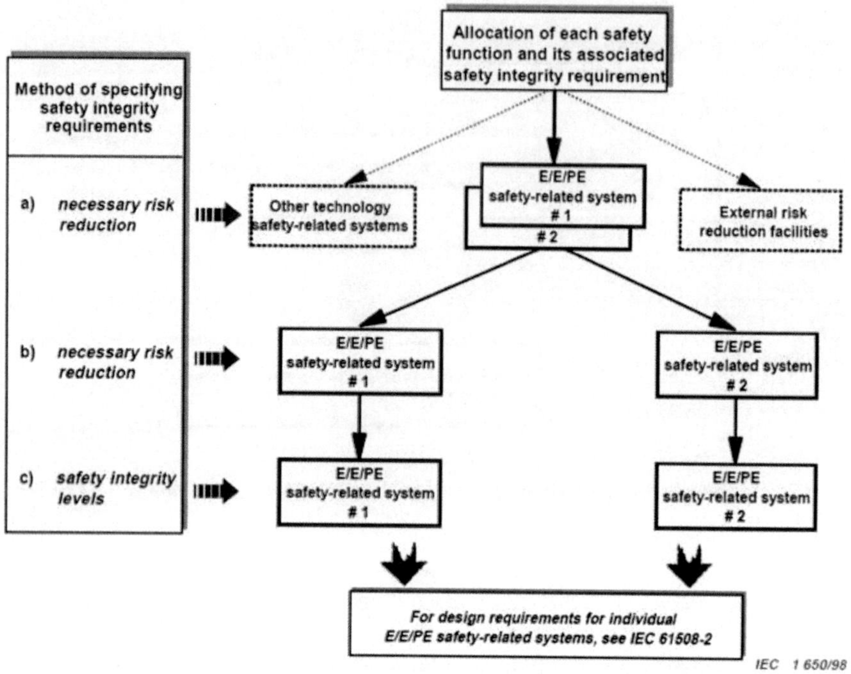

Fig. 2.16 Allocation of safety requirements to the E/E/PE safety-related systems, other technology safety-related systems and external risk reduction facilities. *Source* IEC 61508-1:1998, Fig. 6

Safety integrity level	Low demand mode of operation (Average probability of failure to perform its design function on demand)
4	$\geq 10^{-5}$ to $< 10^{-4}$
3	$\geq 10^{-4}$ to $< 10^{-3}$
2	$\geq 10^{-3}$ to $< 10^{-2}$
1	$\geq 10^{-2}$ to $< 10^{-1}$
NOTE – See notes 3 to 9 below for details on interpreting this table.	

Fig. 2.17 Target failure measures for a safety function operating in low demand mode of operation and the safety integrity level (SIL). *Source* IEC 61508-1:1998, Table 2

The high demand or continuous mode of operations does not consider an error-free EUC and its basic control system. In case the risk event could occur at any time or with a high frequency during operation of the EUC, the reliability and the level of risk reduction must be significant higher. The probability based on the assumption that a fault in the EUC and its basic control system does not occur in the same time, as the EE-system fails to perform the demanding safety function.

Safety integrity level	High demand or continuous mode of operation (Probability of a dangerous failure per hour)
4	$\geq 10^{-9}$ to $< 10^{-8}$
3	$\geq 10^{-8}$ to $< 10^{-7}$
2	$\geq 10^{-7}$ to $< 10^{-6}$
1	$\geq 10^{-6}$ to $< 10^{-5}$
NOTE – See notes 3 to 9 below for details on interpreting this table.	

Fig. 2.18 Target failure measures for a safety function operating in high demand mode or continuous mode of operation and the safety integrity level (SIL). *Source* IEC 61508-1:1998, Table 3

Driving a vehicle in road traffic means already a risk due to the kinetic energy of the accelerated mass of the vehicle. The risk of the occupancies is provided by several protection measures and the rules provided by the road traffic regulations. The other road traffic participants are protected as well by the road traffic regulations and by pedestrian protection measures of the vehicle in case of accidents.

2.4.2 Risk According to ISO 26262

A new revision of ISO 26262 had been published in 2018. The new edition mentioned the main changes to ISO 26262:2011 as follows:

> This edition of ISO 26262 series of standards cancels and replaces the edition ISO 26262:2011 series of standards, which have been technically revised and includes the following main changes:
>
> - Requirements for trucks, buses, trailers and semi-trailers,
> - Extension of the vocabulary,
> - More detailed objectives,
> - Objective-oriented confirmation measures,
> - Management of safety anomalies,
> - References to cybersecurity,
> - Updated target values for hardware architecture metrics,
> - Guidance on model-based development and software safety analysis,
> - Evaluation of hardware elements,
> - Additional guidance on dependent failure analysis,
> - Guidance on fault tolerance, safety-related special characteristics and software tools,
> - Guidance for semiconductors,
> - Requirements for motorcycles,
> - General restructuring of all parts for improved clarity.
>
> *Source* ISO 26262:2018, Foreword.

The standard introduced itself as an adaption of ISO 26262 but with several significant changes which were targeted for the development of electronic system to be integrated in road vehicles. In general, the standard addresses series development parts, which the standard defines as an ITEM.

The new edition of 2018 formulates the "Introduction as follow:

Introduction The ISO 26262 series of standards is the adaptation of IEC 61508 series of standards to address the sector specific needs of electrical and/or electronic (E/E) systems within road vehicles. This adaptation applies to all activities during the safety lifecycle of safety-related systems comprised of electrical, electronic and software components. Safety is one of the key issues in the development of road vehicles. Development and integration of automotive functionalities strengthen the need for functional safety and the need to provide evidence that functional safety objectives are satisfied. With the trend of increasing technological complexity, software content and mechatronic implementation, there are increasing risks from systematic failures and random hardware failures, these being considered within the scope of functional safety. ISO 26262 series of standards includes guidance to mitigate these risks by providing appropriate requirements and processes. To achieve functional safety, the ISO 26262 series of standards:

(a) provides a reference for the automotive safety lifecycle and supports the tailoring of the activities to be performed during the lifecycle phases, i.e., development, production, operation, service and decommissioning;
(b) provides an automotive-specific risk-based approach to determine integrity levels [Automotive Safety Integrity Levels (ASILs)];
(c) uses ASILs to specify which of the requirements of ISO 26262 are applicable to avoid unreasonable residual risk;
(d) provides requirements for functional safety management, design, implementation, verification, validation and confirmation measures; and
(e) provides requirements for relations between customers and suppliers.

The ISO 26262 series of standards is concerned with functional safety of E/E systems that is achieved through safety measures including safety mechanisms. It also provides a framework within which safety-related systems based on other technologies (e.g. mechanical, hydraulic and pneumatic) can be considered. The achievement of functional safety is influenced by the development process (including such activities as requirements specification, design, implementation, integration, verification, validation and configuration), the production and service processes and the management processes. Safety is intertwined with common function-oriented and quality-oriented activities and work products. The ISO 26262 series of standards addresses the safety-related aspects of these activities and work products.

Figure 2.19 shows the overall structure of the ISO 26262 series of standards. The ISO 26262 series of standards is based upon a V-model as a reference process model for the different phases of product development.

Within the Figure:

- The shaded "V"s represent the interconnection among ISO 26262-3, ISO 26262-4, ISO 26262-5, ISO 26262-6 and ISO 26262-7;
- For motorcycles:
- ISO 26262-12:2018, Clause 8 supports ISO 26262-3,
- ISO 26262-12: 2018, Clauses 9 and 10 support ISO 26262-4;

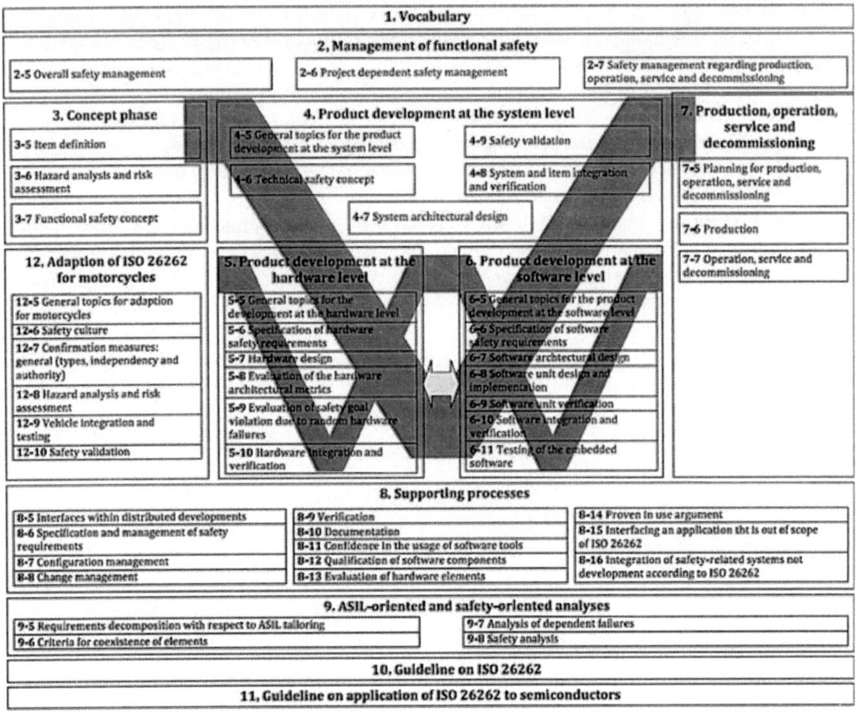

Fig. 2.19 Overview of the ISO 26262 series of standards

- The specific clauses are indicated in the following manner: "m–n", where "m" represents the number

Source ISO 26262:2018, Introduction.

In the scope definition the standard is more precise about the use of the standard, the standard defines its scope as follow:

1. Scope

This document is intended to be applied to safety-related systems that include one or more electrical and/or electronic (E/E) systems and that are installed in series production road vehicles, excluding mopeds. This document does not address unique E/E systems in special vehicles such as E/E systems designed for drivers with disabilities.

NOTE: Other Dedicated Application-Specific Safety Standards Exist and Can Complement the ISO 26262 Series of Standards or Vice Versa.

Systems and their components released for production, or systems and their components already under development prior to the publication date of this document, are exempted from the scope of this edition. This document addresses alterations to existing systems and their components released for production prior to the publication of this document by tailoring the safety lifecycle depending on the alteration. This document addresses integration of existing systems not developed according to this document and systems developed according to this document by tailoring the safety lifecycle. This document addresses possible hazards caused by malfunctioning behaviour of safety-related E/E systems, including interaction of these systems. It does not address hazards related to electric shock, fire, smoke, heat, radiation, toxicity, flammability, reactivity, corrosion, release of energy and similar hazards, unless directly caused by malfunctioning behaviour of safety-related E/E systems. This document describes a framework for functional safety to assist the development of safety-related E/E systems. This framework is intended to be used to integrate functional safety activities into a company-specific development framework. Some requirements have a clear technical focus to implement functional safety into a product; others address the development process and can therefore be seen as process requirements in order to demonstrate the capability of an organisation with respect to functional safety. This document does not address the nominal performance of E/E-Systems. This document specifies the requirements for functional safety management for automotive applications, including the following:

- project-independent requirements with regard to the organizations involved (overall safety management), and
- project-specific requirements with regard to the management activities in the safety lifecycle, i.e. management during the concept phase and the product development phases (at the system, hardware and software level), and regarding production, operation, service and decommissioning.

Source ISO 26262:2018, Scope.

The standard recommended in a note that other safety standards could be complemented with ISO 26262 and vice versa. The standard goes not detailed into a specific direction, which standards for which topic should be preferred. It also does not address harmonised standard such as the machinery guideline from the European Union.

The new edition from 2018 provides in the informative part 10 of the standard the differences between IEC 61508 and ISO 26262:

4.1 Functional safety for automotive systems (relationship with IEC 61508[1])

IEC 61508, Functional Safety of electrical/electronic/programmable electronic safety-related systems, is designated by IEC as a generic standard and a basic safety publication. This means that industry sectors will base their own standards for functional safety on the requirements of IEC 61508. In the automotive industry, there are a number of issues with applying IEC 61508 directly. Some of these issues and corresponding differences in the ISO 26262 series of standards are described below. IEC 61508 is based upon the model of "equipment under control", for example an industrial plant that has an associated control system as follows:

(a) A hazard analysis identifies the hazards associated with the equipment under control (including the equipment control system), to which risk reduction measures will be applied. This can be achieved through E/E/PE systems, or other technology safety-related systems (e.g. a safety valve), or external measures (e.g. a physical containment of the plant). The ISO 26262 series of standards contains a normative automotive scheme for hazard classification based on severity, probability of exposure and controllability.
(b) Risk reduction allocated to E/E/PE systems is achieved through safety functions, which are designated as such. These safety functions are either part of a separate protection system, or can be incorporated into the plant control. It is not always possible to make this distinction in automotive systems. The safety of a vehicle depends on the behaviour of the control systems themselves.

The ISO 26262 Series of Standards Uses the Notion of Safety Goals and a Safety Concept as Follows:

- A hazard analysis and risk assessment identifies hazards and hazardous events that need to be prevented, mitigated, or controlled;
- At least one safety goal is associated with each hazardous event that has been classified as ASIL A, B, C or D;
- An Automotive Safety Integrity Level (ASIL) is associated with each safety goal;
- The functional safety concept is a statement of the functionality to achieve the safety goal(s);
- The technical safety concept is a statement of how this functionality is implemented on the system level by hardware and software; and
- Software safety requirements and hardware safety requirements state the specific safety requirements which will be implemented as part of the software and hardware design.

EXAMPLE the Airbag System.

- One of the hazards is unintended deployment.
- An associated safety goal is that the airbag only deploys when a crash occurs that requires the deployment.
- The functional safety concept can specify a redundant function to detect whether the vehicle is in a collision.
- The technical safety concept can specify the implementation of two independent accelerometers with different axial orientations and two independent firing circuits. The squib deploys if both are closed.

IEC 61508 is aimed at singular or low volume systems. The system is built and tested, then installed in the plant, and then safety validation is performed. For mass-market systems such as road vehicles, safety validation is performed before the release for volume (series) production. Therefore, the order of lifecycle activities in the ISO 26262 series of standards is different. Related to this, ISO 26262-7 addresses requirements for production. These are not covered in IEC 61508. IEC 61508 does not address specific requirements for managing development across multiple organizations and supply chains. Because automotive systems are produced by vehicle manufacturers themselves, by one or more suppliers to the manufacturer or by collaboration between manufacturer and supplier(s), the ISO 26262 series of standards includes requirements to explicitly address this issue, including the Development Interface Agreement (DIA), (see ISO 26262-8:2018, Clause 5). IEC 61508 does not contain normative requirements for hazard classification. The ISO 26262 series of standards contains an automotive scheme for hazard classification. This scheme recognises that a hazard in an automotive system does not necessarily lead to an accident. The outcome will depend on whether the persons at risk are actually exposed to the hazard in the situation in which it occurs; and whether the involved people are able to take steps to control the outcome of the hazard. An example of this concept, applied to a failure which affects the controllability of a moving vehicle, is given in Fig. 2.20.

NOTE: This concept is intended only to demonstrate that there is not necessarily a direct correlation between a failure occurring and the accident. It is not a representation of the hazard analysis and risk assessment process, although the parameters although the parameters evaluated in this process are related to the probabilities of the state transitions shown in the figure.

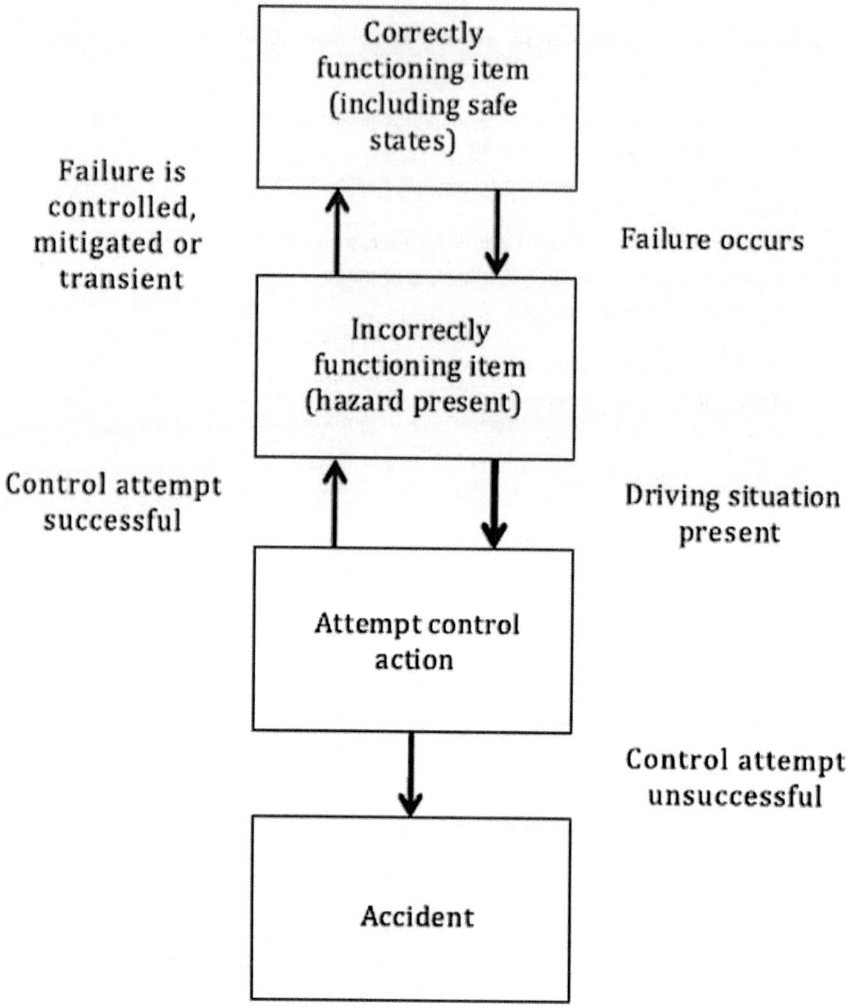

Fig. 2.20 State machine model of automotive risk

The requirements for hardware development (ISO 26262-5) and software development (ISO 26262-6) are adapted for the state-of-the-art in the automotive industry. For the methods listed in the ISO 26262 series of standards specific goals are provided. To achieve these goals, the provided methods can be applied or a rationale that alternative methods can also achieve the goal is provided. Safety requirements in the ISO 26262 series of standards are assigned an ASIL (Automotive Safety Integrity Level) rather than a SIL (Safety Integrity Level). The main motivation for this is in IEC 61508 is stated in probabilistic terms (see IEC 61508-1:2010, 3). IEC 61508 acknowledges that qualitative judgement is often required in respect of systematic safety integrity while requiring quantitative techniques for hardware safety integrity. An ASIL in ISO 26262 is primarily concerned with requirements for achieving systematic safety in the system, hardware and software; however, there are probabilistic targets associated with compliance to the requirements of an ASIL with respect to random hardware failures.

Source ISO 26262:2018, Part 10, Clause 4.1

ISO 26262 defines the relation of risk, danger and safety integrity differently. ISO 26262 does not explain their understanding of safety integrity even that the standard uses the abbreviation: Automotive Safety Integrity Level (ASIL). The principle using the vehicle as an item or applying the idea of "Equipment under Control, EUC" is not considered by ISO 26262. EUC could be explained as "device or system, which should be controlled by safety-related functions". Under certain limiting conditions, ISO 26262 admits to develop a desired vehicle function that is safety related on its own. In this case, the system does not receive safety through EUC itself. Technically, according to IEC 61508, EUC and the safety functions have to cause an error at the same time in order to create a hazardous situation. Considering a hydraulic braking system as a EUC, which in its function can be monitored by an EE-system, errors of the hydraulic systems could be avoided or controlled by the EE-system.

In ISO 26262, part3, Annex B1 ISO 26262 explains the principles of the methodology for the "Hazard Analysis and Risk Assessment".

The standard start as follow:

Hazard Analysis and Risk Assessment

General

This annex gives a general explanation of the hazard analysis and risk assessment. The examples in B.2 (severity), B.3 (probability of exposure) and B.4 (controllability) are for information only and are not exhaustive.

For this analytical approach, a risk (R) can be described as a function (F), having three parameters: The frequency of occurrence (f) of a hazardous event, the controllability (C), i.e. the ability to avoid the specific harm or damage through timely reactions of the persons involved, and the potential severity (S) of the resulting harm or damage:

$$R = F(f, C, S) \tag{B.1}$$

The frequency of occurrence f is, in turn, influenced by two factors. One factor to consider is how frequently and for how long individuals find themselves in a situation where the aforementioned hazardous event can occur. In ISO 26262 this is simplified to be a measure of the probability of the operational situation taking place in which the hazardous event can occur (exposure, E). Another factor is the occurrence rate of faults in the item. This is not considered during hazard analysis and risk assessment. Instead the ASILs that result from the classification of E, S, C during hazard analysis and risk assessment determine the minimum set of requirements on the item, in order to control or reduce the probability of random hardware failures, and to avoid systematic faults. The failure rate of the item is not considered a priori (in the risk assessment) because an unreasonable residual risk is avoided through the implementation of the resulting safety requirements.

The Hazard Analysis and Risk Assessment Subphase Comprises Three Steps, as Described Below.

(a) Situation analysis and hazard identification (see 6.4.2): the goal of the situation analysis and hazard identification is to identify the potential unintended behaviours of the item that could lead to a hazardous event. The situation analysis and hazard identification activity requires a clear definition of the item, its functionality and its boundaries. It is based on the item's behaviour; therefore, the detailed design of the item does not necessarily need to be known.

EXAMPLE Factors to Be Considered for Situation Analysis and Hazard Identification Can Include:

- vehicle usage scenarios, for example high speed driving, urban driving, parking, off-road;
- environmental conditions, for example road surface friction, side winds;
- reasonably foreseeable driver use and misuse; and
- interaction between operational systems.
- T&B base vehicle, vehicle configuration and vehicle operation.

(b) Classification of hazardous events (see 6.4.3): the hazard classification scheme comprises the determination of the severity, the probability of exposure, and the controllability associated with the hazardous events of the item. The severity represents an estimate of the potential harm in a particular driving situation, while the probability of exposure is determined by the corresponding situation. The controllability rates how easy or difficult it is for the driver or other road traffic participant to avoid the considered accident type in the considered operational situation. For each hazard, depending on the number of related hazardous events, the classification will result in one or more combinations of severity, probability of exposure, and controllability.

(c) ASIL determination (see 6.4.3): determining the required automotive safety integrity level.

Source ISO 262622018, Part 3, Annex B1.

The description of the methodology the standard addresses as an example.

- Vehicle behaviour and related scenarios,
- Environmental conditions,
- Reasonable foreseeable misuse,
- Interactions between operational systems and vehicle configurations and operation, etc.

However, it provides no guidance nor requirements how to avoid for example systematic failure out of the scope of the ITEM, as defined by ISO 26262.

The automobile industry relies here on other technology and engineering of the electronic safety system will be considered as a fail-safe or a degrading system. They also referencing directly to state of the art of software and hardware development.

The Requirements for Hardware Development (ISO 26262-5) and Software Development (ISO 26262-6) Are Adapted for the State-Of-The-Art in the Automotive Industry.

The state-of-the-art from 2018 does not necessarily represent the state of the art of the future. Especially in the field of electronic hardware, the technology will continue to develop significantly. Typical architecture for future safety systems which provide proven technology from other industries is whether considered in the hardware nor in the requirements for software development.

The following principles are not addressed by the standard:

- Systematic failure out of the scope of the ITEM, whether between ITEMS, or on vehicle level, nor in relation to vehicle external system, e.g. charging systems for E-mobility.
- Other risk beside malfunctions failure, etc., of EE-systems
- Combined EE-system and mechanical safety mechanisms
- Protection mechanism
- Active safety systems
- Voting architecture
- Close-loop safety-related control systems (or any interface to real world or other technology, also not failure of other technology)
- Basic control system by hydraulic, pneumatic, other technology and EE-systems for risk reduction.
- Risk derived from vehicle design, or risk derived from other material, poisoning, chemical reactions, etc.
- Risk due to human factors outside and inside of the vehicle (accept controllability of EE-failure).

The standard addresses for example thermal incidents of EE-system elements or foreseeable misuse, but does give no specific requirements how to deal with such kinds of risks.

ISO 26262 provides a safety lifecycle which considers the management and development activities around the ITEM, but even the context of the system is only addressed as part of the ITEM Definition (Fig. 2.21).

The Safety Lifecycle according ISO 26262 defines a basic process framework in their chapters 2 till 7. The purpose of the Safety Lifecycle understands the standard as follow:

> The ISO 26262 reference safety lifecycle encompasses the principal safety activities during the concept phase, product development, production, operation, service and decommissioning. Planning, coordinating and monitoring the progress of the safety activities, as well as the responsibility to ensure that the confirmation measures are performed, are key management tasks and are performed throughout the lifecycle. The safety lifecycle may be tailored (see Clause 6).
>
> *Source* ISO 26262:2018, Chapter 2, 5.2.1

In Part 7, operation, service and decommissioning are addressed, but all the requirements are dealing with the ITEM, the operation of the vehicle is not addressed in the standard at all.

ISO 26262 mentions normative methods that describe a systematic derivation of the potential risk, which may originate from the investigated of the considered Item (vehicle system), based on a hazard analysis and risk assessment. Hazard or risk analyses are not normatively defined in other safety standards. Either the requirements for these methods are listed or the method itself is exemplarily described (Fig. 2.22).

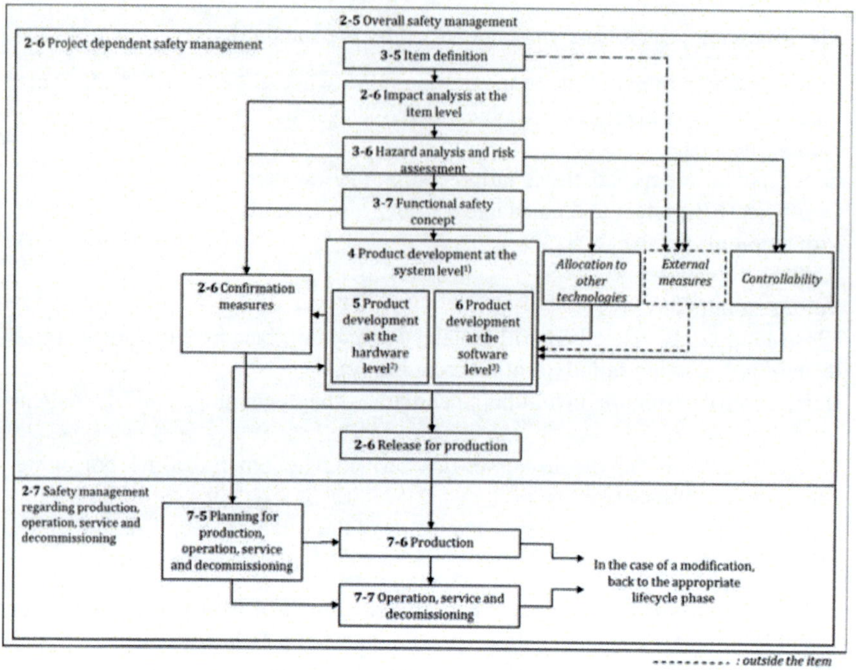

Fig. 2.21 Safety lifecycle according ISO 26262:2018. *Source* ISO 26262:2018, This chapter, Fig. 2.2

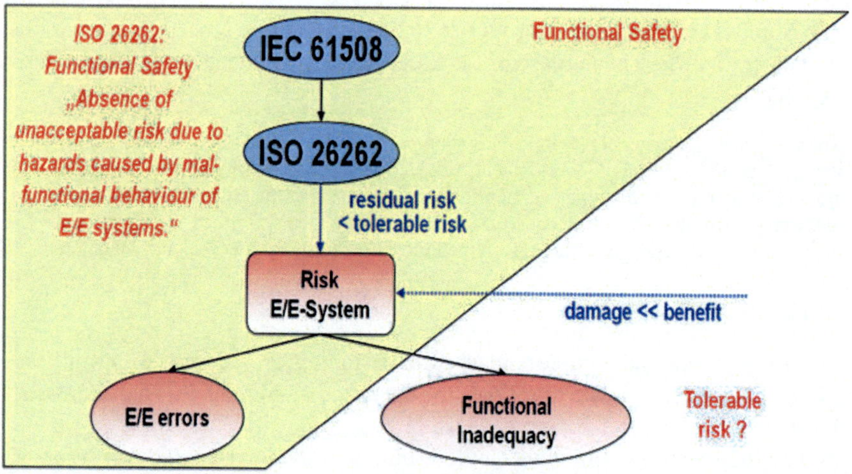

Fig. 2.22 Distinction of hazards, based on correctly functioning systems. *Source* unpublished research project

The reduction of risk cannot be achieved with the activities and methods mentioned in ISO 26262 if a function is not suitable, inadequate suitable, inadequate or falsely indicated safety-related functions. This represents a special challenge, considering that ISO 26262 does not directly addresses a Equipment under Control (EUC, e.g. a system, machinery or vehicle, which should be controlled safety related) or the distinction between safety functions of designated safety requirements for on-demand (low demand) or continuous mode (high demand) safety systems. How is it possible to find out whether or not reactions of a vehicle system or certain measurements are sufficient, tolerable or safety-related appropriate?

2.5 Critical Infrastructure

Legislation of Critical Infrastructure Protection began to be actively addressed in Europe in 2005, when the European Union issued a Call for a European Programme for Critical Infrastructure Protection (further only CIP) that first drew attention to the fact that several sectors and their elements were significant for today´s society—and that their failure could have serious impacts on national security, the economy, public administration and the basic vital functions of Member States´ populations. Three years later, the EC issued European Council Directive 2008/114/EC, which requires Member States to identify the European Critical Infrastructure Elements.

Critical infrastructure is a term used by governments to describe assets that are essential for the functioning of a society and economy—the infrastructure. Most commonly associated with the term are facilities for:

- Shelter; heating (e.g. natural gas, fuel oil, district heating);
- Agriculture, food production and distribution;
- Water supply (drinking water, waste water/sewage, stemming of surface water (e.g. dikes and sluices));
- Public health (hospitals, ambulances);
- Transportation systems (fuel supply, railway network, airports, harbours, inland shipping);
- Security services (police, military).
- Electricity generation, transmission and distribution; (e.g. natural gas, fuel oil, coal, nuclear power)
- Renewable energy, which is naturally replenished on a human timescale, such as sunlight, wind, rain, tides, waves and geothermal heat.
- Telecommunication; coordination for successful operations
- Economic sector; goods and services and financial services (banking, clearing).

2.5.1 *Organisation for Critical Infrastructure*

In order to manage and maintain critical organisation, a dedicated organisation is unambiguous. The target of the organisation is to provide the intended service and adequate protection of the critical infrastructure (Fig. 2.23).

The ETH Zurich proposes the following targets in their four-pillar model from critical infrastructure:

- Prevention and early warning
- Detection
- Reaction and
- Crisis Management.

For that they propose the organisational measure for a cooperation model:

> Essential Partners: The Cooperation Model
>
> Given its involvement across the four pillars, and the challenges they pose, the CIIP unit has to have different specialized competencies, implying a large and complex organization. However, the CIIP unit can be streamlined without loss of capacity by establishing cooperation among different partners who are best qualified to cope each with one of the tasks. If each partner concentrates on their competency, existing know-how can be applied most efficiently, saving costs, as well as manpower. According to the tasks of each of the four pillars, CIIP requires different organizational, technical, and analytical competencies. The CIIP unit should ideally include three partners:
>
> - A governmental agency, providing strategic leadership and supervision (the Head of the CIIP unit),

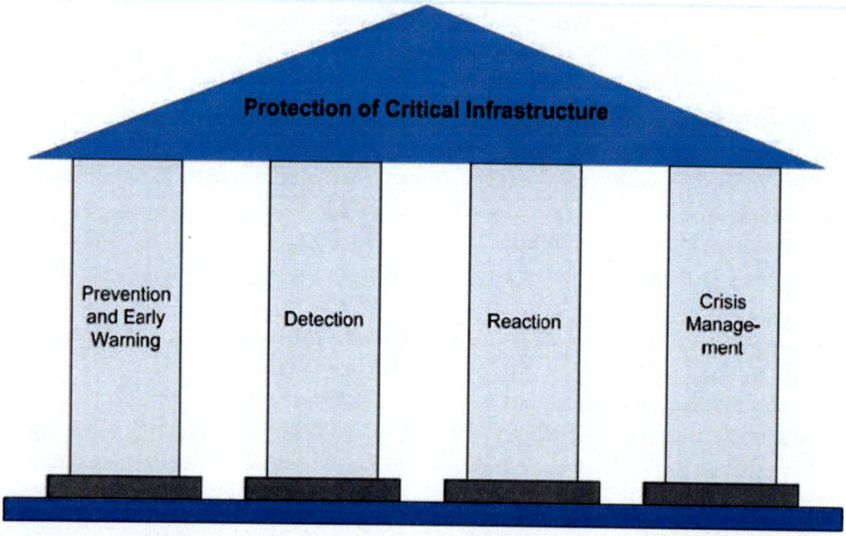

Fig. 2.23 Illustration of the four-pillar model. *Source* Four Pillar Model, from "A Generic National Framework. For Critical Information Infrastructure Protection (CIIP)", ETH Zurich, August 2007

- An analysis center with strong linkages to the intelligence community (the Situation Center),
- A technical center of expertise, usually consisting of staff members of a national CERT (the CERT Team).

This Section Describes These Partners and Examines Which Partner Should Provide Which Services to Form an Effective CIIP Unit.

Source Cooperation Model, from A Generic National Framework For Critical Information Infrastructure Protection (CIIP), ETH Zurich, August 2007.

Such a management of control systems for critical infrastructure requires a competent organisation where the cooperation with governmental authorities and collaboration with infrastructure providers and certification bodies play a major role.

2.5.2 *Cloud Computing*

When you think of a "cloud," do you think of public clouds? You're not alone—many organizations think that "heading to the clouds" means deploying a public cloud, but you can get the same agility and flexibility in your own data centre with a robust private cloud. But what is the impact of safety and security for a public cloud, and what are the benefits for safety, security and privacy with a private cloud? (Fig. 2.24).

A cloud system could be distinguished by the following questions:

- Who manages the infrastructure?
- Who owned the infrastructure?
- Where is the infrastructure located?
- Who has access rights or the right to consume data?

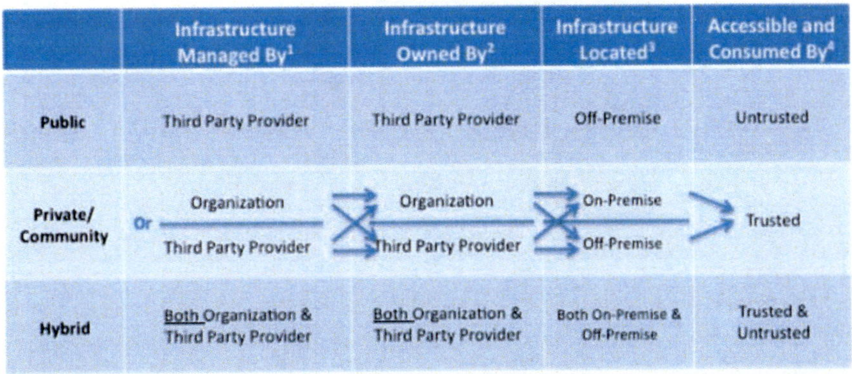

Fig. 2.24 Source Security guidance for critical areas of focus in cloud computing V3.0, 2017

This often differs for

- Public cloud systems
- Private and Community owned or
- Hybrid cloud systems.

The owner must be primarily responsible for the services and also for the necessary protective mechanisms. He can delegate this task to an operator or service provider. For such a system, certain rules must be defined, from which the guarantees and obligations for the users are then derived. In particular, internationally operated systems are used by operators in one country and users in another. Users, consumers and all those who have access to the system must abide by common rules. Even a public cloud system is subject to at least the legal requirements of the country in which the service is offered. Often the legislator will prohibit the use of a public cloud service for critical infrastructure.

A very interesting model for the assessment is the "Jericho Cube Model" (Fig. 2.25).

The Cloud Cube Model illustrates the many permutations available in cloud offerings today and presents four criteria/dimensions in order to differentiate cloud "formations" from one another and the manner of their provision, in order to understand how cloud computing affects the way in which security might be approached (Fig. 2.26).

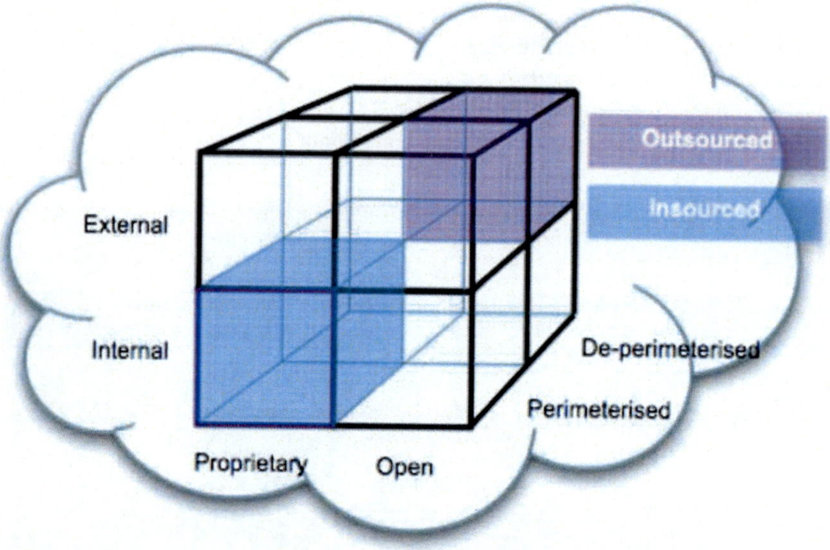

Fig. 2.25 Jericho Cloud Cube Model. *Source* Security guidance for critical areas of focus in cloud computing V3.0, 2017

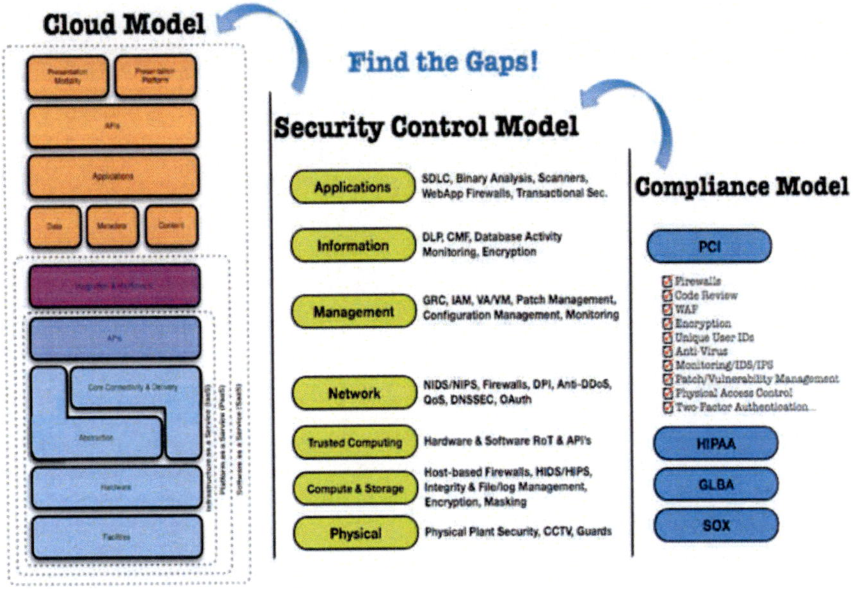

Fig. 2.26 Mapping the cloud model to the security control and compliance. *Source* Security guidance for critical areas of focus in cloud computing V3.0, 2017

The cloud model provides a generic architecture, how to understand the cloud architecture including all mayor functional units. A generic risk analysis provides a security model with adequate measures for the following area:

- Application
- Information
- Management
- Network
- Trusted computing
- Computer and storage as well as
- The underlying physical system.

The compliance model provides support for the measures for the assurance case. The Cube model does not address other safety domains, such as functional safety, privacy.

2.5.3 Transport and Traffic and Critical Infrastructure

Transport and traffic belong also to critical infrastructure. The transport and traffic sector includes:

- Air transport
- Maritime transport

- Inland waterways transport
- Rail transport
- Road transport
- Logistics.

Transport and traffic belongs in any case to the critical infrastructure, the reason, why derive from various risks related to transportation and traffic systems. Typical risks are:

- Unavailability of transport systems
- Blocking of emergency ways
- Accidents (independent if caused by systems or humans due to misuse or active sabotage etc.)
- Environmental damages, etc.

Stakeholders manage risk to critical infrastructures because it is simply infeasible to protect all infrastructures against all threats. In order to be used more effectively and efficiently, limited governmental and private sector resources must be applied in a manner that reflects risk. Resources should be focused on what is critical, vulnerable and facing the most consequential threats. Once risks are assessed, strategies and priorities for managing these risks, including mitigation, transfer, and, in some cases, acceptance, can be established.

And since risk management is a continuous process, rather than an end state, strategies for mitigating risk must be employed across all phases of the risk management process: specifically, prevention, preparedness, response and recovery. The detailed risk in the various segments of traffic requires different approaches for the risk management. A typical sequence for the risk analysis as proposed in various guides is based on the following steps:

- Determining risk management Scope
- Identifying critical information infrastructure functions
- Analysing critical function value chain and interdependencies
- Assessing critical function risk
- Prioritizing and treating critical function risk.

These steps are working more on a user view and evaluate the risk in operation scenarios. The result of the analysis is the risk assessment and necessary protection mechanisms.

In a more data integrity approach for technical systems, the steps could be as follows:

- Identify assets and identify which are most critical
- Identify, characterise and assess threats
- Assess the vulnerability of critical assets to specific threats
- Determine the risk (i.e. the expected consequences of specific types of attacks on specific assets)
- Identify ways to reduce those risks
- Prioritize risk reduction measures based on a strategy.

The terminology derives more from the cybersecurity domain and focusses more on the evaluation of necessary security mechanisms.

In order to develop the right protection system for infrastructure-based traffic systems, the various different purposes of the applications, the dedicated risk scenario and the resulting hazardous situations need to be analysed. If the intended purpose and the usability is not understood during development, safety, security and protection measures, the measures could not be systematically validated and not monitored during operation.

Chapter 3
Automation in Mobility

Future mobility seems leaving the idea of individual mobility; the key motivation seems to be autonomous or at least automated driving. Machine learning and artificial intelligence get their revival, but the automotive industry could not find the right context yet. All that cyberart and cyberchallenges sound similar to the first automation ideas more than 100 years ago. In the late fifties of last millennium, the new science became famous and called "cybernetics". The area of success for cybernetics was telecommunication and missile control, which lead to our space programmes. Landing on the moon and return by humans was one of the biggest and remarkable early steps. In plant automation, cybernetics became more "measurement and control", and in the late eighties, fuzzy control was well-established technology. Machine-learning algorithms implemented on the "Distributed Control Systems, DCS" provided the same results just by a software program. In automotive industry, we need to obviously make now our own experiences, so that we understand how human driving works, before we could start with the automation of the driving task.

3.1 Human Driving: a Closed-Loop Control System

Automated driving is a well-discussed topic, before we start to provide solutions to support or assist driving or even compensation or replacing of human functions by automation systems.

The questions like the following arise:

- How does human driving work?
- What does the driver do during driving?

Any kind of automation deals with human interaction. On the one hand, we provide assistance system, which should support a human intend, or we try to automate a function, which replaces the human work.

H.-L. Ross, *Safety for Future Transport and Mobility*,
https://doi.org/10.1007/978-3-030-54883-4_3

When it comes to automation of driving functions, a cybernetics approach and the targets get more and more in the focus. Physics is concerned with the material of the world, and energy matters somehow. Cybernetics abstracts from the material and concentrates on control, communication, information of these actions, behaviour or required behaviour or actions.

The philosophe Schopenhauer provides the following formula for what exists:

the World = will + representation

Since we consider this from an observer's point of view, the following formula could be considered:

the world = behaviour + representation

All players or any device that is not fixed moves somehow. Humans move due to a motivation or their will, or they could move intentionally but also unconsciously or unwillingly. A technical system could be moved by providing energy. Goods, artefacts, things or any other devices only move due to certain kind of energy if there is no energy impact they remain where they are on earth due to the gravity. Whenever things and humans move, it leads to an action, which could be observed. These actions could be described as behaviour from an observer's point of view.

3.2 Human Driving

A driver is a person that operates a technical system called vehicle. The behaviour of the driver and the resulting behaviour of the vehicle could be observed and described.

In case of driver assistant systems, we need to capture the human intend and amplify the human action. If you want to automate human functions or actions, we need to understand the objective of the human activity. If you want to automate human functions, the question arises whether you can implement the functions in a system or whether you consider the driver's goal and should achieve the goal with other technical solutions.

Driving of a vehicle based on specific actions of the driver, which is based on "driving", it includes selecting the driving direction, steering and braking. Further secondary actions in addition to basic driving are indicating driving direction to other road traffic participants with the turn indicator, showing others the position or presence by switching the light on or the brake intention by brake lights. In order to improve the driver's ability to observe the road traffic environment, the headlights could be switched on. In order to protect occupancies, several measures are considered like using a safety belt or doing some test before driving like testing the airbag functionality.

3.2.1 Driver's HMI

The driver's HMI is not just a display. The driver controls a vehicle with nearly all his senses. In Germany, there used to be a TV-programme called "Der siebte Sinn" which means "The Seventh Sense". Here, it was pointed out repeatedly to foresighted driving and experience values from accident research, which one cannot perceive with the eyes alone. Pedals, steering wheel, all switches and levers give feedback or indicate with their position or attitude to the driver. The driver for the respective driving strategy uses all this information subconsciously depending on the driving experiences (Fig. 3.1).

The driver gives control commands with

- steering wheel,
- gear switch,
- drive pedal and
- brake pedal.

The commands of the driver affect the propulsion, steering and brake systems to perform the required action. The driver receives feedback from his commands by observation of the road environment. The key information the driver perceives is

- road condition,
- weather condition,
- how good the lane and trajectory could be followed,
- the behaviour of other traffic participants,
- obstacles or people near or in the driving corridor or lane,

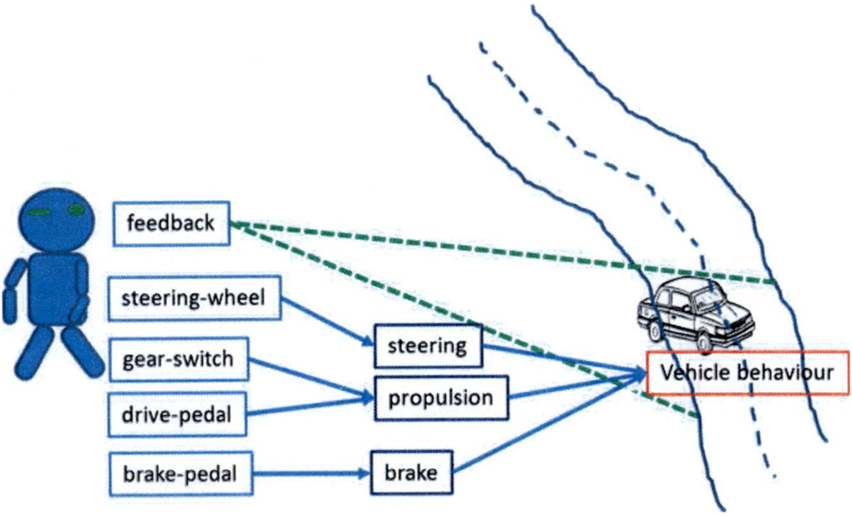

Fig. 3.1 Manual driving on roads

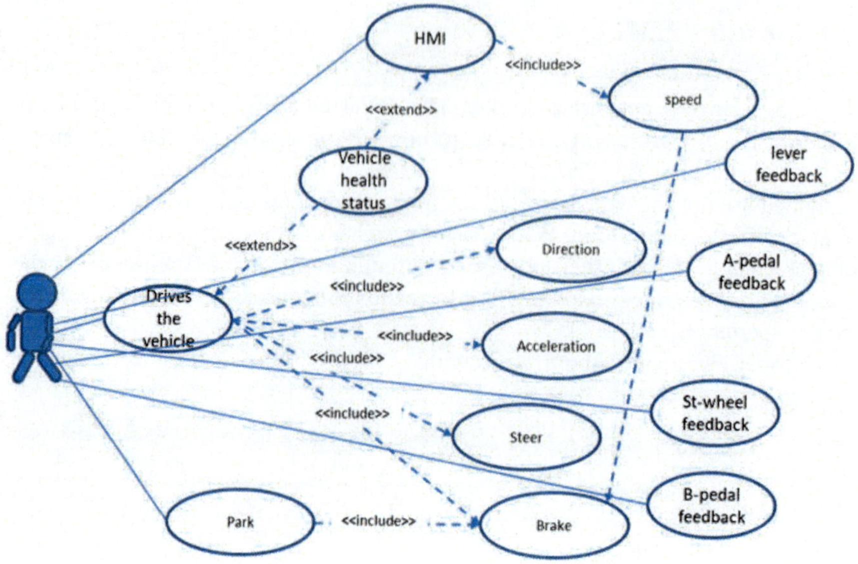

Fig. 3.2 Typical driver use cases

- road markings,
- traffic signs, etc. (Fig. 3.2).

Driver's HMI is a bidirectional interface, and it is based on the typical indicators like

- speed metre,
- gear selector indication,
- light information,
- vehicle health status like ABS warning lamb, which indicates a failure in the electronic brake system.

In the other direction, the driver gets feedback by the steering wheel, pedal feeling and also the position of switches or handles in the vehicle.

Based on all this information and feedbacks, the driver commands all the vehicle behaviour. In a very basic car, no electronic device is involved at all in that control process.

3.2.2 Driver in the Loop

During the driving education, the driver learnt that he needs to slow down in cases

- where he could not observe the driving corridor within the stopping distance or
- could not control the vehicle due to his physical abilities (Fig. 3.3).

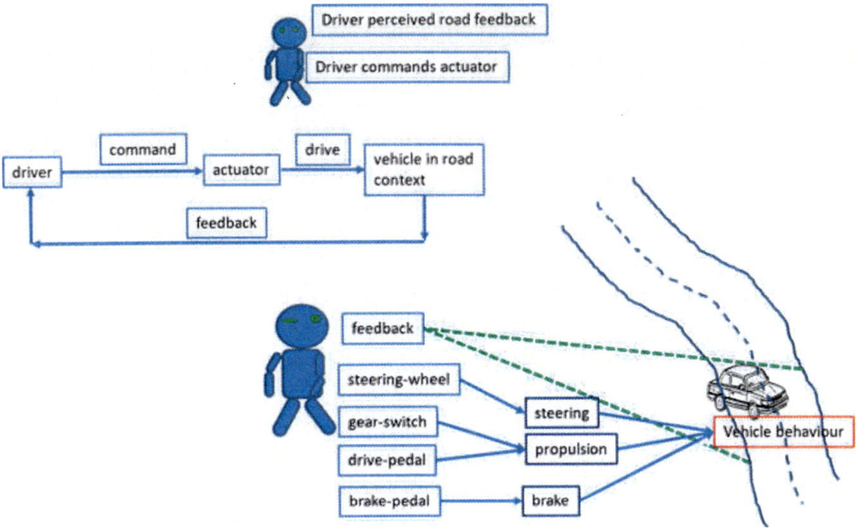

Fig. 3.3 Driver as the controller in a closed-loop system

The driver could be considered as a controller in a closed-loop control system. The equipment under control (EUC; see also IEC 61508) is the vehicle in its road traffic environment or the context of the "operational design domain, ODD". The ODD is the vehicle in the intended context. Consequently, the vehicle needs to be designed in a way that it could be safely operated in the intended operation context or environment, and such a vehicle fulfils the requirements for roadworthiness.

3.2.3 Human Driving Journey

As long as society sees no advantage in automating individual mobility and the modes of transport are shown to offer only minor safety benefits, it will be difficult to obtain funding for automating individual mobility. Consequently, automated protection systems will be increasingly demanded. The approaches for air taxis are becoming more and more mature. Even the aviation authorities such as EASA and FAA are already working on the necessary certification standards. While driving, he is requested to adjust his speed, considering the traffic situation, weather and road conditions. He should behave in a manner appropriate to the traffic conditions, not endanger or hinder others as possible. He is requested to use the right lane and to drive largely in its middle. Furthermore, he should drive with foresight and conservatively in the interest of the environment and safety. In general, the experienced driver orients himself to the right edge of the road. As long as there are no reasons to change his speed, he drives at a constant speed or with sufficient distance to the vehicle in front in the lane provided for this purpose. The mission and the map of the journey are stored somehow in the brain. People driving supported by a navigation system could very often not explain others how they reach their target. In a professional driving training, you learn how to plan the course of the road with your eyes; i.e. you look ahead as far as you can see and recognise the course of the road, the signs, crossings and critical situations and areas. The complete planning of the driving style for the straightforward section of the road is of course done in the brain. Nevertheless, such single sections are not stored in the long-term memory. If a person drives to his destination with the help of a navigation system and is asked about the chosen route, he will often not be able to name the concrete route. The driver has concentrated purely on the information provided by the navigation system and has already forgotten the previous road at the next junction. He will have forgotten all the insignificant information and events that have influenced his speed or most of his activities that influenced the driving behaviour at all. All road unevenness or the compensation of crosswind, etc., will probably have been performed by him alone as a "spinal cord function", and he will not be able to consciously describe or explain it in any comprehensible way. However, this also means that most of the functions take place in the driver's subconscious while driving. The driver continuously learns and improves these functions, by daily driving. Every new situation, as well as every condition and the possible combinatorics must be learned. This begins in childhood and ends with the driver saying goodbye to the traffic. If a driver finds himself in an unknown traffic situation, he has to adapt his driving style accordingly conservatively. In the end, the inexperienced driver will always be asked to drive at a correspondingly slow pace.

3.3 Human Control Mechanism

In order to control a vehicle in the road traffic environment, people need sensors and actuators.

The main sensors are:

- eyes
- ears
- mouth and tongue
- nose
- skin.

The main actuators are:

- voice
- physical movement
- liquid secretes (Fig. 3.4).

The tongue as a taste sensor is today not important for driving a vehicle, but it is also the basis for speaking. The skin our biggest sensor secretes liquids if we sweat, and the liquid could influence our ability to detect information or limit the

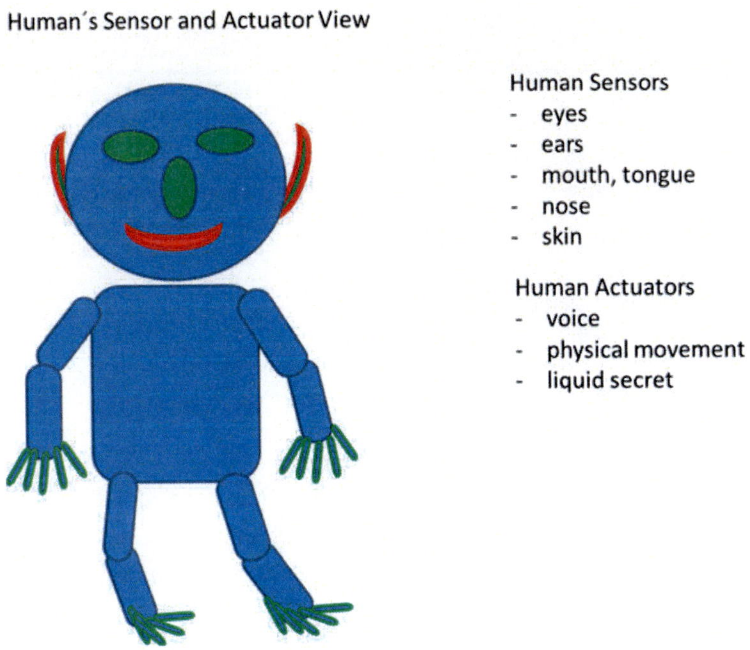

Fig. 3.4 Human sensors and actuators

grip at the steering wheel. That could influence sensors like the capacitive sensors, which monitor the driver activities but also lead to wrong driver impacts in sudden situations. The skin, tendons and the nervous systems are configurable pressure or torque sensors, which could also directly initiate actions.

For any action, the information and the energy to perform the action need to be available at that point in time, for those humans have a very complex network for communication and control.

Organs and mechanisms inside human bodies like

- brain and their sectors,
- spinal cord,
- tendon,
- nervous system,
- sense of balance,
- muscles,
- blood track,

have multiple tasks, together with the human sensors and human actuators are the human-system. They transport information and energy but also doing control. This could be local control functions like reflexes, but also decentral tasks like thinking in the various segments of the brain (Fig. 3.5).

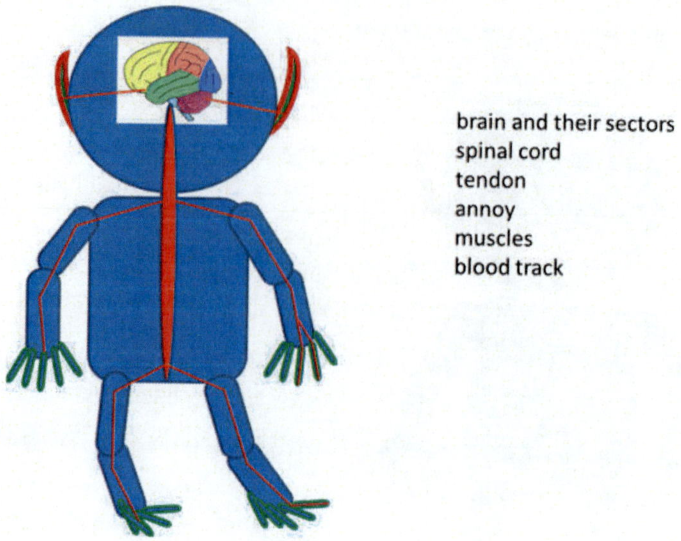

Fig. 3.5 Human control and communication devices or the human system

For example, the human driver does not control the uneven road with his brain, but with reflexes or trained motions based on the feedback he gets from the steering wheel. This way he sometimes grips the steering wheel tighter or he lets the steering wheel slide loosely in his hands.

Nervous system, tendons and the spinal cord, as well as the veins that supply the human system with blood, are linked to the different parts of the brain through a flexible logic system. Every nerve cell or the synapses in the brain are programmable logic units that can perform their own operations, but also adapt information or simply forward them. Say every single nerve cell or synapse is a configurable logic element.

Tendon and nervous system are not just communication systems; they provide also the energy supported by the muscles. For some functions, they could be compared with wired systems or hydraulic control systems. Reflexes are sometimes actions, which are based on inherent experiences from the early beginning of humans. The sense of balance is an essential organ to be able to orientate yourself in the environment.

Those reflexes are in dangerous situations, very important safety mechanism. One of the challenges of automated driving leads to the question how could these reflexes implement and how could you validate those?

3.4 Human Behaviour and Context

Human awareness or more the psychological consciousness is a very complex topic. In order to automate human behaviour or even to implement autonomous driving functions, we need a model about human and "driving" in a specific context. This citation from "Heinz von Foerster" should show a bridge, without diving too deep into psychological theories.

> Significance. It may be strange in times like these to stipulate autonomy, for autonomy implies responsibility:
>
> If I am the only one who decides how I act then I am responsible for my action. Since the rule of the most popular game played today is to make someone else responsible for my acts - the name of the game is heteronomy - my arguments make, I understand, a most unpopular claim. One way of sweeping it under the rug is to dismiss it as just another attempt to rescue "solipsism", the view that this world is only in my imagination and the only reality is the imagining "I". Indeed, that was precisely what I was saying before, but I was talking only about a single organism. The situation is quite different when there are two, as I shall demonstrate with the aid of the gentleman with the bowler hat (Fig. 20) (Fig. 3.6).
>
> He insists that he is the sole reality, while everything else appears only in his imagination. However, he cannot deny that his imaginary universe is populated with apparitions that are not unlike himself. Hence, he has to concede that they themselves may insist that they are the sole reality and everything else is only a concoction of their imagination. In that case,

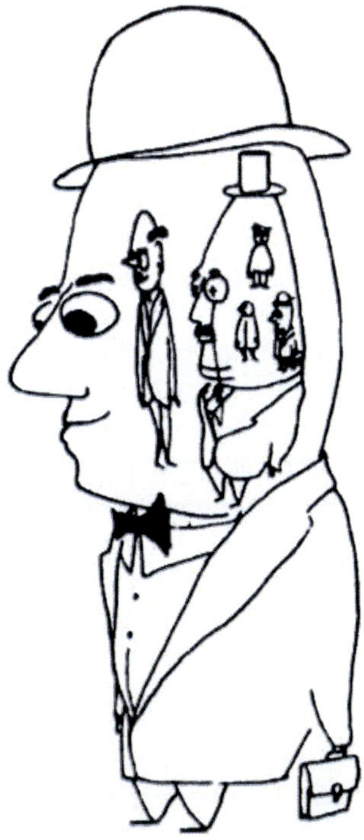

Fig. 3.6 Heinz von Foerster, man with the bowler

their imaginary universe will be populated with apparitions, one of which may be he, the gentleman with the bowler hat.

Source 1 On Constructing a Reality, Heinz von Foerster,

The focus of the entire paper is very similar to that what we intend to do with autonomous driving. These ideas are already hundreds of years old, but in laboratory environments we can already implement such functions. Bringing such kind of technology in mass mobilisation is still science fiction, but not impossible.

Making smaller steps in the way like automation of human driving instead of autonomous driving could provide more efficiency with resources and safety or even social security for road traffic.

3.4.1 Observer of the Observer

In engineering science, you always observe the world, from an engineering point of view. In psychology, the psychologist tries to put himself in the patient's place so that he can understand his problems. In the same way, when a person is thinking about decisions, he puts himself in the position he thinks of and observes himself in the environment he thinks of. However, if he sees himself or his environment wrongly, then he will also estimate his position and situation differently. For each observation, it is important to know the perspective and position, i.e. the context, of the observer, so that one can assess the observation. Nevertheless, even this observer will only ever be able to perceive and describe the observation from his perspective (Fig. 3.7).

In a court case, the judge will have a different perspective on the facts of the case than the prosecutor, the defence counsel or the expert witness. The accused will have a completely different position here than the defendant. This position is not always driven by the case facts, how the damage occurred or what the cause of the damage was. In order to be able to understand the expert witness, the defender and the prosecution often commission different experts, who of course each take the different views of the accused or the defendant. The judge has to evaluate the expert opinions from the point of view of the case law. For this purpose, he has to know and be able to assess the individual perspectives of the different experts' witness.

This is the same issue with any technical specification, and even an independent observer has a certain context, which he provides based on the professional experience, motivational context, etc. Without a context description, also a so-called neutral observer witness or observation leads to a one-sided perception. Unless the context is available, any other stakeholder could understand the point of view or draw his own conclusion. In software technology, there is a method that the observer pattern (also called listener pattern) is a design pattern from the field of software development. It

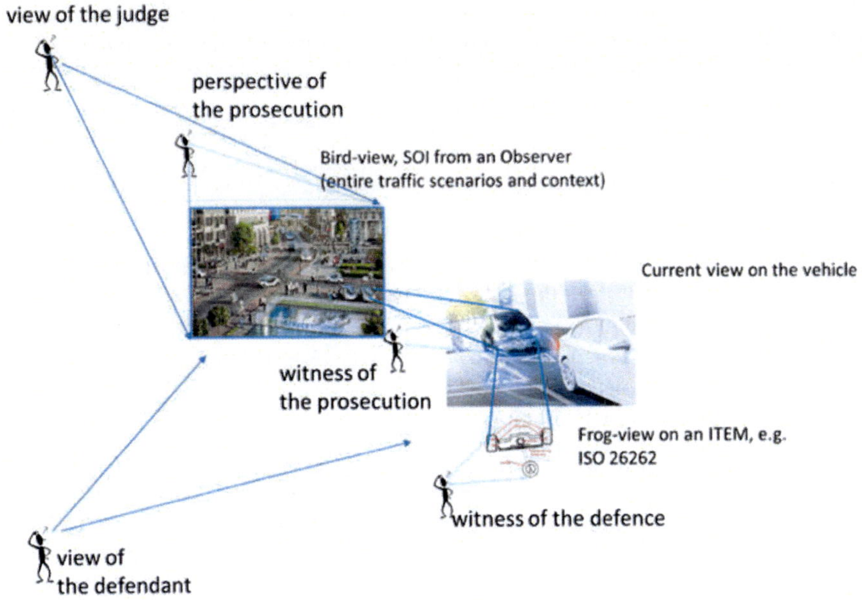

Fig. 3.7 Viewpoints and perspectives in a court case

belongs to the category of behavioural patterns and serves to pass on changes to an object to structures dependent on this object.

In addition to typical publish-subscriber pattern, the observer pattern is an enhanced developed with the signal-slot concept. Observer is in that concept, a behavioural design pattern that lets you define a subscription mechanism to notify multiple objects about any events that happen to the object they are observing.

Besides such specific tools, the respective comparison between the intended goal and the observable behaviour will be an essential aspect of the development process. Again, there is the point of view to give a behaviour during development and the expected behaviour in the real world. In safety analyses, one often considers the behaviour and conditions that can be observed during development and makes estimates of how the behaviour will show up in the real world. Especially when technical systems have to cooperate very closely with humans, it is necessary to consider how the technical system behaves together with humans. The perspective of intended use case will also be different from the perspective in case of occurring damages.

A very typical approach is also the comparison of customer specification and performance specification. The customer specifies his requirements based on the needs for the integration into his professional intend or his wider product scope. The customer describes his tasks and problems in a so-called "Functional Specification", which should not restrict the solution space for the suppliers.The supplier specifies his capability, what ideas the supplier has to provide a solution for the customers' problem. The customer offers his performance within his performance specification. The customers know that the suppliers write their performance specification on different contexts, which lead also quite often to different solutions for the same customer problem. This is another example, where a different context could provide very different results in the observation.

3.4.2 *Industry Automation*

In any industry, we made the experience; whenever we try to automate human labour, we could not just copy the human functions. The McKinsey Quarterly, "Where machines could replace humans—and where they can't (yet), July 2016" shows many examples where human functions could not automate. For some of the examples, you could easily find solutions, which could make the human labour obsolete. If huge forestry machines are a good solution for our environment and should be supported by society or politics is a secondary question, but not negligible (Fig. 3.8).

For any automation task, the operational intend in the intended context has to be specified. Without that any product description, rights for the product like patents but most likely, any burden for the product could be defined. Consequently, the product is not sellable.

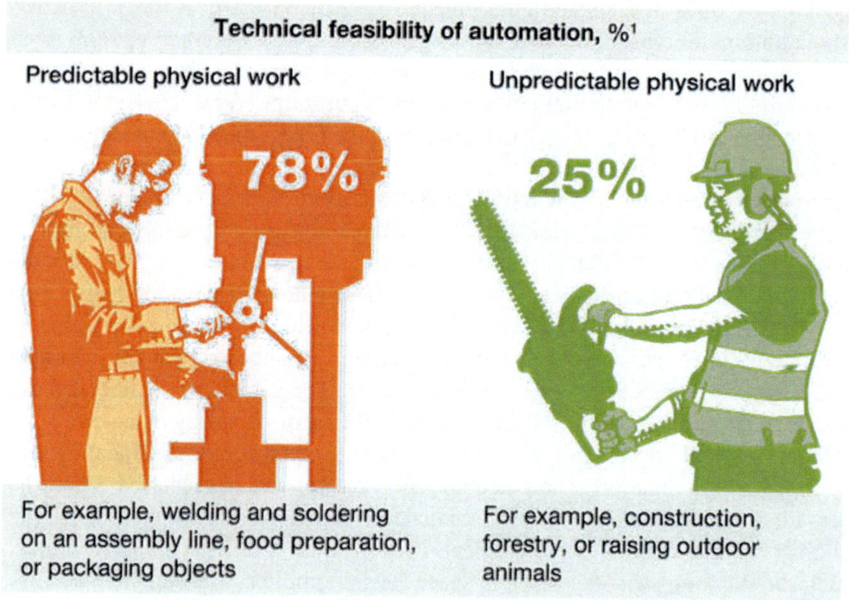

Fig. 3.8 Where machines could replace humans, McKinsey. *Source 2* McKinsey Quarterly, where machines could replace humans—and where they can't (yet), July 2016

Today, people often talk about robot taxis; the term robot is no longer associated with the original idea of creating a human-like machine. Even though the word "robot" has Czech roots, where it was originally derived from the term for "artificial human" or automaton. It was soon discovered that the robot could serve humans better if it is optimised for its intended use. If, for example, the robot only uses paths that are intended for it, then the paths should be designed in such a way that the robot can use them as efficiently as possible for its purposes. This means designing the intended range of motion for the robot so that it can perform at its best. However, even the typical robot is not anymore just created in the image of a human being and it can be seen that the robot works better on wheels than with legs. It also shows that the robot is easier to realise with wheels. Of course, the developer of the robot must also make sure that the robot does not do things that the people around it do not want or even endanger. Everyone probably knows the stories and films where robots become independent. This means that a space is always created for a robot in which it can function optimally; these functions could lead to a hazard in another space. The automated vehicle is used not only by the driver, but also by the people who are transported from A to B inside, or by the logistics company that has its goods transported from C to D. However, the space, the roads and the traffic environment themselves are usually not designed for robots, but for vehicles driven by a human driver.

In any scene or situation, these roles can have different requirements. Currently, there is no such clear assignment of legal requirements for Automated Driving.

The scene can be compared to a backdrop in a theatre play. The stage set including the artefacts such as walls, furniture and pictures forms the static traffic environment. The script describes the dynamic behaviour of the actors and the effects, which are initiated by the stage designers. The effects are often natural in the traffic environment, as if weather conditions or pollution, dirt, etc., are blown by wind onto the road. With rain, such a dirty road surface quickly becomes a greasy and slippery affair for traffic. A similar risk means water on the road at low temperatures, which leads to black ice. Only the overlapping of the stage set, with the moving elements and the actors, leads to the play, which the spectator, or the observer of the scene, perceives (Fig. 3.9).

Driving in various scenarios leads also to very many actions required by the driver and many possible impacts to perceive. The driver at least has the following basic tasks:

- determining the lane or trajectory,
- monitoring the traffic environment and
- safely controlling the vehicle.

Approaching another scene, the transition to another scenes and the new scene itself could provide many different impacts. Typical situation is approaching a road crossing. The driver enters a new scene, so other actors and a new scenery for all actors come into play. This also changes the necessary role behaviour in the new scene. The transitions between the scenes always mean an increased risk, because all actors have to be synchronized with each other in the new context. And of course the road type could change, and new regulations for the traffic need to be applied (Fig. 3.10).

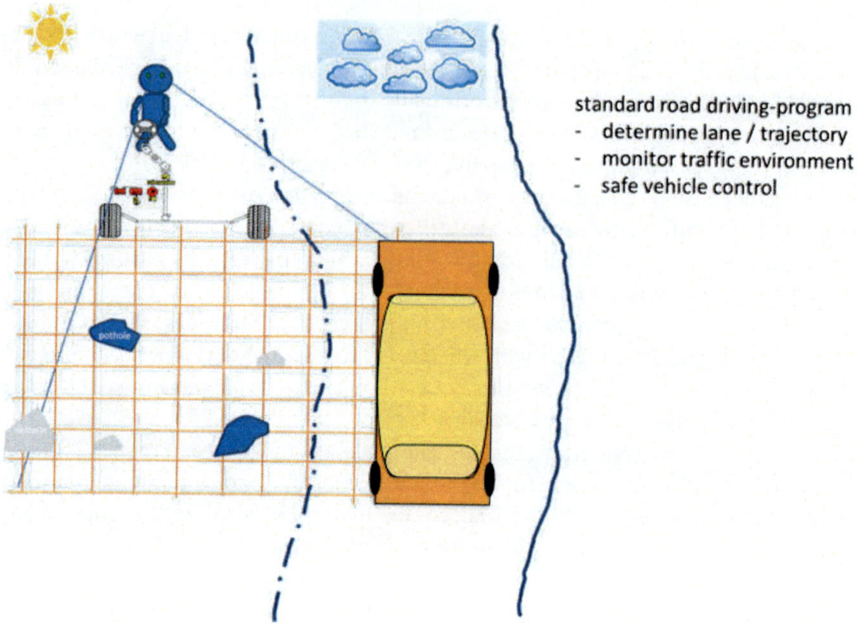

Fig. 3.9 Scene of steady driving in road traffic

An actor on stage knows how to behave in the new scene or with a new set in the background. The scriptwriter has anticipated all these scenes and situations; the producers and directors continuously interpret during rehearsals and the performance whether the actors are following the guidelines. Thus, based on his driving school and driving experience, the driver should also know what he has to pay attention to, what risks he has to calculate with and how he has to adapt his driving behaviour to the new scene or traffic environment.

3.4.3 Multiple Contexts for Different Layers of Abstraction

The context of the driver in a vehicle is limited. The driving seat, the position in the seat, the segment of the windscreen, etc., are designed in a way, so that drivers could see, what they need to see or what they should see to fulfil the road traffic regulations. What the driver could see or perceive, from the physical ability, from the position of the driving seat and from the entire ergonomic and design of the vehicle is very different. We need to ask the question what the driving system needs to perceive to achieve safe driving (Fig. 3.11).

We need to consider a multi-context for the "driving system". With today's vehicle architecture, the driver could only influence the car, what he is driving. All the approaches for autonomous driving follow the same principles. For the road

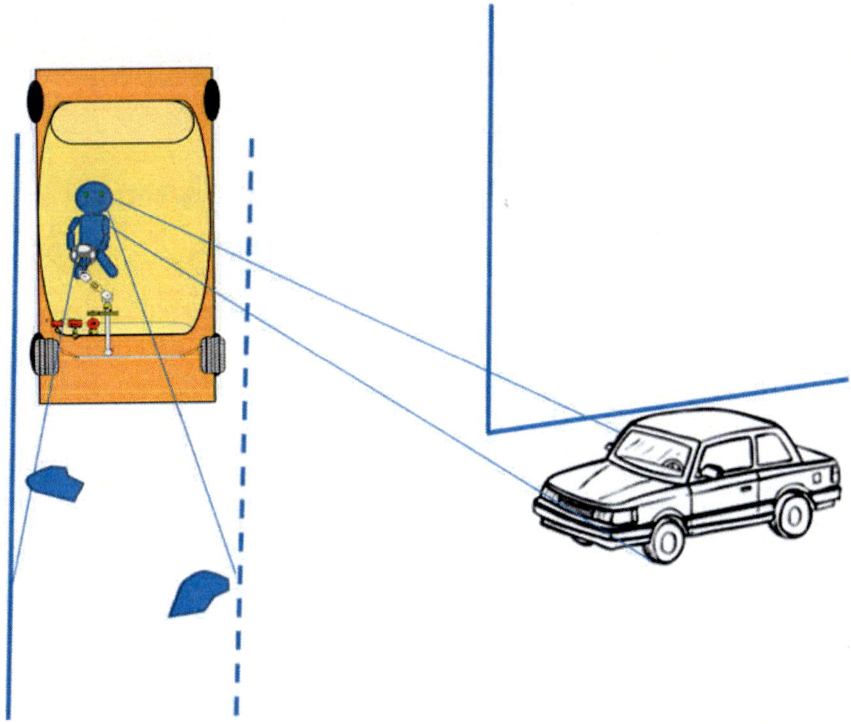

Fig. 3.10 Driver in the context of vehicle and road traffic, new scene approaching a crossing

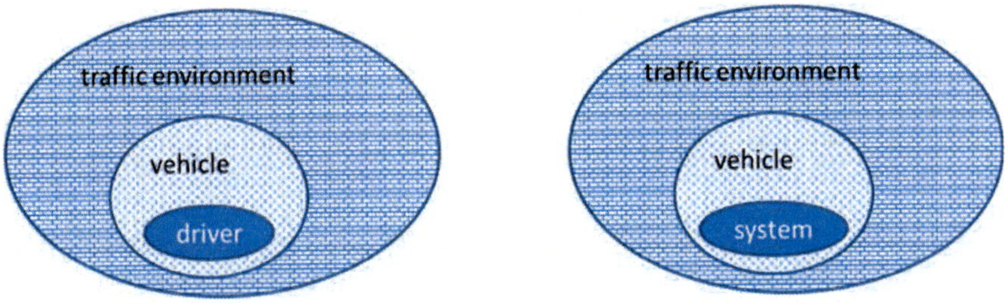

Fig. 3.11 System in the context of vehicle and traffic

traffic context, consequently the autonomous driving system could only influence the vehicle in a way that its behaviour follows the rules given for the vehicle from the road traffic context. Most of the requirements to be considered for the driving system are derived from the road traffic regulations like

- road design and behaviour under certain conditions,
- vehicle design and principles of operation,
- drivers' capability and expected behaviour,

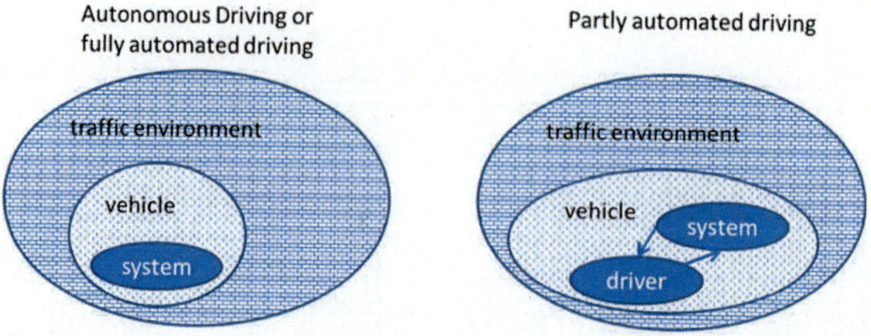

Fig. 3.12 HMI, system and driver in the context of vehicle and traffic

- behaviour of other road traffic participants,
- facts of life, like weather, how to respect weaker road participants, expected coexistence of humans, etc.,
- penalty and liability, etc.

For pure autonomous driving, the context seems to be equal to fully automated driving. In case of partly automated driving, we need to consider the drivers' interfaces in all conditions and situations with the context of road and vehicle (Fig. 3.12).

The positive aspect of these complex scenarios is that their Road Traffic Act defines most of the correct behaviour of vehicles, drivers and any other road traffic participants in nearly all countries of the world. The interpretation of all that rules and regulations for road designer, vehicle manufacturer, drivers and rule makers is much unconsolidated, which is shown by current publications.

3.5 Communication and Interfaces

Communication is closer to understanding than to speaking. Heinz von Foerster tries to make transparent in his publications that communication is dominated by the receiver of the information. Unless the sender of the information could not communicate verbally and bring his words in the right context for the receiver, communication could fail. The receiver could only understand the words from the sender unless they speak the same language. Several tests have shown that human stores information in clusters. Depending on the receivers' background, the receiver stores the information in certain categories. If sender and receiver have a different background, like a baker and a hunter, the same words could lead to a different understanding of the information (Fig. 3.13).

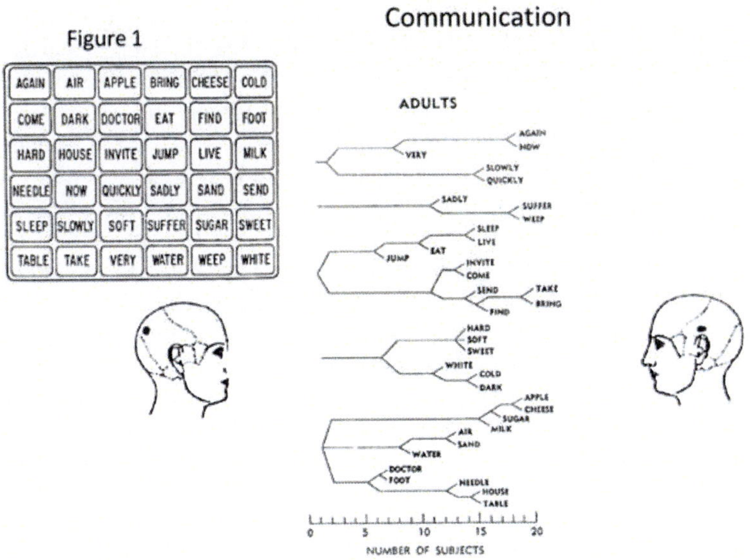

Fig. 3.13 Free interpretation of Heinz von Foerster, Understanding Understanding. *Source* 3 Understanding Understanding: Essays on Cybernetics and Cognition, Heinz von Foerster, 2003

In the referenced publication from Heinz von Foerster, children would have clustered the information completely different. The same is true for a technical communication, unless you bring information from sender to the receiver not in the same context, communication could fail (Fig. 3.14).

If a controller commands an actuator, the reaction depends on the context of the receiver. "Go left" could be understood for example

- from the sender perspective or from the receivers' perspective,
- from the context of the receiver (e.g. going on the left side of the road).

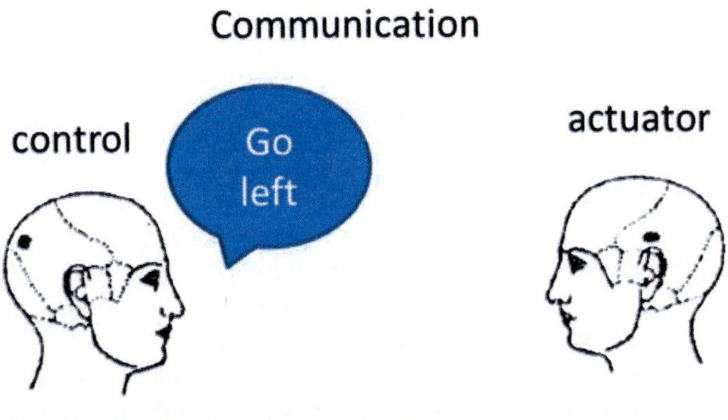

Fig. 3.14 Communication in a specific context

Those assumptions seem to be true for any interface, and not only the dedicated function or information needs to be communicated, also the context needs to be consistent. Communication requires also the synchronisation of the context, this could happen by many kinds of measures, and for example the context could be part of the message, could be communicated separately or could be provided by the design of the interface. For technical communication systems, it is necessary to bring the communication partners into the same context. Unless the context is not synchronised and harmonised, the communication partners will not understand each other. A word is only an odour or noise, which can be understood by the listener in its word content, if he speaks the same language. Thus, a byte is also only a set of electrical impulse series that must be generated by the transmitter according to a scheme known by the receiver so that the receiver can initiate the appropriate action or simply receive the byte as such.

The same challenge concerns the developer who wants to develop an automation function. If he does not meet the requirements of his stakeholders, it will not be a successful product. If it does not meet the expectations of its users, the product will not be used. If, for example, a driver assistance function is developed, the driver will have to be able to see a benefit from the function. If he does not recognise it, he will not be able to implement the function correctly, i.e. the benefit of the function will not be able to reach the user of the function. For this reason, some form of communication must be established between the developer and the user, which puts the developer in the context of the user's expectations. If the developer works in a different application context, he can also endanger the user or people in the environment. These are essentially the aspects of user safety or safety-in-use that ensure the safe handling of the products or working in a safe context of a technical product or functionality. In case of mass production products, the user manual is often used to put the user in the right context. For a passenger car, the manuals are already so huge that a normal driver or owner cannot read them completely or even understand them. This means that the driver needs technical aids so that he can be put into the right context for the various automation functions. Of course, these functions must be active while the vehicle is in operation, or it must be possible to activate them in the relevant situations.

3.6 Human Driver as a Control System

Humans work with feedback, and they learn with feedback. Even perceiving is more a kind of recognition. Some events, scenes or objects, humans learn to perceive, by training. Even before birth, embryos learn the voice of their parents. Even the own image and the own voice are learning processes for a baby. Some of the learning seems to be also in our genes so that we can inherent effects from our parents (Fig. 3.15).

Through the lenses in our eyes, we see the whole world upside down. The two eyes provide us with two images that are adjusted and stored in our brains, or we compare them with something familiar to classify what we saw. In long-term memory, we

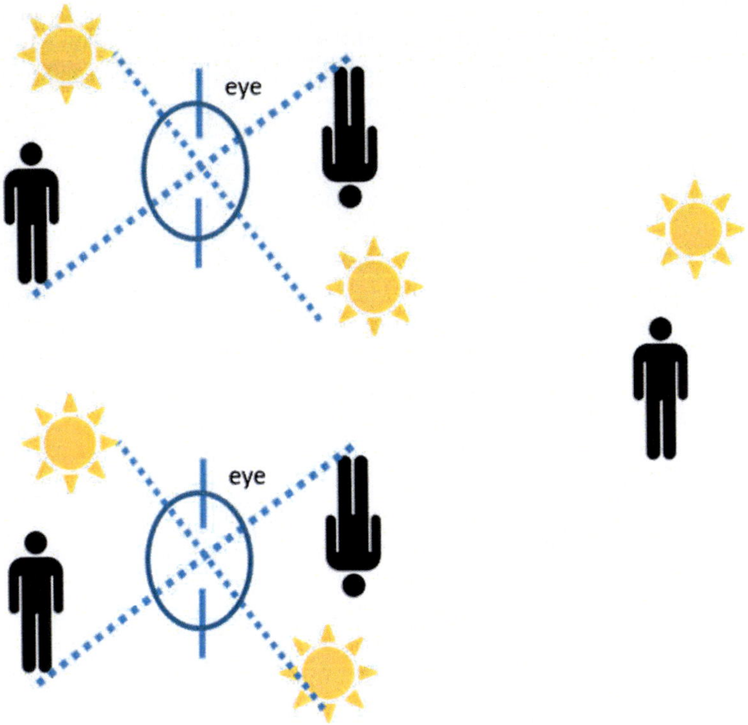

Fig. 3.15 Human perception with the eyes

probably only save the class. The older we get, the more scenarios and classes we deposit in our store. We must practise all reflex-type exercises like dancing regularly. Certain control algorithms such as cycling are probably stored as a pattern in our nervous system, and the control parameters are probably in the brain. The sense of balance is a typical feedback sensor for the information we perceive with our eyes (Fig. 3.16).

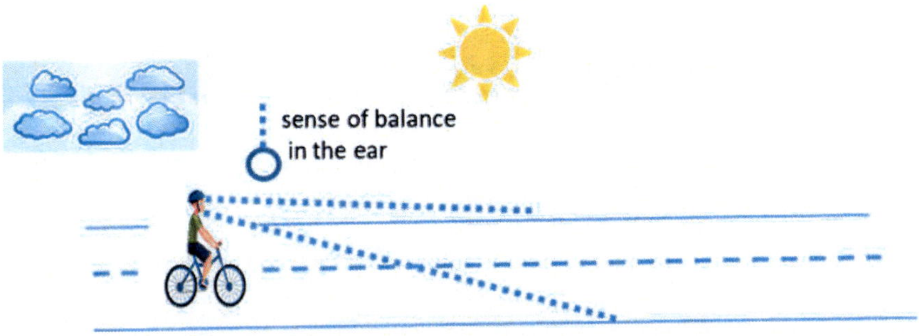

Fig. 3.16 Bike cycling as control loop

However, during cycling not only the eyes and the balance sensors are involved in cycling. Our nervous system learnt how the pedals feel, how much force we need in the legs and how the entire body balances disturbances due to road and weather. Another control loop is about speed. In order to cycle stabile, a certain speed is required. Measuring sufficient speed with scanning the environment is a learning issue. Learning cycling requires to estimate the necessary speed and to compensate wind to keep the trajectory. In case of side wind, it could be a serious challenge to balance wind during driving over a bridge or if a bus overtakes a cyclist. Driving uphill or downhill requires also estimations about needed pedal force to maintain sufficient speed for balancing, and during downhill braking needs to control the speed in order to be able to react on events like obstacles on the road or sharp curves. Of course, during downhill the cyclist needs to know the brake performance of the bike, so that he could stop or reduce speed if required. Besides management of the sensors and actuators, a huge synchronisation work of all the information from sensors, actuators and their feedback is necessary. The impact of the pedals and handles and their effect on the bike need to be trained, so that the control pattern remains permanent in the memory mechanism of the brain and the nervous system (Fig. 3.17).

A typical safety-relevant control system requires two sensor units for the safe detection of the real world. Two eyes do not necessarily offer the necessary different perspectives. A comparison with the expected context from the environment can improve the content of the sensor information and its correctness. For the actuator feedback is required for the correct movement of the actuator itself and the feedback for the entire control circuit (Fig. 3.18).

During driving, we read back the torque, for example, from a motor in order to get feedback about the correct controlled torque. The driver receives visual feedback on the quality of the lane keeping and compensates via the steering wheel, for example uneven roads. In a technical control system, sensors and the reference variable usually have to be measured or determined. The behaviour of the actuator is ensured by an internal control loop. The resulting behaviour of the vehicle and the compensation of the environmental influences are controlled by further independent feedback loops. If the control loops and the feedback loops are not independent, there can be very large

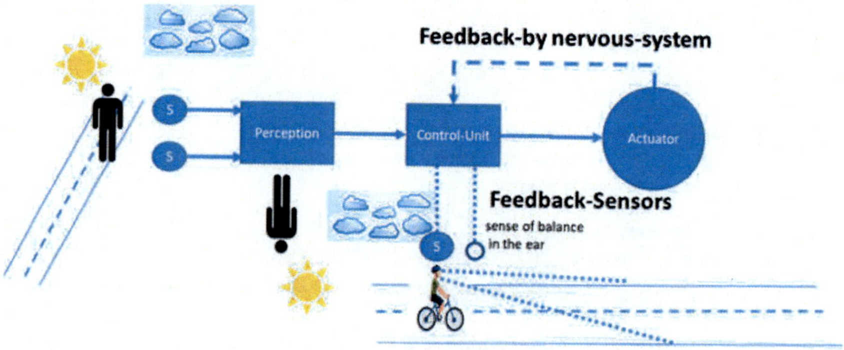

Fig. 3.17 Feedback systems

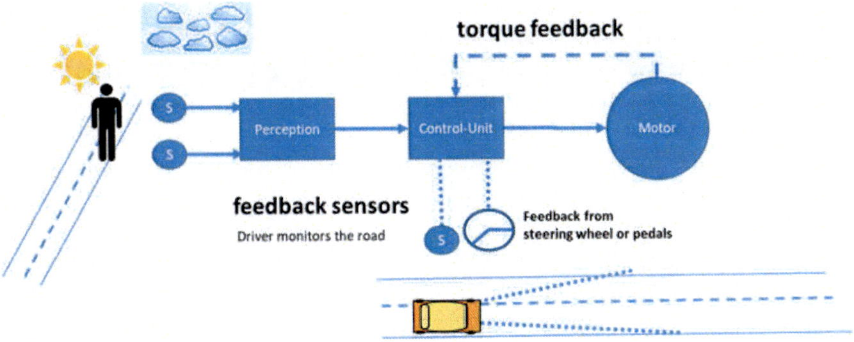

Fig. 3.18 Technical and human feedback systems

disturbances; the system cannot follow the reference variable nor can it compensate for any disturbances without error. If several control loops are implemented in one microcontroller, technical influences and runtime variations are a challenge for the control quality (Fig. 3.19).

There are in today's control system various control loops, and the feedback loops are based, for example, on:

- Microcontroller including application software, basic software, hardware–software interface to the output pin of the microcontroller. In safety-related application and adequate controller, it is possible to read back the data of the pin registers.

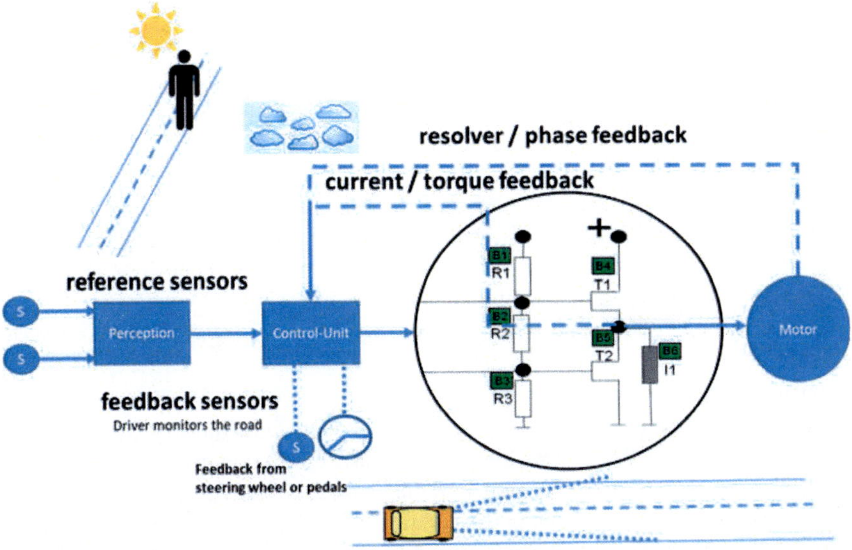

Fig. 3.19 Control cascade in vehicle systems

The application software controls the output based on the feedback from internal or controller external information.
- Transistor power driver uses high-side and low-side drivers as electrical redundancy. In case of motor drivers, the current feedback could be also providing an equivalent torque feedback calculated from the power current.
- Resolver could provide phase position feedback or feedback about slip in an asynchrony motor.
- Wheel speed sensors, acceleration sensors or gyros could provide feedback about the vehicle behaviour and speed and compensate weather and road condition impacts.
- Steering wheel or pedal feeling provides the driver feedback about road or weather condition.
- The driver monitors usually the lane trajectory and corrects it by continuous steering. Minor influences caused by uneven roads, wind, weather or system friction, he compensates in the subconscious.

Vehicle behaviour reduces on two major degrees of freedom, the longitudinal and the lateral axle. In today's driver assistant system, the driver is the closed-loop controller for the prevailing vehicle behaviour.

3.6.1 Human Communication

Human communication only works if the communication partners speak the same language. The typical elements of a communication are:

- sender,
- receiver,
- communication media,
- information transport media.

The human voice is just noise for an alien. A common language uses the same vocabulary and the same syntax and semantics. Signs could base on voice or on letters. The syntax of a language (a sign system) describes the rules according to which the language constructs (signs of the sign system) are formed. The semantics of a language (a sign system) describes the meaning of the language constructs (signs of the sign system). Thus, the word "sun" is a noun and the word "shine" is a verb (Fig. 3.20).

If a person speaks to another person the following sentence: "The sun is shining", he produces a set of noise, sounds or signs. The sender of the information produces a sequence of sounds that follow the agreed rules. If both communication partners understand the same rules in the same way, the receiver will understand the sentence. If the sender does not send the sounds according to the agreed rules, no matter why, the receiver will have difficulty understanding the sentence. The communications

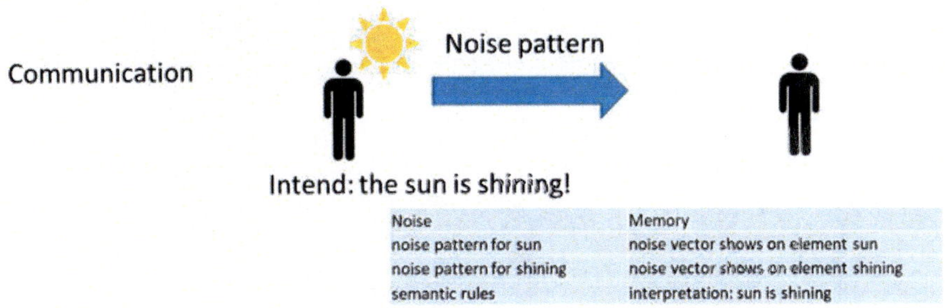

Fig. 3.20 Example for human communication

work if both have the ability to produce sets of signs, sounds or noise according to the same rules.

It doesn't matter, if the communication media is voice, written letters or information on digital communication systems, the rules between sender and receiver need to be unique or even unambiguous.

3.6.2 Human Perception

Human perception is a kind of communication or a part of it. The recipient wants to grasp an element, a property or a behaviour in nature or in the real world and store it in his memory for further actions (Fig. 3.21).

Usually humans

- see with their eyes,
- hear with their ears,
- feel with the skin or the nerves in it,
- smell with the tongue, etc.

In case of verbal communication, humans convert the sound pattern into set of images in the brain. In case of optical perception, we can compare it with railway signals. Free travel or "go" means the signal arm is raised and horizontal signal arm means "stop". The train driver perceives the position of the signal arm and decides if he brakes the train or continues driving. This case is a very simple example of perception, since the train driver is educated in the rules and there are only two possible conditions of the signal arm (Fig. 3.22).

In case of the perception of the shining sun, we could reduce it also to two basic conditions: the sun is shining or not. However, people perceive with their whole body. Even the different parts of the brain react very differently to the different senses. The overall reaction of the person to the different perceptions through their different senses forms a learned combination of functions that can be very different depending on the context. When we perceive loud noises in the dark, the whole body adjusts to danger. We begin to sweat, our hair stands up, and our muscles pretension

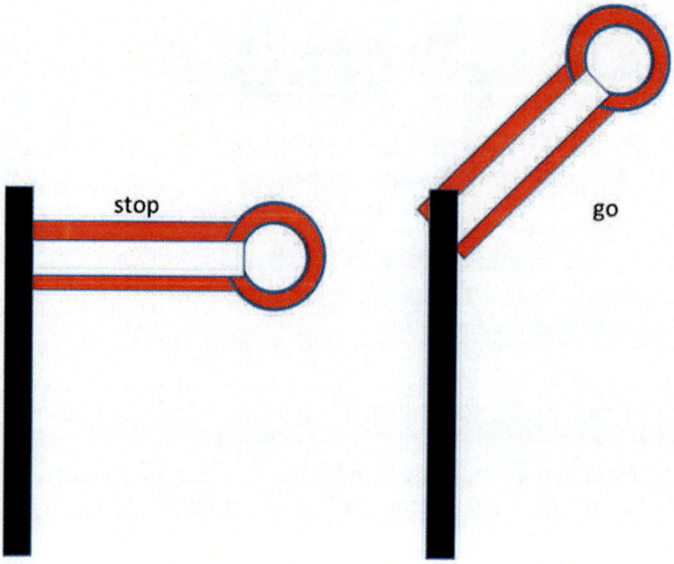

Fig. 3.21 Optical signals and perception

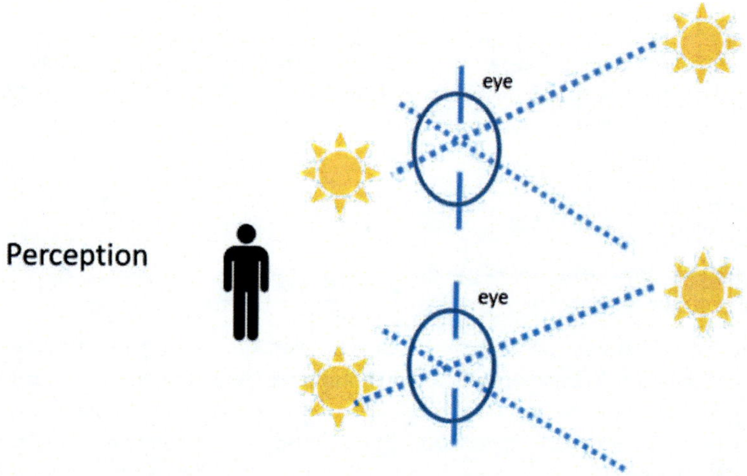

Fig. 3.22 Human perception

to run away quickly. Reflexes react to other stimuli that are perceived by other senses and are processed in other parts of our brain. Thus, the right and left hemispheres of the brain are different in their perceptual mechanisms. One is systematic and fact-based and the other more image-oriented. We often hear that artists perceive more in pictures, and scientists are more fact-oriented. Either heredity or learned

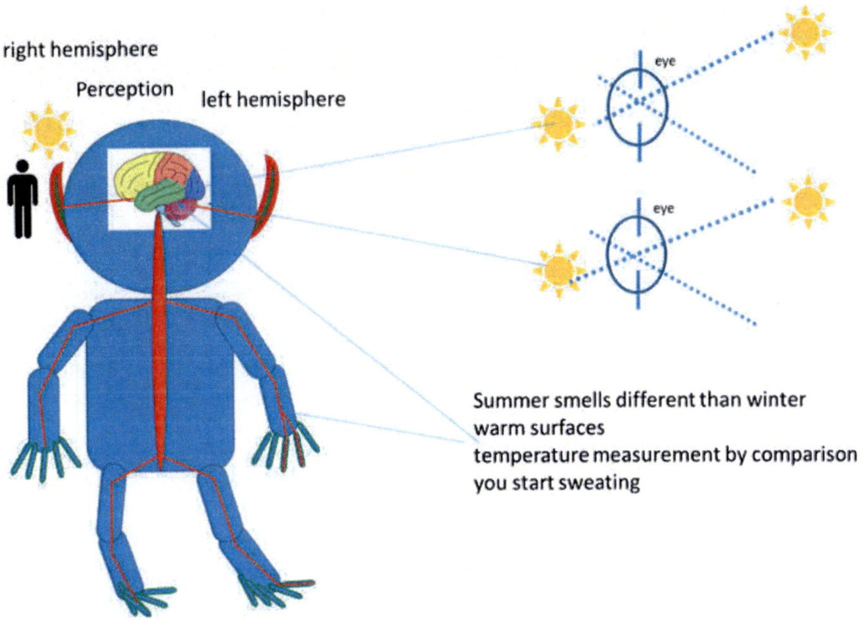

Fig. 3.23 Human perception

behaviour could lead, for example, to anxious or factual person. Nevertheless, even very objective people can be anxious (Fig. 3.23).

Living beings perceive with the whole body or their entire entity. The example with the sun shows that one does not necessarily have to understand the detailed wording to understand the phrase; "the sun is shining". The glare of the sun from above, the warmer surfaces, the skin transpiring, the smell of nature in heat and so on are enough to understand the statement. Even the two eyes are just two sensors. Three-dimensional perception is not based on measuring the focal points on the iris in the eye, but on experienced images that have learned the third dimension.

Children are not so good at judging the distances of moving cars, but adults also perceive distances and speeds in the dark much more uncertainly.

Reflex-like movements are often primordial instincts that become dulled in the course of life, and one becomes more relaxed. You learn to assess such situations better. Of course, such serenity has not only cost jungle hunters their lives.

The capability of humans is not limited to the perception of real-world issues, and it is more how people could react on environmental events or stimuli, some response impulsive, some sporadic, some after a while of thinking or balancing of the consequences of the reaction. Sometimes, it is the key to response on stimuli at the right moment in the right way. It is not only a matter of surviving of the strongest or the most lucky one; it is often a matter of assessing the situation and risks and weighing up which decision is the right one at the right moment in time within in the right situation.

3.6.3 Technical Control System Comparison

John McCarthy provides the following definition for "*artificial intelligence (AI)"for layman or students in the first semester: "It is the science and engineering of making intelligent machines, especially intelligent computer programs. It is related to the similar task of using computers to understand human intelligence, but AI does not have to confine itself to methods that are biologically observable*".

He provides a list of branches of AI:

Logical AI: What a program knows about the world in general the facts of the specific situation in which it must act, and its goals are all represented by sentences of some mathematical logical language. The program decides what to do by inferring that certain actions are appropriate for achieving its goals. The first article proposing this was [McC59].[McC89] is a more recent summary.[McC96b] lists some of the concepts involved in logical aI.[Sha97] is an important text.

Search: AI programs often examine large numbers of possibilities, e.g. moves in a chess game or inferences by a theorem-proving program. Discoveries are continually made about how to do this more efficiently in various domains.

Pattern recognition: When a program makes observations of some kind, it is often programmed to compare what it sees with a pattern. For example, a vision program may try to match a pattern of eyes and a nose in a scene in order to find a face. More complex patterns, e.g. in a natural language text, in a chess position, or in the history of some event are also studied. These more complex patterns require quite different methods than do the simple patterns that have been studied the most.

Representation: Facts About the World Have to Be Represented in Some Way. Usually Languages of Mathematical Logic Are Used.

Inference: From some facts, others can be inferred. Mathematical logical deduction is adequate for some purposes, but new methods of non-monotonic inference have been added to logic since the 1970s. The simplest kind of non-monotonic reasoning is default reasoning in which a conclusion is to be inferred by default, but the conclusion can be withdrawn if there is evidence to the contrary. For example, when we hear of a bird, we man infer that it can fly, but this conclusion can be reversed when we hear that it is a penguin. It is the possibility that a conclusion may have to be withdrawn that constitutes the non-monotonic character of the reasoning. Ordinary logical reasoning is monotonic in that the set of conclusions that can the drawn from a set of premises is a monotonic increasing function of the premises.

Circumscription is Another Form of Non-Monotonic Reasoning.

Common sense knowledge and reasoning: This is the area in which AI is farthest from human-level, in spite of the fact that it has been an active research area since the 1950s. While there has been considerable progress, e.g. in developing systems of non-monotonic reasoning and theories of action, yet more new ideas are needed. The Cyc system contains a large but spotty collection of common sense facts.

Learning from experience: Programs do that. The approaches to AI based on connectionism and neural nets specialize in that. There is also learning of laws expressed in logic.[Mit97] is a comprehensive undergraduate text on machine learning. Programs can only learn what facts or behaviors their formalisms can represent, and unfortunately learning systems are almost all based on very limited abilities to represent information.

Planning: Planning programs start with general facts about the world (especially facts about the effects of actions), facts about the particular situation and a statement of a goal. From these, they generate a strategy for achieving the goal. In the most common cases, the strategy is just a sequence of actions.

Epistemology: This is a Study of the Kinds of Knowledge that Are Required for Solving Problems in the World.

Ontology: Ontology is the study of the kinds of things that exist. In AI, the programs and sentences deal with various kinds of objects, and we study what these kinds are and what their basic properties are. Emphasis on ontology begins in the 1990s.

Heuristics: A heuristic is a way of trying to discover something or an idea imbedded in a program. The term is used variously in AI. Heuristic functions are used in some approaches to search to measure how far a node in a search tree seems to be from a goal. Heuristic predicates that compare two nodes in a search tree to see if one is better than the other, i.e. constitutes an advance toward the goal, may be more useful.[My opinion].

Genetic: programming Genetic programming is a technique for getting programs to solve a task by mating random Lisp programs and selecting fittest in millions of generations. It is being developed by John Koza's group and here's a tutorial1.

Source 9 https://jmc.stanford.edu/articles/whatisai/whatisai.pdf, John McCarthy, download 2020.

Artificial intelligence is not assigned to a system, and it is often considered as the appearance for an observer. Machine learning is often described of learning in which machine can learn by its own without being explicitly programmed. It is a subset or an application of AI, which provides systems the ability to automatically learn and improve from experience.

In system engineering, system architecture and design provide the basis for technical functionality. If we jump from artificial intelligence or machine learning to human behaviour, we always forget the impact of the human system. If we want to provide human functionality, it is not only a matter of software algorithm; we also need a system, which provides the required feature, performance and capabilities.

The brain is a kind of high-performance computer system with functional units like

- a huge memory,
- monitoring centre and
- central intelligence with distributed intelligent I/O system (Fig. 3.24).

The high-performance controller is shown redundantly, and it could be the equivalent for various parts of the brain, which operate separately from each other but it

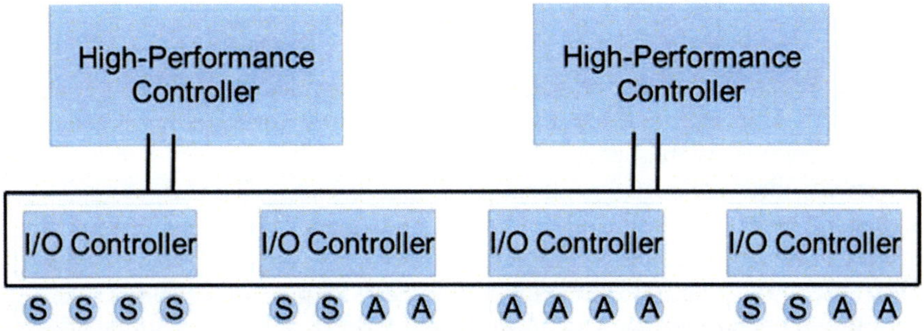

Fig. 3.24 Human system as a distributed control system

could be also just redundant resources to host various function groups of the brain. Similar like the brain, separate independently operating resources should be available to manage the input from the sensors with a different focus or a different configuration. An example could be the theories about the left and right part of the brain, the hemispheres. If you are turning clockwise, you prefer your right brain to be more creative, intuitive and emotional, and that is where images predominate. If it rotates counterclockwise, your left brain dominates, making you more of a structured, logical–analytical thinker. Another example is that in right-handed people, predominantly the left hemisphere controls all linguistic functions. Calculation and logical thinking also take place here. As the largest commissure (connection in the brain), the corpus callosum connects the two cerebral hemispheres. The right hemisphere could be also seen as a part, which monitors from a context point of view, and how pictures are in its environment, and the left hemisphere records and controls the specific actions. Reading and listening are managed, for example, in different hemispheres.

The autonomic (vegetative) nervous system regulates the processes in the body that cannot be controlled by the will. It is constantly active and regulates, for example, breathing, heartbeat and metabolism. It receives signals from the brain and sends them to the body. It provides all the information and controls all the necessary functions for utilities and basis functioning of the human system.

The big miracle addresses consciousness versus conscience, which seems to be not a real issue for the automation of human transportation systems. It is very questionable if road traffic systems should make their own ethical decisions.

It is important to understand the human functions, how we use them to mobilise or operate the actuators and how we get the feedback, the measurement and control system behind. If these functions and interactions are known, it is possible to ensure a safe coexistence between humans and mobile transport systems. Furthermore, the human system seems to demonstrate efficient resource utilisation and can implement fast-cascaded control loops with high reliability even by using machine learning or any other artificial intelligence.

3.6.4 Control System Architecture in Comparison with the Human Control System

Engineers from plant automation will immediately recognise the distributed control system (DCS) from their industry. Even today, the high-performance buses between those units are well established (Fig. 3.25).

There are local control loops, distributed control loops, cascades, systems for data history, preventive maintenance systems and far more functions that humans also need to control a vehicle (Fig. 3.26).

The I/O controller needs to be flexible so that today's typical I/Os could be adapted fast to the I/O controller but also all I/Os need to made available with dedicated

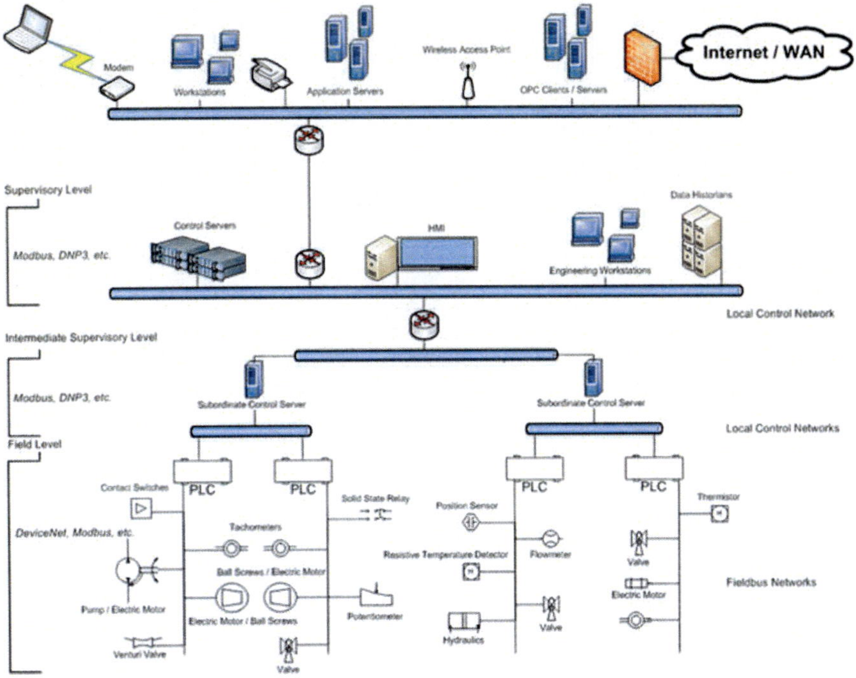

Fig. 3.25 Typical distributed control system. *Source* 4 download from: figure/Typical-architecture-of-a-Distributed-Control-System_fig1_242457425

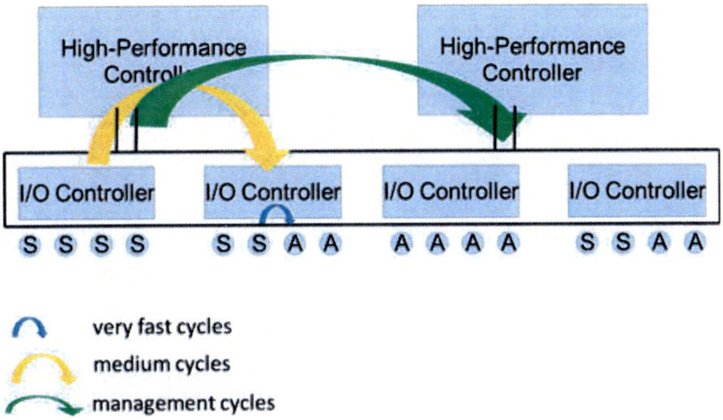

Fig. 3.26 Distributed control system

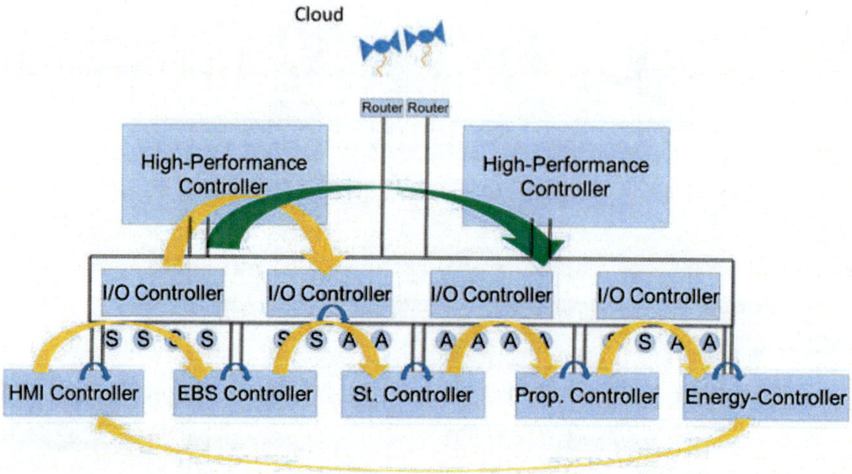

Fig. 3.27 Extendable distributed control system

identifier to all other controller units. In case of redundant Ethernet backbone, any extensions are possible (Fig. 3.27).

Router could extend the network with cloud systems to provide services like remote control, diagnostics, flashing-over-the-air, etc.

Existing control devices with redundant Ethernet communication are

- HMI controller,
- electronic brake controller,
- steering controller,
- propulsion controller,
- energy management controller or any other utility controller.

The control loops could be implemented on any control level, and the loops could implement various cascades in various system states. Re-configuration or re-flashing during operation could be implemented due to existing redundancy concepts and central diagnostic or predictive health management in control devices within the network.

Similar to much enhanced actuators, also human drivers perform closed-loop control just with the actuator. With pedal feeling and the feeling of the steering wheel, the driver perceives the frequency to control vehicle vibrations, road unevenness, etc., and these are just reflexes or other local control functions by the nervous system. When driving over a pothole, you drivers learn to grab the steering. A steering actuator increases in some conditions the current due to road feedback, and no other sensor needs to be involved for such closed-loop control. Typical valve control automatically increases the current in case of higher feedback pressure. With just a one-bit operation command, the function could be inverted (e.g. decrease in case of higher feedback pressure). Such very fast closed-loop principles drive the design of brake and steering actuators, and any kind of cyclic control system by using microcontroller functions could never reach the performance of such actuator design principles. ASICs and optimised hybrid transistors could perform very fast cycles in the order of microseconds. Those fast control loops operate on the basic electro-technical principles like voltage, current or frequency controls. The same electro-technical principles use a human driver with his skin, tendon and alloy to actuate his muscles. Even high-speed graphic cards, many core architecture or so-called General Purpose Units (GPUs) are slower if they control actuators based on sensors or if they work on cyclical processes.

3.7 Cyclic Data Processing Versus Distributed Event-Driven Data Processing

Most processing units follow even today the so-called von Neumann architecture (Fig. 3.28).

The von Neumann architecture is the idea of a computer architecture based on a 1945 description by the mathematician and physicist John von Neumann and others in the First Draft of a Report on the EDVAC (Electronic Discrete Variable Automatic Computer). Parallel processing or using configurable subprocessor networks as humans provide with their nervous systems requires other architectures. FPGAs are quite close to the idea, but they still work digitally with memory and not on learning object-based structures. Maybe configurable semaphores could provide similar processing functions as human systems could do (Fig. 3.29).

McCulloch defines a neuron as smallest unit of a nervous system.

> A neuron is more than just a switch; it is at the same time battery and generator; besides, it provides for the maintenance of its three parts; cell body, dendrites and axon.

This understanding of a neuron as technical elements leads to a very heterogeneous unit, with many technical functions. At least it is information management, logical

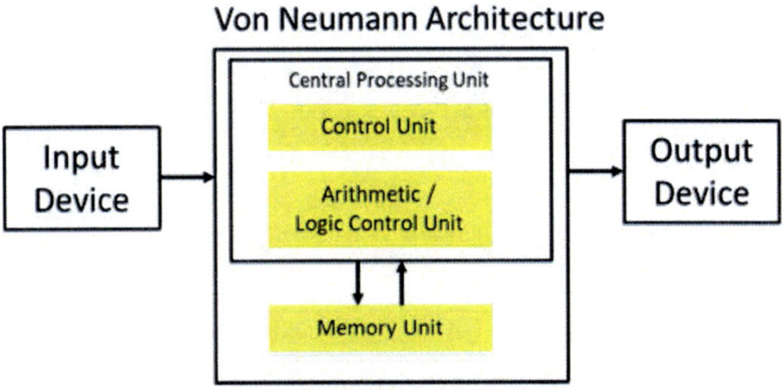

Fig. 3.28 Basic principle of the "von Neumann architecture"

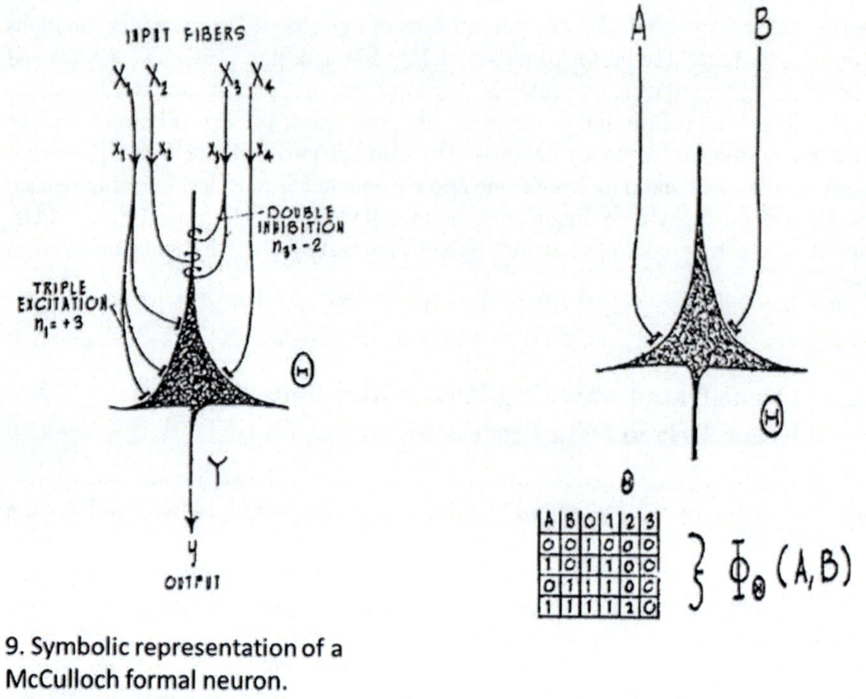

Fig. 3.29 Basic principle of formal neuron, McCulloch. *Source 3* Understanding Understanding: Essays on Cybernetics and Cognition, Heinz von Foerster, 2003

unit and energy management in one system element. McCulloch shows in various diagrams the huge amount of logical combinations of a neuron (Fig. 3.30).

An ASIC with some DSP functions could combine various inputs already in many options. Some peripheral elements from microcontroller could be configured based

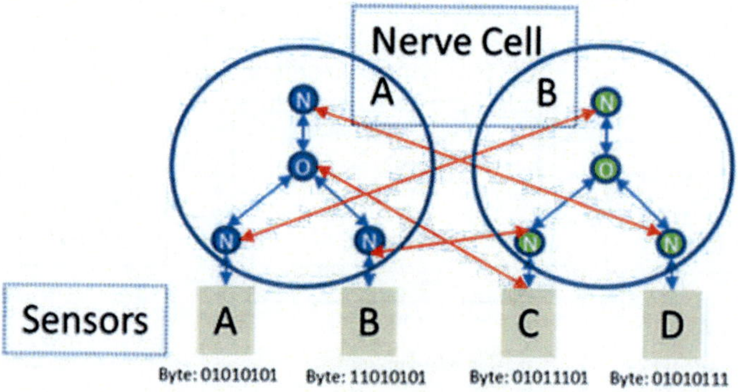

Fig. 3.30 Basic idea of a logical unit similar to a neuron from multiple sensor inputs

on the same hardware as input or output. Providing, distributing, management of energy on the same device is still a challenge realising that in the size of a neuron by this energy efficiency.

Many neurons could be in various combinations and cycles with other neurons, and so sequence pattern leads always to another combination of results. Today machine learning quite of defines by such pattern of sequence a new structure of information (Fig. 3.31).

However, a neuron seems to have more capability than just the above DSP function and it could have also in combination with other cells different states. Due to multiple states, it could reach depending how transitions are from a state to another possible various different action networks could occur (Fig. 3.32).

Due to various combinations and many cycles, a stochastic action network could lead to many combinations of transition and sequences (Fig. 3.33).

Due to feedbacks on one neuron or multiple neurons, any state could lead to many combinations of information. Such states and transitions could be a memory cell, an instruction or a program control function and that in many patterns like semaphores, type repeater or whatever logical possible.

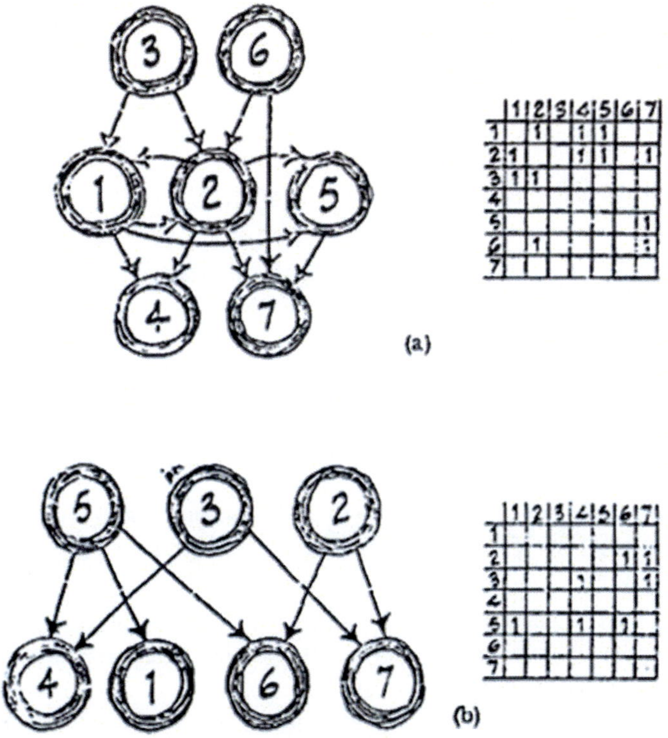

Fig. 3.31 Hybrid network (**a**) and action network (**b**), Fig. 3.2. *Source* 3 Understanding Understanding: Essays on Cybernetics and Cognition, Heinz von Foerster, 2003

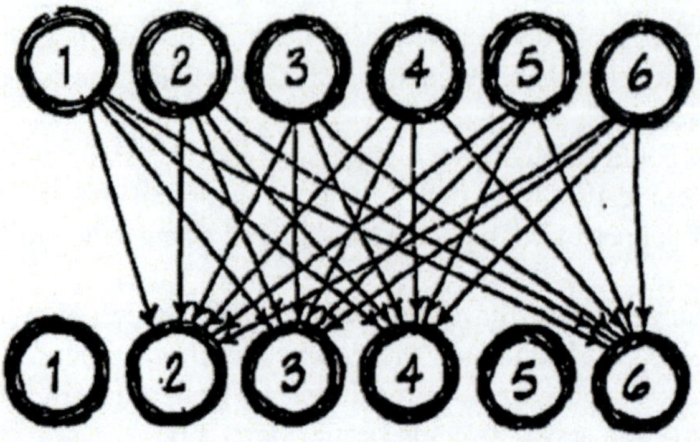

Fig. 3.32 Stochastic action network, Fig. 6. *Source* 3 Understanding Understanding: Essays on Cybernetics and Cognition, Heinz von Foerster, 2003

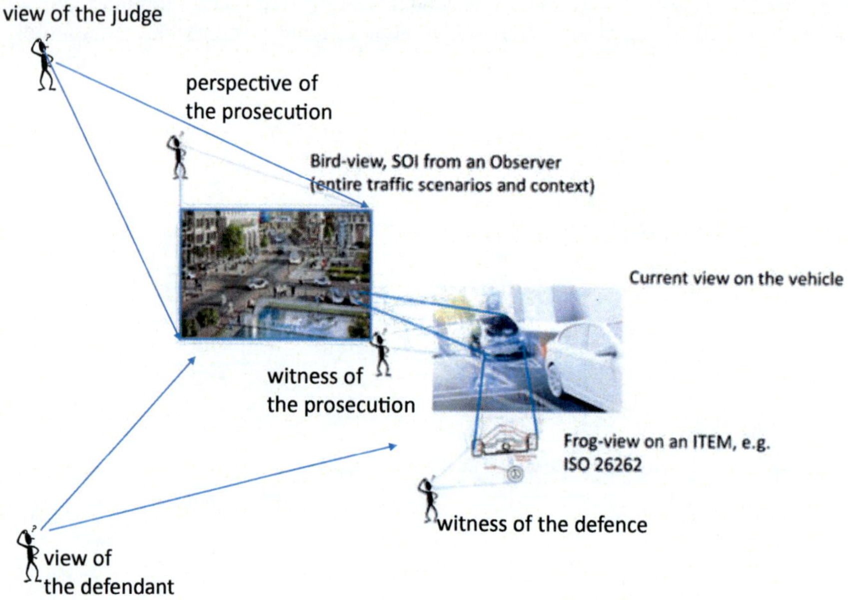

Fig. 3.33 State transition diagram, Fig. 7. *Source* 3 Understanding Understanding: Essays on Cybernetics and Cognition, Heinz von Foerster, 2003

It seems so that the human system is still far ahead of our technical capabilities. Maybe in some laboratories, such prototypes are working, but for mass mobilisation, it would be a long way to go.

The following example is how early people already saw the potential of distributed control.

3.7.1 Distributed Control

Benjamin Peters published a book, "How Not to Network a Nation: The Uneasy History of the Soviet Internet", 2016, the first ideas of the Internet in Russia. It describes the story of "ОГАС", "National Automated System for Computation and Information Processing"; it was a Soviet project to create a nationwide information network, which starts in the early sixties.

In this book, one of the targets of the project was to find solutions to overcome the bottleneck of the "von Neumann architecture" and the opportunities of an open data world. Quite early, it had been recognised that there are also risks

- The access to the data from various parties and the challenge of access and authorisation.
- How could it be limited to Russia?
- Is the financer of the project also the owner of the resulting benefits?
- How could the authorities control local developments?

These are all valid points of today's Internet risks and the new challenges and opportunities of 5G technology.

In 1960, the Russian scientists developed the idea of "Cybertonia", a virtual land ruled by a robot council. Cybertonia then organised regular recreational activities in Kiev and Lviv, conferences and children's parties, published leaflets, its own currency and even a draft for a cybernetic constitution. In 1965, for example, children were given the passport of a country that had a robot on its cover that played saxophone: a symbol of American music, as Benjamin Peters notes in his book: "How to network no nation. The turbulent history of the Soviet Internet" (Fig. 3.34).

Adhesive effects often determine the behaviour of social groups. Such remanence causes certain effects to occur late; i.e. it is known that they occur, but not when. The challenge and the chance of the future technology are probably to understand the inner forces of groups and to intervene in control, so that a benefit for these people is created and the interventions do not endanger people. External influences could affect the intensity of remanence, the kind of activated inner forces or dependencies for the behaviour of social groups. The ontology of their environment is another important factor. Understanding those dependencies and influences seems to be a good task for artificial intelligence and particularly for machine learning algorithms. Establishing frameworks, solid barriers or deterministic structures and architectures are necessary. Those frameworks should be transparent and valid for any user or stakeholder of such system to establish the necessary acceptance by the society. Implementing such a framework by artificial intelligence would lead to the horror scenarios, as we know from science fiction movies.

Some statements from one of the scientists, Viktor Mikhailovich Glushkov behind Cybertonia:

- "No doubt, it's impossible to program a machine so that it resolves in all unforeseen situations."

– “Body withstands enormous air pressure (about 20 tons) which is be balanced by the same from the inside. We take circa half-litre air when breathing while the total capacity of lungs is nearly 4 L. Human kidneys conduct 1700 L of blood per day while only 4% of his nervous system are reasonably used”.

Some targets of team had been published on the following (source 8) web page: https://aeon.co/essays/how-the-soviets-invented-the-internet-and-why-it-didn-t-work:

> Glushkov’s teams proposed ‘macro-piping processing’ modelled after the simultaneous firings of many synapses in the human brain. In addition to countless mainframe computer projects, other theoretical schemes included automata theory, the paperless office, and natural language programming that would let humans communicate with computers semantically, not just syntactically as programmers do today. Most ambitiously, Glushkov and his students theorised ‘information immortality’, a concept we might call ‘mind uploading’ with Isaac Asimov or Arthur C Clarke in hand. On his deathbed decades later, Glushkov comforted his grieving wife with a resonant reflection: ‘Be at ease,’ he soothed her. ‘One day the light from our Earth will pass by constellations, and on each constellation we will appear young again. Thus we will be together forever in the eternities!’.

It shows that those people had many ideas from today’s digital millennium in their mind, but also all the concerns and limitation.

Fig. 3.34 Passport of a citizen of Cybertonia. *Source* 7 Benjamin Peters, how to network no nation. The turbulent history of the soviet internet 1965

3.7.2 Configuration of Human Control System

Architecture and design of a human system represent always the static part of a system. Without the configuration and calibration data, no electronic system could word. Even architecture and design could never provide any dynamic if not an impulse releases energy so that an observer could perceive a behaviour. The human system provides resources with certain properties. These are used for different tasks with different frequencies, duration and intensity. To initialise the system, to start certain synchronisation points and to achieve transitions to certain states in the defined or intended context, data are necessary. If new life is created from an egg cell through cell division, data are inherited that determine but also restrict the space and abilities. The genes form the coding data for cell growth, and the human system is based on these data. This means that the genes are the blueprint, the program plan and the inherited configuration or calibration data. One can imagine that the genes are data patterns that are used by different stimuli or functions and can control certain functions or also restrict and limit their function. How these functions form new functions or variants of these functions with the configuration data (e.g. the human genes), depends on the respective context, the intensity, timing and many combinations of all of these. Active configuration data a function at different times the same data can lead to different perceptible functions. If a function is switched off in time, the body parts can be protected from unwanted damage; if it is activated too long or at the wrong time, it can lead to unwanted or even dangerous behaviour. Genes inherit and change through inheritance. If the frequency is requested too often or too intensively by another stimulus, it can lead to damage. Characteristics could be filtered, bypassed or systematically weakened by external influences; the observable behaviour can be described as inertia. In the case of intensive demand, it could be perceived as hot-blooded behaviour during appropriate education or training. Monozygotic twins may also have a weak or a strong heart with the same inherited genes, which may be influences during pregnancy, childbirth or growth. It can be attributed to food, stress or exercise, as well as disease. Such effects naturally lead to statistical similarities to genes and their coding, but the influencing factors over time are so heterogenic that selective manipulation of the genes cannot lead to excessive systematic influences. Today, it is possible to influence, manipulate or even switch off certain genes. Their presumed effect on the body by means of gene shears could lead to short-term benefits. The long-term impacts and what the deviation means for humans in the long term can be very different. Breeding intelligent people or breeding particularly efficient people can perhaps last as long as faithful shepherd dogs. If certain genes are identified as the cause of disease, it will certainly make sense to eliminate or evacuate the genes. Using configuration data to increase the performance of a weak system will only lead to short-term success because long-term robustness could not be achieved.

3.8 Cybernetics and Control

Cybernetics is evident in all aspects of nature and technology. It occurs when a device, systems or an organism uses feedback loop to generate a certain performance. Usually, closed-loop concepts are considered, but any of such system requires energy to provide the intended performance.

The diagram below shows a simple first-order cybernetic control system in the P3 management context (Fig. 3.35).

Such closed-loop control systems are considered as closed regarding the cyclic behaviour provided by the feedback. Usually, such a control algorithm requires energy to assure required action and resulting behaviour of the considered system (Fig. 3.36).

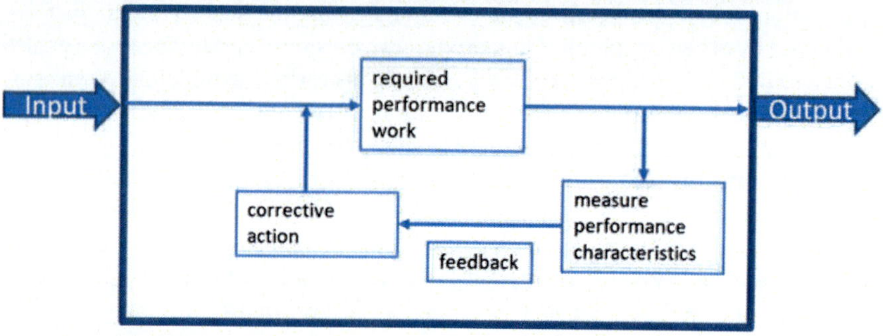

Fig. 3.35 Closed-loop control

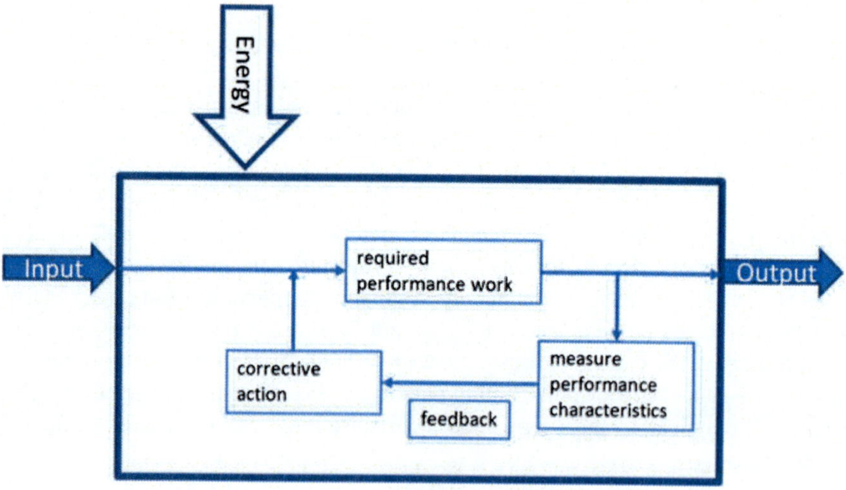

Fig. 3.36 Energy provision for closed-loop system

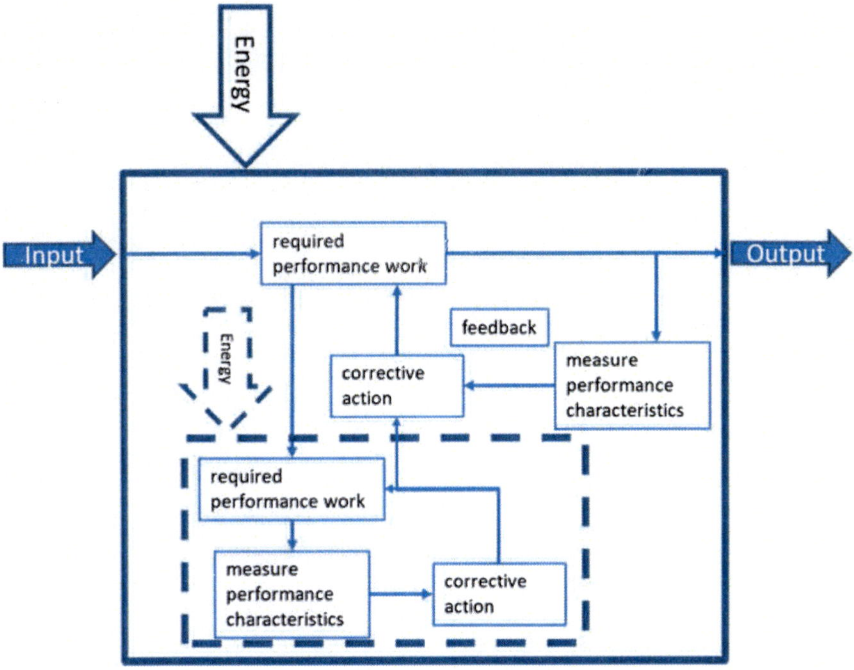

Fig. 3.37 Control cascade

The energy needs to be provided in order to perform any control measure in the system. In case the control action means a rectification, also this energy would be required. In case of a cooling system, maybe energy could be delivered, e.g. as warmer cooling water.

Such a system considered as a whole leads to control cascades (Fig. 3.37).

Control cascades could be very complex devices. Nearly any kinds of physical coupling and dependencies are possible. Understanding the control strategy in order to understand the design of the feedback loops would be essential to achieve stabile behaviour.

The basic control loop could be a machine and the secondary control loop the cooling system for the machine. Unless we consider pure technical systems, understanding is obvious (Fig. 3.38).

People could be in various roles in such a control cycle. They could be in a monitoring role like monitoring the road characteristics and correct steering as required for safe driving. They could be in a secondary position like providing more energy for a combustion engine, for example, by pressing the drive pedal. In typical assistance systems, people are monitored in order to support human actions and provide, e.g., amplification of the desired action. Humans could be in the role of driver, occupancies, pedestrians, cyclists or driver of other vehicles. A typical control task for a vehicle driver is to adequately passing tasks pedestrians with adequate distance and

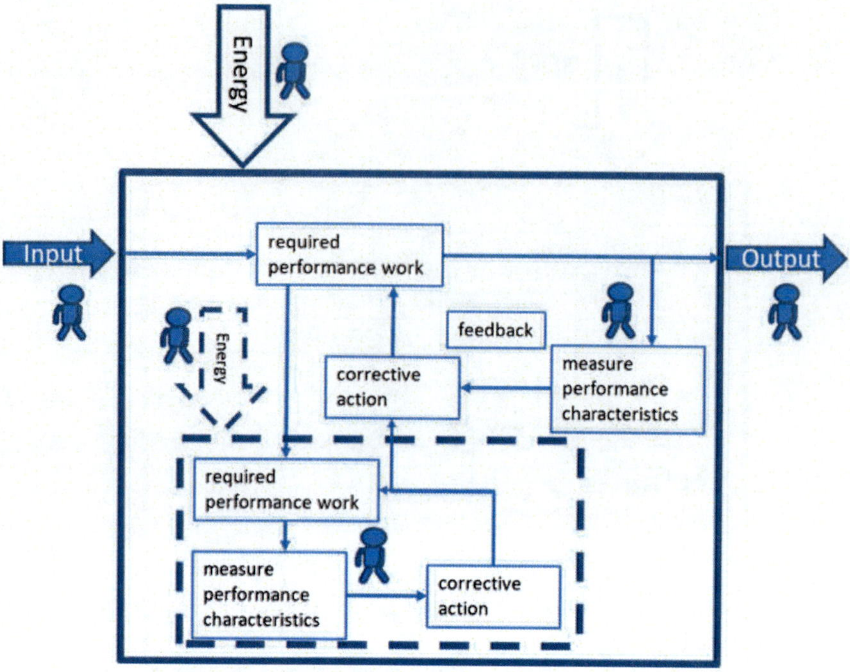

Fig. 3.38 Human in technical control systems

speed. In this case, at least to control outputs need to be controlled the speed of the vehicle and the distance to the pedestrian.

The options and the possible impact of humans within the various roles and the interdependency of all control cycles become easily very complex.

In case of static driving on a highway, we need at least to synchronising two actuators, the steering system and the propulsion. The speed controller is depending on vehicle inner resistance and outer resistance like road friction, wind, etc. The same impact is true for the steering system, but the impact to keep the target value in focus is far more performant. Wind, road friction and inner resistance affect speed moderately and the feedback more or less steady or even constant. For steering it is the other way round, wind and road affect the steering function with high frequency and significant amplitudes.

The impact and the effect of humans in various roles like the driver, occupancies, pedestrians, cyclists, driver of other vehicles are very heterogeneous. They can act willingness and reflexing; they could pursue or break them. All those options need to be considered.

3.8.1 Digital Twin

The approach with the redundancy in the brain with right and left hemisphere is one aspect. Research about twins is a huge field to understand consciousness. Oliver Sacks, "twins", had described another interesting approach. Handicapped twins developed a huge way of communication, so that they could calculate prime figure better than any cyclic computer even could do today. They developed a kind of communication, where one twin gave the basis for the calculation of the other twin and vice versa. The approach of mathematical coprocessing was famous in the nineteens with the Intel 80,286 and its coprocessor 80,287. A coprocessor could perform floating-point arithmetic, graphics, signal processing, string processing, cryptography or I/O interfacing with peripheral devices (Fig. 3.39).

Another approach is based on the theory that twins are somehow equal and could communicate in their under-consciousness, and that a digital model becomes a copy of the considered reality. "Vehicles Capable of Dynamic Vision" from Ernst Dickmanns describes a 4D model, which is also a dynamic 3D model of the environment, and the force dimension is the time (Fig. 3.40).

Dynamic machine vision considers three major areas of activity:

- object detection (central arrow upwards),
- tracking and state estimation (recursive loop in lower right),

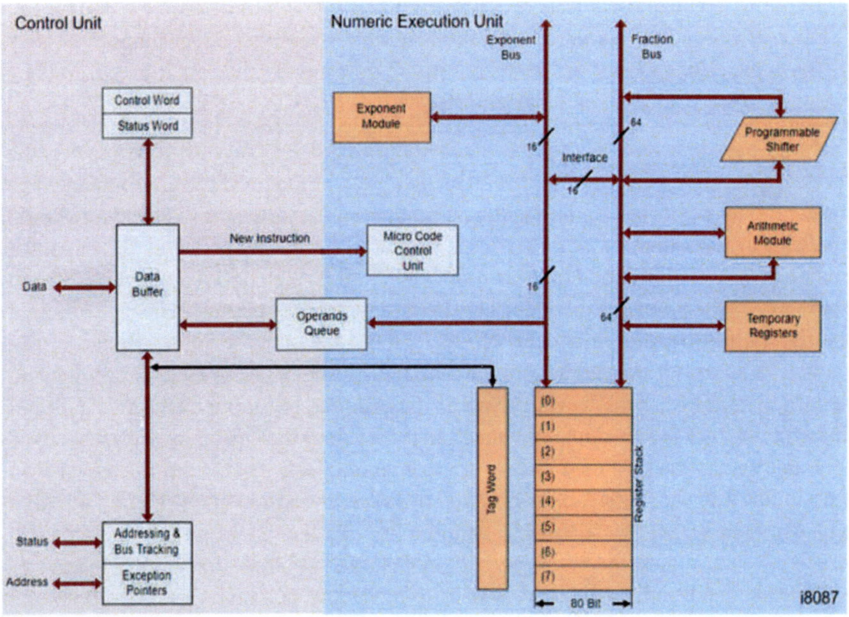

Fig. 3.39 i8087 and i80287 microarchitecture. *Source* 5 Intel manual for 80xx architecture

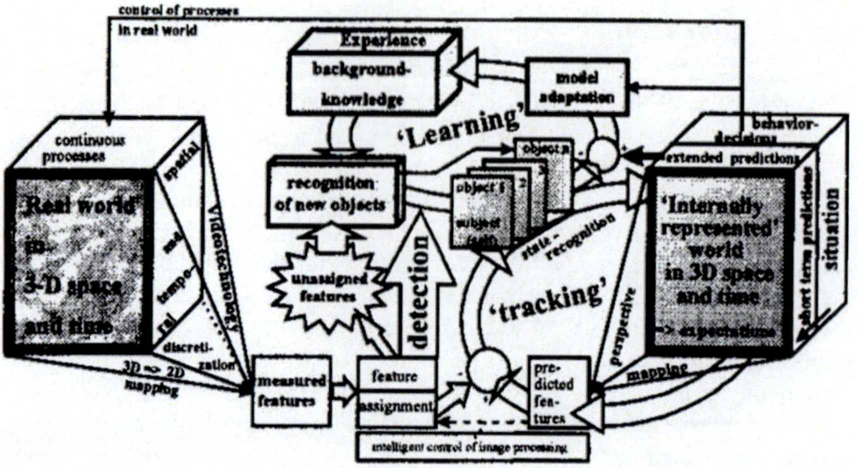

Fig. 3.40 Survey on the 4D approach. *Source* 6 vehicle capable on dynamic vision, Ernst Dickmanns, 2004

- learning (loop in centre top), the latter two being driven by prediction error feedback (Fig. 3.41).

Model-based validation is based also on the idea of a digital twin. There are at least two principles,

- The real validation candidate could operate in a simulated environment.
- Use a behaviour model of the system to be developed, and test it in a simulated environment or in the real environment.

Model-based validation requires to develop a model during development of the product under development or the dedicated context or the product environment. The context model needs to be validated against the reality in which the product should be used. For a product model, it is important that the model provides the expected behaviour, which is intended for the product. Of course, it is required that the intended behaviour of the intended product had been developed upfront. After a sufficient maturity during development, the product model becomes a digital image of the product. After an adequate development and realisation of the dedicated product and the sufficient and successful validation of the product, the model becomes a digital behaviour-twin of the product. The question is now does the "digital twin" only "live" in the laboratory environment during development or during the entire lifecycle? In some industries, the digital twin remains in order to test change impacts, possible enhancements, e.g. for performance, material improvements or for fault injection test to see the reaction and the resulting behaviour of the product. Especially in case of field events, the digital twin could be used to reproduce accidents and their potential cause. The digital twin provides traceable and reproducible behaviour, even if the product had been damaged completely. In automation for transport systems, it is

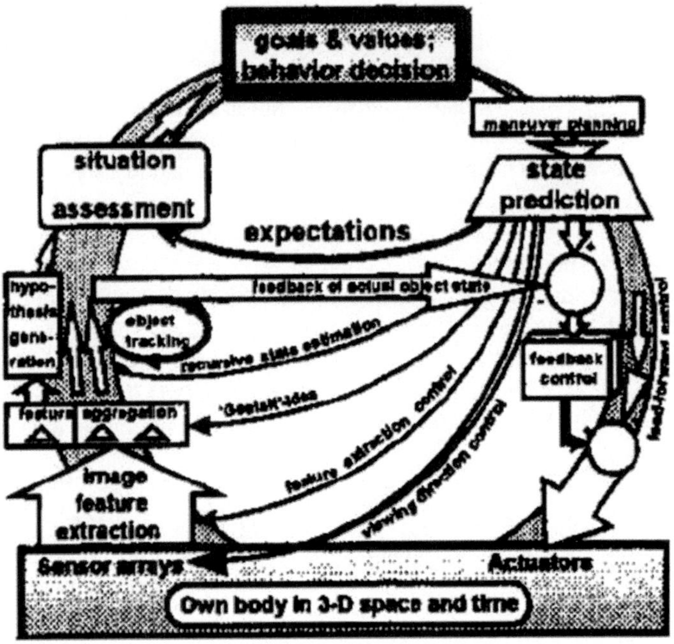

Fig. 3.41 Own body in 3D space and time. *Source* 6 vehicle capable on dynamic vision, Ernst Dickmanns, 2004

meaningful that the twin could be implemented in the transport system; in that case, the twin could do many meaningful activities during the operation of the transport system.

- In high available systems like airplanes, the digital twin could be available in case parts of the operation system failed; the twin could calculate data like a coprocessor.
- The twin could take over control for a certain time, and the operation system could recover, e.g. performing a reset in case of significant software failure.
- Validation measures could be performed during operation, e.g. if new external events affecting the product.
- In case of incidents, if the product shows unexpected behaviour, the cause inside of the product, but also the impact of the environment in that moment could be repeated and becomes traceable.
- The digital twin could calculate load or tension scenarios before entering a critical situation, so that the performance capability could be forecasted.
- Certain stress scenarios could be simulated on the twin, and test vectors for the operating product could be generated so that predictive health management could be supported.

There are many more advantages for using a digital twin, especially a digital twin for the logic control part. For hot standby systems, the digital twin philosophy more and more replaces the ideas of diverse systems in order to avoid incidents due to systematic failure. The twins test during operation permanently themselves vice versa with stress and fault infections during operation and prevent sudden shutdown situations. The healthy twin operates the actuator; the hot standby twin performs tests or even runs self-tests and evaluates the test results. In case of system weaknesses, the stress for the system or the system performance could be reduced, so that sudden critical situations could be avoided.

Of course, the same approach could be applied for the operation environment; in case of full available system performance, the transport system could operate more offensive, operate closer to the risk limits or run longer in a higher de-rating.

The question is do we need to implement the "digital twin" in the transportation system in order to achieve the same advantages? Could we not install the twin in the cloud?

If the heart of a person fails, a few minutes later, the brain-dead of the person enters.

When the spine is severed, the spinal cord can no longer transmit information to the limbs from the brain. Consequently, the actuators in the remote-operated transportation system need some self-intelligence when the connection to the control centre is interrupted. How long the remote unit has to function autarkic, and how performant the separated systems need to be depends on the required or necessary performance of the systems in the current operating environment.

Today, we still could not transfer energy wireless, except sunbeams. So that, the control and availability of energy need to be assured within the automated transportation system. Unless we could accept an energy-less state as a safe state, it would be fine. A train that stops on the tracks could be considered to be a safe stop, unless the railway control system assures no other train in that segment. That it leads to a complete failure of the entire operation of the railway is just a commercial issue and leads of course to many angry customers.

The energy problem could be solved by sufficient redundancy in the transportation system. The necessary limp home or degraded functionality needs to be planned based on the system requirements and the intended operation environment.

Whenever modelling, it is always a matter of abstraction. Using the entire environment and trying to simulate complete environment lead to too many degrees of freedom. Testing, validation and assessments are only possible in a reduced view. The artwork behind modelling identifies the right level of abstraction and finding a level, which allow communicating the result.

3.8.2 Cybernetics About Behaviour and Control

Cybernetics deals with regulatory systems, their structures, constraints, possibilities and controls and resulting observable behaviour. The essence of this approach is to understand the functions and processes of systems capable of receiving, storing and processing information and then using it for its own control or to control machines in close cooperation with humans. Cybernetics try to explain humans' behaviour and the intrinsic motivation behind and describe it similar to system behaviour and their inner structure by providing a dependency of observable points by mathematical transition functions.

Maxwell Maltz defined even psycho-cybernetics in 1960. Maltz became interested in why setting goals work. He learned that the power of self-affirmation and mental visualisation techniques used the connection between the mind and the body.

Norbert Wiener, the founder of cybernetics, did come more from the physics and mathematics. Many publications from Heinz von Foerster try to find the link from biology to human control, and how the human control system works. He did define bionic control, which seems to be still a wide field for machine learning, artificial intelligence and how to communicate with other systems and humans.

In automated driving, we could better describe not only the interface to a human driver, and with help of the cybernetics we also could better describe the behaviour of other road traffic participants like pedestrians, cyclists, drivers of other vehicles and other road traffic participants.

The challenge with observable fix-point is that any uncertainty in the observable fix-point increases or even potentiates the uncertainty in the resulting target, e.g. the correctness of the trajectory. Adequate feedback loops could lower the impact of those uncertainties in the observable fix-point or in the influence to the resulting target. Multi-step approaches like Kalman filter could reduce the uncertainty.

A very new approach is the Cramer–Roa bound. The Cramér–Rao bound (CRB), Cramér–Rao lower bound (CRLB), Cramér–Rao inequality, Fréchet–Darmois–Cramér–Rao inequality or information inequality expresses a lower bound on the variance of unbiased estimators of a deterministic (fixed, though unknown) parameter. The methodology goes back to Harald Cramér, Calyampudi Radhakrishna Rao, Maurice Fréchet and Georges Darmois all of whom independently derived this limit to statistical precision in the 1940s.

Such methods like CRLB use so-called square filter or Kalman filter, and they tell us the best at least squares filter can do, e.g. by the Riccati equations with zero process noise and infinite initial covariance matrix. There are challenges behind like:

- Knowing the best a filter can do does not tell us how to build the filter so that it will work in the real world and in real time.
- Generally, building a filter with zero process noise is a bad idea because the filter stops paying attention to the measurements.

Rao has defined a new paradigm and defined the following formula about uncertainties:

$$\begin{aligned}&\text{Uncertain Knowledge} + \text{Knowledge of the Amount of Uncertainty in It}\\&= \text{Usable Knowledge}\end{aligned}$$

The formula is not wrong but too general, and the relation between

- time dependent and time independent (static, dynamic, real-time, reference system),
- direct information or indirect information (e.g. measurement of real effects and indirect measurement, or observed malfunction and approximated malfunctions),
- probability and statistics,
- random process and estimated theory,

needs to be evaluated based on the context in that a control task should be considered.

The frequency domain approach of the Wiener filter (1949) has been enhanced to the natural time domain approach by the Kalman filter. The challenge seems to be to realise an adequate electronic to run such filter in real time.

The issue with the uncertainties

- of demand,
- of sufficient perception,
- of relevance of impact on technical systems and
- behaviour of humans in handling, reaction or coexistence with technical systems or equipment or,
- impact on individuals and groups of people, for example,
- the different reactions of groups of people with technology and between each others,
- etc.

leads to huge field of risk for automation projects. This shows that the classical idea of safety integrity is just a small segment of system safety.

The of safety integrity statement:

> If you control, mitigate or avoid the failure impact or their occurrence, the system could be considered as safe.

Safety Integrity is just one type of measure to reduce the risk below a tolerable level accepted by the society. Any kind of misuse scenarios, insufficient regulations or inadequate technology and their possible affect to safety requires other measures.

3.8.3 *Cybernetics and Sufficiently Perceiving the Environment*

Cybernetics is as good as we could explain the behaviour with models and mathematical formulas. If we could sufficiently explain with the models and formulas the necessary behaviour and feedback of the real world as better, we could control things in this world.

A great tool for understanding the real world and how much, what and how good we need to perceive from the real world is the receiver operating characteristic (ROC) methodology or the ROC curve.

ROC means "receiver operating characteristic", and it is the visual representation of performance of binary classifier. "False-positive rate" versus "true-positive rate" visualise the understanding of the classifier's performance.

AUC means "area under the curve", and it is a numerical representation of the performance of binary classifier.

When using normalised units, the area under the curve (AUC) is equal to the probability that a classifier will rank a randomly chosen positive instance higher than a randomly chosen negative one.

- True positive (TP), predicted positive, and it is true.
- True negative (TN), predicted negative, and it is true.
- False positive (FP), predicted positive, and it is false.
- False negative (FN), predicted negative, and it is false.

The "predicted values" could be in the condition as Positive or Negative, and the "actual value" could be True or False (Fig. 3.42).

Recall: Out of all positive classes, how much are predicted correctly.

Precision: Out of all positive classes predicted correctly, how many are actually positive.

Accuracy: Out of all classes, how much are predicted correctly (Fig. 3.43).

The ROC provides following information or statements:

- A true positive is a true observation correctly predicted to be true.
- The true-positive rate (TPR) or "recall" is a ratio between 0 and 1.
- A false positive is a false observation incorrectly predicted to be true.
- Specificity is one minus the false-positive rate (FPR; it is a factor to assess the correctness of the specification).
- The best point to be on the ROC chart is the top left corner, with 100% TPR, sensitivity, or recall, and 0% FPR, or a correct specification "specificity".
- The best feasible points are on the ROC curve. As you move from left to right, you decrease your threshold to classify an observation as positive. You get more true positives, but also more false positives.

The ROC curve visualises the set of feasible solutions, as you vary the classification threshold, implicitly varying the cost of false positives relative to false negatives.

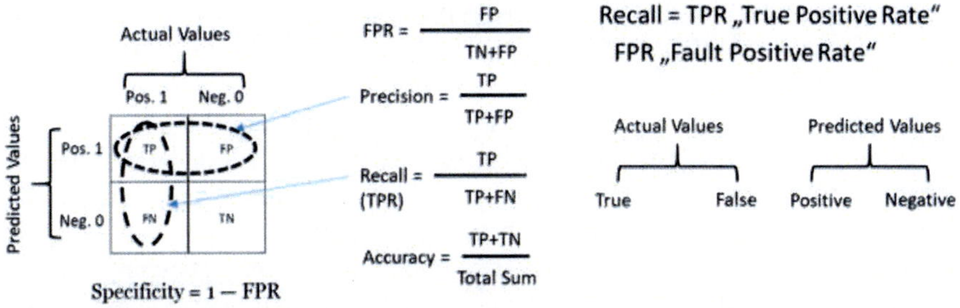

Fig. 3.42 Values and rates for ROC

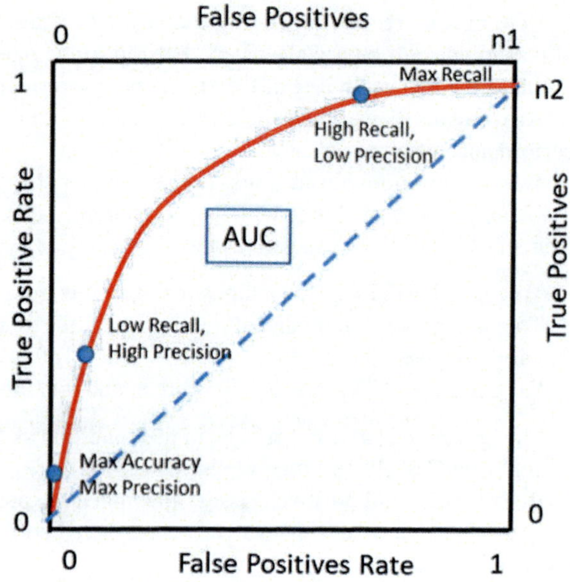

Fig. 3.43 ROC and resulting AUC

The slope of the ROC curve reflects how many true positives you gain for each false positive you accept.

ROC is often used to test and validate models in order to answer the questions like

- How good is the model or is the model sufficient?
- How the parameter needs to be tuned to find a sufficient result?
- How much or what kind of redundancy is necessary to perceive the real world sufficiently?

The method is often used also in medical and social prediction. In some cases, a static sensor perceives the real world better, because the observation point is always in the same position related to the object or area under observation. Perception with an approaching vehicle could perceive with just one sensor an obstacle sometimes better, due to the movement of the vehicle any snapshot from the sensors provide a different perspective. For example a lidar or radar observes the object from a different perspective at different time slots. If the perceiving system could record the and process the data sufficient deterministic under given real-time aspects, requires high performing controllers in order to compensate, vibration, noise, temperature impacts from the driving vehicle, as well as road and weather conditions.

Providing services or assistance for humans requires a good understanding of human needs and behaviour. As better, the designer of the control system understands all the relation as more benefit could be provided for humans.

The ROC or fault-negative curve (FN curve) has been particularly established in test processes. End of last millennium, JOHN S GERO and UDO KANNENGIESSER published various ideas how to systematically perceive and understand the world, and how to make it suitable for system technology. The "function–behaviour–structure (FBS) ontology" provides a methodology to represent processes despite its original focus on representing objects. The FBS ontology provides a uniform framework for classifying processes and includes higher-level semantics in their representation.

In the following referenced paper, they introduce the framework. The abstract of the paper gives the following objective:

> This paper extends the Function-Behaviour-Structure (FBS) framework, which proposed eight fundamental processes involved in designing. This framework did not explicitly account for the dynamic character of the context in which designing takes place, described by the notion of situatedness. This paper describes this concept as a recursive interrelationship between different environments, which, together with a model of constructive memory, provides the foundation of a situated FBS framework. The eight fundamental processes are then reconstructed within this new framework to represent designing in a dynamic world.

In this paper, they recognise that artificial intelligence (AI) in design research has increasingly focused on developing agent-based design systems.

> They claimed: Yet, many of these approaches have shown only limited success in supporting conceptual designing since they ignore one of conceptual designing's (as opposed to routine designing's) most distinguishing features. As not all, the requirements are known at the outset of a design task, conceptual designing involves finding what is needed and modifying it again during the process.
>
> *Source* THE SITUATED FUNCTION - BEHAVIOUR – STRUCTURE FRAMEWORK, JOHN S GERO AND UDO KANNENGIESSER, Australia, 2000.

They introduce the "situated function–behaviour–structure" or "situated FBS framework". It is based on three different stages of consideration:

- the external world, as we observe the "real world",
- the interpreted world, as we understand what has been observed and
- the expected world, as we need them to make them usable for humans.

The FBS ontology supports a situated view of processes based on a model of three interacting worlds. The "situated FBS framework" provides a methodology to describe the situated design of processes.

The situated FBS framework, based on cyclic experiences with the real world in order to generate a model of systems, that could beneficial cooperates with the real world. With the FBS, the framework supported a systematic approach in order to specify the system and necessary interfaces.

We need to understand many of the real-world impacts and the way that humans intend to work and even live with digital technology. If we do not understand the relation, we could cause more damages than providing benefits for humans and our society.

References and Sources:

Source 1: On Constructing a Reality, Heinz von Foerster, This is an abbreviated version of a lecture given at the opening of the fourth International Conference on Environmental Design Research on April 15, 1973, at the Virginia Polytechnic Institute in Blacksburg, Virginia. 2 Brown, G. S., Laws of Form. New York, Julian Press, page 3, 1972.

Source 2: McKinsey Quarterly, Where machines could replace humans—and where they can't (yet), July 2016

Source 3: Understanding Understanding: Essays on Cybernetics and Cognition, Heinz von Foerster, 2003

Source 4: Download from September 2019, https://www.researchgate.net/figure/Typical-architecture-of-a-Distributed-Control-System_fig1_242457425

Source 5: Download from September 2019, Intel manual for 80xx Architecture, https://wikimili.com/en/Coprocessor

Source 6: Download from September 2019, https://pdfs.semanticscholar.org/99b4/ec66e2c732e4127e13b0ff2d90c80e31be7d.pdf, Vehicle Capable on Dynamic Vision, Ernst Dickmanns, 2004

Source 7: Benjamin Peters, How to network no nation. The turbulent history of the Soviet Internet 1965

Source 8: Download from November 2019, https://aeon.co/essays/how-the-soviets-invented-the-internet-and-why-it-didn-t-work

Source 9: Download from February 2020, http://jmc.stanford.edu/articles/whatisai/whatisai.pdf, John McCarthy

Source 10: THE SITUATED FUNCTION - BEHAVIOUR – STRUCTURE FRAMEWORK, JOHN S GERO AND UDO KANNENGIESSER, Australia, 2000

Chapter 4
System Safety Engineering

System engineering is much more than a method to describe a technical product. With the idea of automation, we need to ask the question, what is the purpose of a product, how does it provide a benefit for the user and what risk does it mean humans and the environment. Mobility means using machines to transport people or goods from one place to another place. The "automobile" should perceive the intend of the user and safely fulfil the desired actions. The question arises if it is possible to fulfil all stakeholders' expectations, without violating the society tolerance level and finding acceptance by users, owners and other road traffic participants. How can we manage the complexity for future automation?

The target of the chapter should provide a rational for a structured system safety approach. It proposes reasoning and proposals how automation functions could be realised for a specific vehicle or automation scenarios of a road traffic mobility concept with various vehicles.

In automation of human driving functions, the view just on the system under development is insufficient, and consequently the following sentence is an error:

"*If all devisable mistakes/errors/faults of a system are brought under control, the system is regarded to be safe*."

The context and the interactions with the nature, physical ontology, environment and humans or with the so-called real world in many kinds of roles provide further challenges, which could lead to further accident scenarios, that need to be considered.

4.1 Observer of a System

A system is always depending on the context within which it should be used (Fig. 4.1).

Current approach of ISO 26262 considers only the safety integrity of a so-called ITEM, which could be a steering system in the vehicle context. Typical requirements for the item, like the intended functionality or preliminary assumption derive often from existing vehicles, which should be modified.

H.-L. Ross, *Safety for Future Transport and Mobility*,
https://doi.org/10.1007/978-3-030-54883-4_4

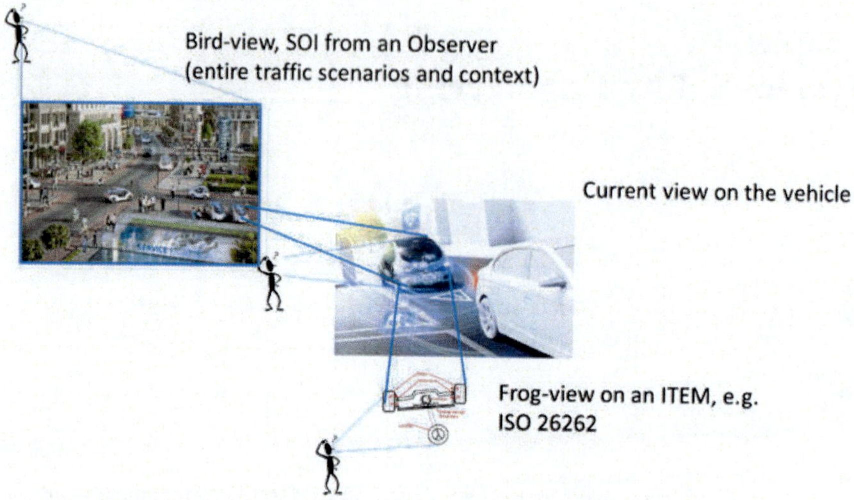

Fig. 4.1 Typical perspective and viewpoints from a position of different observers

Current legislation for homologation considers a view on the vehicle. Regulations like UN ECE R79 for steering systems define their requirements how a steering system should operate in a standard vehicle. Typically, a M1-class vehicle is considered where we need to understand, the viewpoint from the people that have established such regulations. We need to understand the viewpoint of the people that formulate legal requirements, and the viewpoint of the lobbyist from the motor industry, that try to protect their business. If we describe a vehicle, we need to understand from what viewpoint the technical observer defines and specifies the requirements and constraints.

The birds view considers an environmental-centric approach, where the vehicles and other road traffic participants are dynamic objects in various road traffic scenarios. The current view is the typical "normal" view of a vehicle manufacturer. The entire vehicle design, their feature and the way how the benefits of the vehicle could be communicated are in the focus. Safety would be more communicated as a user benefit and not in the focus of vehicle safety, for example the homologation. Therefore, NCAP recommendations have a higher priority as, for example, the legal requirements. The frog view is the typical TIER1 view, they see the steering, brake or propulsion system in their focus and quite often is the scope of delivery just a control unit for the entire system.

4.1.1 Viewpoints and Perspective

There are always many stakeholders for a product. During a systematic development, those different stakeholder views need to be considered and also the impact of those stakeholder requirements. Such requirements, feature, functions, constraints, design limits and target use cases could effect on any level of abstraction the architecture and the resulting design. Ideally, we could abstract certain information from the total description model according to the profile of stakeholders. A classification of the stakeholder requirements leads to a huge matrix in what characteristic of the product they could be relevant. We do not have a standardised version of stakeholders and their often-varying interests and besides that also a basis data model, which would be able to consider entire world and its coherences.

Social system engineering could help to classify such stakeholder requirements, but the impact to the system is always a matter of an analysis. Even such genius data management and information systems such as Google, Wikipedia or an artificial intelligence algorithm would be stretched to their limits. Norbert Wiener popularised the social implications of cybernetics, drawing analogies between automatic systems and humans. The term "sociocybernetics" was coined. Sociocybernetics can be described—in a nutshell—as systems science in sociology and other social sciences. In addition to fundamental theoretical, epistemological and other philosophical questions, the field of work of sociocybernetics extends to various topics of applied research, empiricism, methodology and ethics. What unites them is not so much a specific unified field of research, but rather a common orientation towards specific theoretical and methodological basic assumptions and a focus on problems of complexity in society. Sociocybernetics is less interested in the isolated analysis of specific causal relationships, but rather in the mutual influence of dynamic self-regulating systems. It follows the claim, formulated by W. Ross Ashby as early as 1956, not to focus on the special properties of the object of research, but on its forms of operation: "It does not ask 'What is a thing', but 'What does it do?'" (Ashby 1956). System processes should provide the link between technical systems and the human actors. Adequate analysis should provide concepts how the technical system should behave and operate in conjunction with users, or people in the operation environment.

Anybody who has ever tried to build a house knows a construction drawing. The aim of this construction drawing is of course to show the later owner how the house will look once it is finished. Often, particularly for houses that are built by a building contractor, there are also construction specifications, but the average person is unable to read this without the help of an attorney.

The construction drawing often shows a front, back and various side views and vertical sections in order to see the arrangement of the different stories as well as horizontal sections to see the arrangement of doors, for example. However, the drawing always shows us the same house. Our expectation is that the different views are consistent and that we can see the entrance door in the front view on the same place in the house, as we would expect to see it from the horizontal view.

Nevertheless, different stakeholders of architectural company need to be identified and all of them of course only want to see the view of the architecture that interests them. Therefore, if we send the carpenter the plan for the inside doors, he will be interested in knowing the height of the floor fill but not the allocation of door lintels and how much iron was used for ultimate load. Actually, here we would need to consider the perspective of the finance controller and the project manager since the resources used already determine the safety sufficiency.

Towards the end of the 1960s (67–69), Phillipe Kruchten described his four views, which lead to the following (here compared to UML) 4 + 1 views:

- The logical view describes the functionality of a system for the end-user. Logical elements are used in order to show different dependencies of elements. Class diagrams, communication diagrams and sequence diagrams can be used as UML diagram.
- The development view or implementation view describes the system from the viewpoint of the developer. Component diagrams or package diagrams can be used as UML diagrams.
- The process view (behaviour or functional view) describes the dynamic aspect of systems as well as the behaviour of elements at their intersections to each other and in a defined environment. Relations could be any kind of communication (technical but also man–machine communication, etc.) time behaviour as well as allocation and structure aspects such as parallelism, distribution, integration, performance and scalability. Activity, sequence or timing diagrams can be used as UML diagrams.
- The physical view or the deployment view describes the system from the point of view of the deployment or rather the manager of deployment. It should include the allocation of components, modules or electric components and elements that have to be obtained and deployed for the communication among each other (e.g. cables, bus, plugs). Distribution diagrams can be used as UML diagrams.
- The scenario view is the additional view, and it describes the planned use cases of application, possible configurations and behaviour versions. This can be the basis for the planned behaviour of elements among each other. The architecture verification later forms the foundation for integration tests. Use case diagrams can be used as UML diagrams (Fig. 4.2).

A more UML oriented representation could be using UML language artefacts to represent the 4 + 1 views from Kruchten (Fig. 4.3).

- The scenario view could be considered as the stakeholder view and could represent the use cases and user stories. The various benefits, constraints, expectations and requirements, etc., could describe the various stakeholder perspectives. In addition to the stakeholder view, a neutral observer view could be defined which could, e.g. describe the entire lifecycle of the product or the system that should be under development.
- The process view could be considered for the conceptual activities. The information flow and expected communication with the various elements are relevant for the development. In that view, the scope of development could be identified. By boundary analysis, interfaces could be identified of elements, which are in or out scope. The communication to technical or functional elements as well to logical elements could be defined. Behaviour models could describe the interaction of humans and technical elements. Another important aspect is the evaluation of the environmental impact to the considered development item. A product could be used in various use case scenarios or like a vehicle in various operational scenarios.

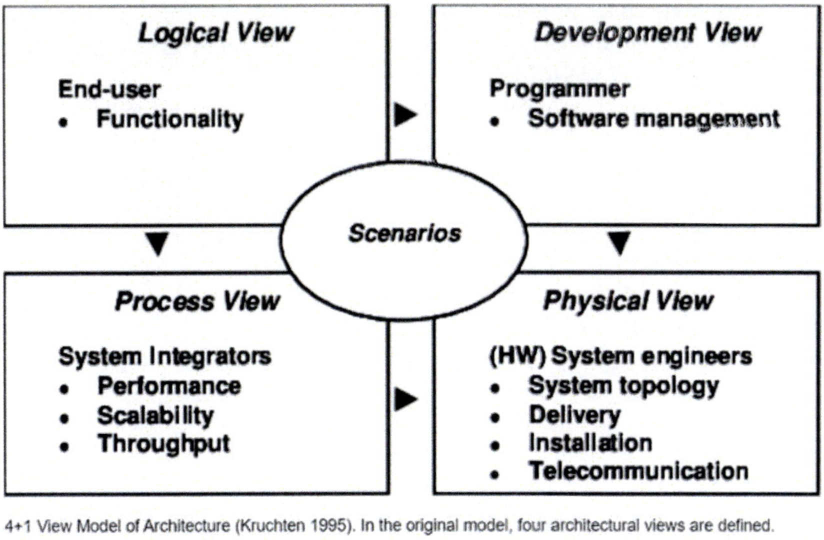

Fig. 4.2 4 + 1 view model of architecture. *Source* Kruchten 1969

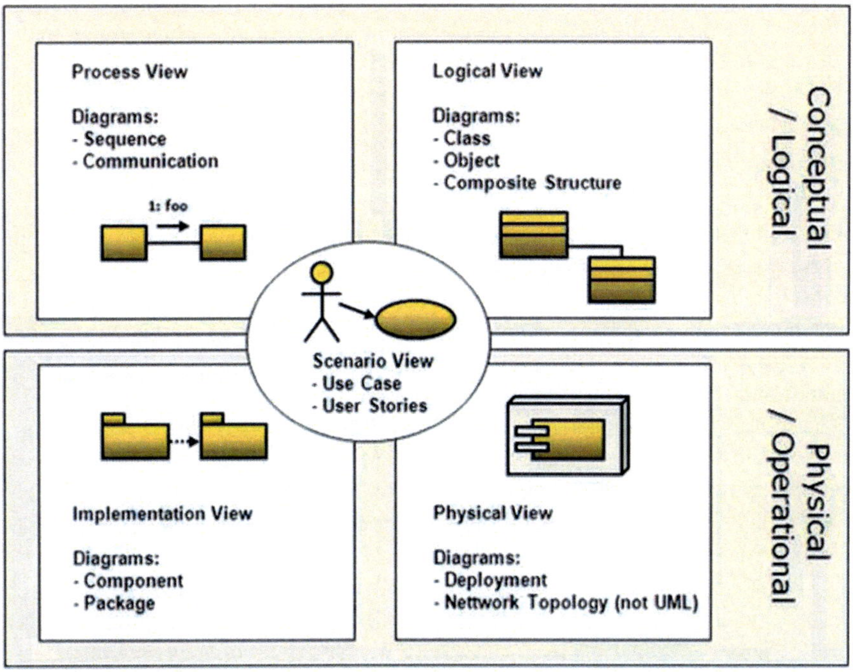

Fig. 4.3 4 + 1 Views from Kruchten in UML artefacts

Typical mobility concepts require heterogeneous operation conditions, situations, scenarios, state and transitions in many different road traffic environment, with many different interfaces to human actors in roles like driver, occupancies, pedestrians, cyclists or services like maintenance, traffic observation (e.g. police), rescue and many more.

- The logical view should describe the logical dependencies and behaviour of the elements itself and their interaction between each other. In cooperative mobility systems, where rules (like road traffic regulations) are compulsory to be considered the UML diagrams are not anymore sufficient for the logical dependencies and their variances and probabilities of occurrence. Class, composite, structure and object diagrams are a very good basis for logical representations, but they could only cover a small segment but they are not adequate to provide a logical view on a mobility scenario with the relevant human impact. For automation projects, the typical closed loop control architecture with all the feedback lines and possible impact to the behaviour need other type of representations.
- The implementation and physical view are more considered for software architecture in UML. The typical component and packaging diagrams for the implementation view or the deployment diagrams for the physical view are addressing more the characteristics of software elements. Even the behaviour and the specification of the hardware–software interface require other rules and representations of the modelling elements. For scheduling and resource usage, other types of diagrams are recommended. Technical communication like a point to point CAN bus communication could be described by UML. Network solutions like multi-layer systems or solutions based on the ISO/OSI layer model require other description methods.

All the different viewpoints provide also new aspects of a system. The people involved in the construction or implementation are more focusing on the technical design aspects and how they could manufacture or produce, a controller would ask the question, is it better to buy or make the product by their own. Lawyer, customers, users and many other groups have other stakes on a product or a system, etc. For all that groups a product could provide different forces or benefits, and they have demands and constraints for their acceptance (Fig. 4.4).

The figure shows how heterogeneous the stakeholder requirements could be in what way those influence a product development and limits the range of use for a product depends on very many aspects. The influence of those factors could be only evaluated if a dedicated context for the product and its properties are defined and analysed by all relevant scenarios for the range of use (Fig. 4.5).

In case of an automation system, any kind of stakeholder requirements could be relevant. Which requirement or which constrains are relevant in what degree depends on many aspects. An analysis from a neutral observer's point of view could balance all the requirements.

As an input of the analysis, the following aspects need to be evaluated:

- What are the intended features or functionality that should be provided as a benefit for what kind of users?

- In what environment, conditions, situations, etc., shall the features or functionality be provided?
- What are the use cases, which are to be considered or how do we expect that the user handles the product?
- How do we intend to realise the product?

Based on these questions, a hypothesis could be formulated which is the basis for the stakeholder analysis. The hypothesis is a viewpoint of the person or group

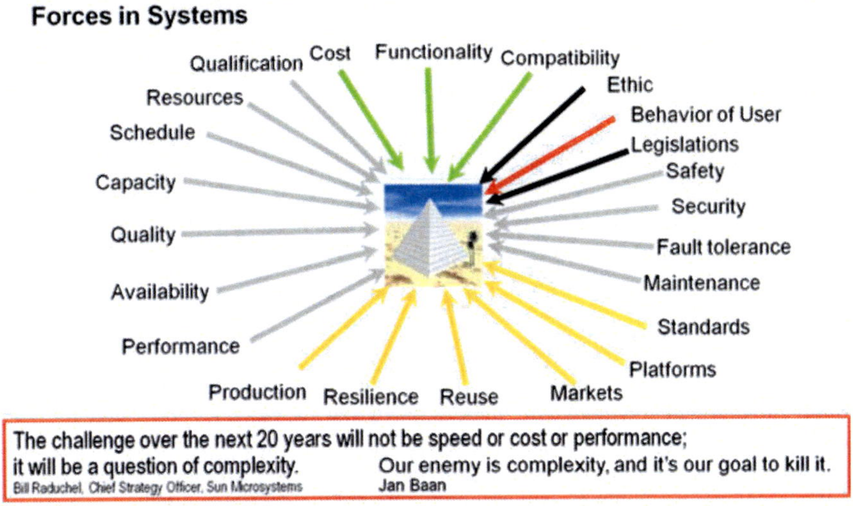

Fig. 4.4 Forces in a system. *Source* IBM

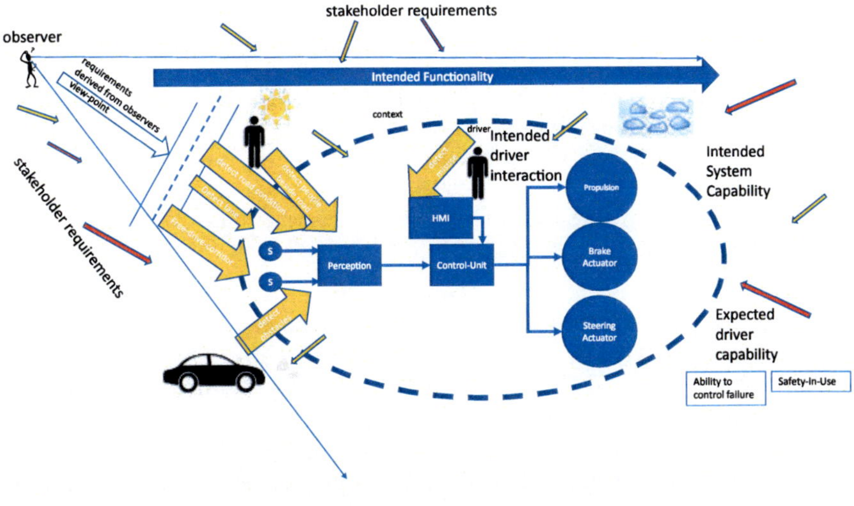

Fig. 4.5 Dependencies in product development process

of people that analyses the stakeholder requirements. We also need to understand their viewpoint and reasons for periodisation. The entire business case and relevant market for the products depend on the result of the analysis. Any kind of product characteristics could be limited in any market different, for example, by legal regulations, the typical handling of the product could be different in the different regions of the world or customer would pay for different feature money or would even ignore the feature. A typical SWOT analysis could lead for any market, use case or area of use to a different result,

- **S**trengths
- **W**eaknesses
- **O**pportunities or
- **T**hreats

lead for any scenario to a different result. Today, we realised that feature like a combustion engine, which was a key selling proposition, became a non-seller. Even the point in time decides about the impact of a stakeholder requirement.

During the development of the international standard ISO 26262, also the automotive industry tried to establish typical views and perspectives of architecture. In order to explain the system safety approach as considered by ISO 26262, the funding project from ITEA called "safe" did provide an enormous framework. "SAFE" is the abbreviation for "**S**afe **A**utomotive so**F**tware archit**E**cture".

One of the key recommendations from ISO 26262 in order to reduce complexity was the hierarchical structuring of the so-called system of interest (SOI). The SEBoK provides two sources of the term:

(1) The system whose lifecycle is under consideration. (ISO/IEC/IEEE 2015)
(2) The system of interest to an observer (Bertalanffy 1968).

Source

(1) ISO/IEC/IEEE. 2015. Systems and Software Engineering—System Life Cycle Processes. Geneva, Switzerland: International Organization for Standardization (ISO)/International Electrotechnical Commission (IEC), Institute of Electrical and Electronics Engineers. ISO/IEC/IEEE 15288:2015. The second definition is an expanded version of the ISO/IEC/IEEE version.
(2) von Bertalanffy, L. 1968. General system theory: Foundations, development, applications, revised ed. New York, NY: Braziller.

Source Guide to the Systems Engineering Body of Knowledge (SEBoK), V2.2, 2020, https://www.sebokwiki.org/wiki/Guide_to_the_Systems_Engineering_Body_of_Knowledge_(SEBoK)

Consequently, it become necessary to introduce a hierarchical model of the vehicle within its context. Such a model provides the necessary abstraction on which the integration elements, or the items as they were named in ISO 26262, should be described with their interfaces (Fig. 4.6).

The automotive industry established very similar typical level of abstraction like in the avionic industry. The highest level for vehicle system is the vehicle level, which

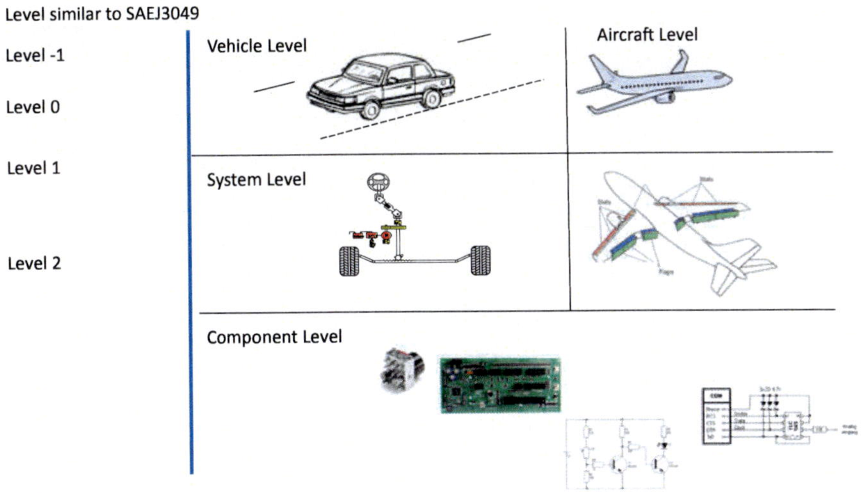

Fig. 4.6 Horizontal level of abstraction

sees the entire vehicle similar like the aeroplane or aircraft level in the avionics industry. In order to find the adequate system level, one or more system levels could be introduced to describe various interfaces before different integration items could be identified, which are real physical units including virtual units like software components.

Based on a funding from German government, the SPES methodology had been established. SPES 2020 stands for "Software Platform for Embedded Systems". The SPES methodology targets on modelling and analysis techniques for software-based systems. Unfortunately, SPES and SAFE are just focusing on a straightforward way to specify the embedded software, so that the system engineering approach in the layer above is very lean (Fig. 4.7).

In theory, SPES provided already a context with separate artefacts for systems and their context and find interfaces to the physical world. In their operational context, also, various roles of humans had been considered and their interfaces, due to the upcoming ideas of "connectivity", so-called cyberspace services and data had been considered.

SPES did provide a matrix with three views and introduced level of abstraction (Fig. 4.8).

The four viewpoints are as follows:

- requirement view
- functional view
- logical view and
- technical view.

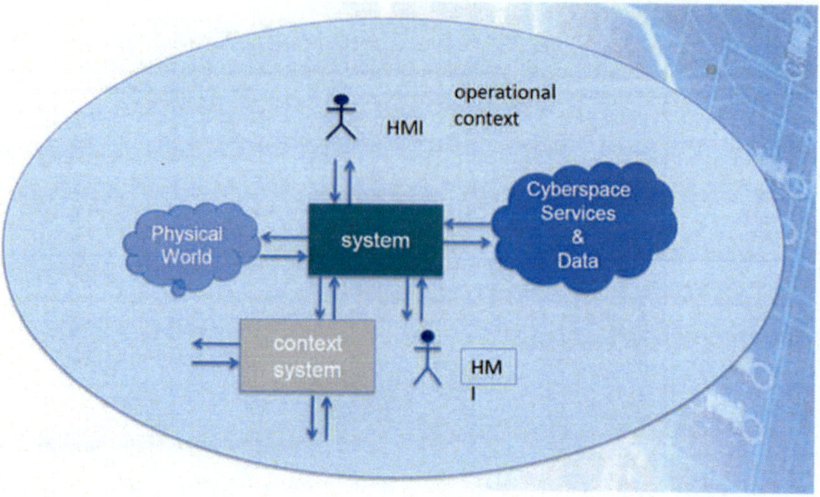

Fig. 4.7 System and their context according to SPES 2020. *Source* Vision for SPES, BROY, 2006

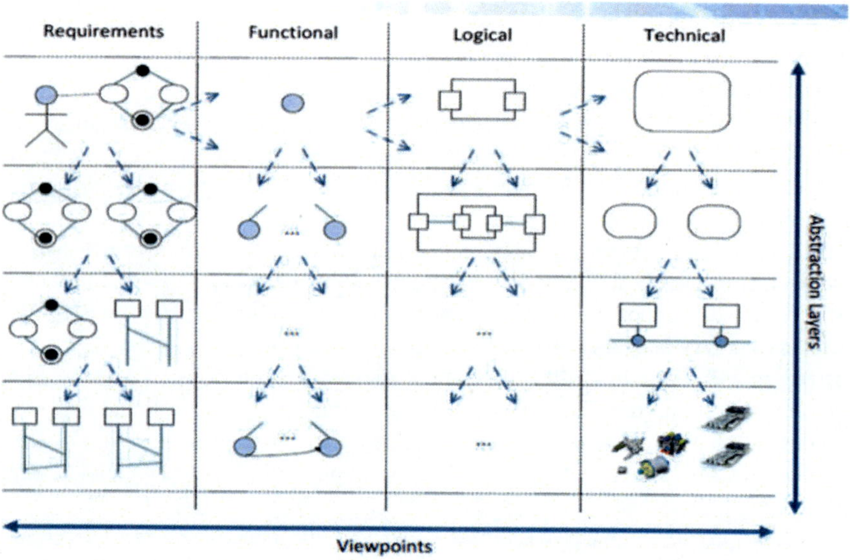

Fig. 4.8 SPES matrix with viewpoints and horizontal layer of abstraction. *Source* Vision for SPES, BROY, 2006

Control circuits and signal chains had been considered as logical views. In many publications, the functional view is only functional dependencies without any assignment to logical or technical elements (Fig. 4.9).

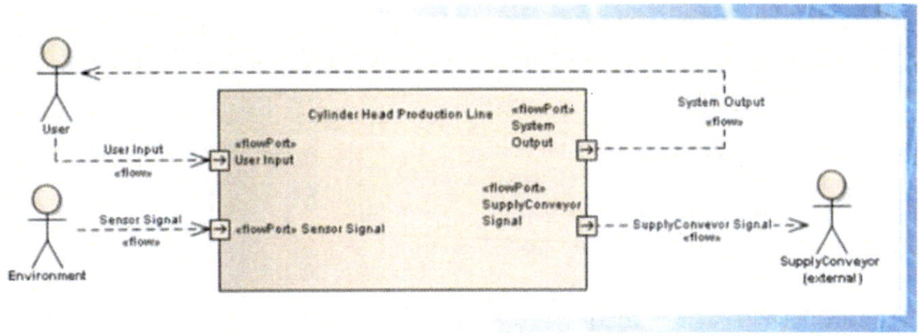

Fig. 4.9 SPES example for HMI. *Source* Vision for SPES, BROY, 2006

A typical example had been provided by SPES how to represent a human–machine interface (HMI) with input from humans and feedback from a system.

The funding project "safe" did provide further views and perspectives (Fig. 4.10). The individual perspectives could be described as follows:

- The operational perspective provides viewpoints on the behaviour interfaces between humans, the considered system within the relevant context.
- The functional perspective provides viewpoints on observable behaviour by functional artefacts.
- The variability perspective provides viewpoints on feature supported by the considered system, including various stakeholder viewpoints and variations in implementation or realisation.

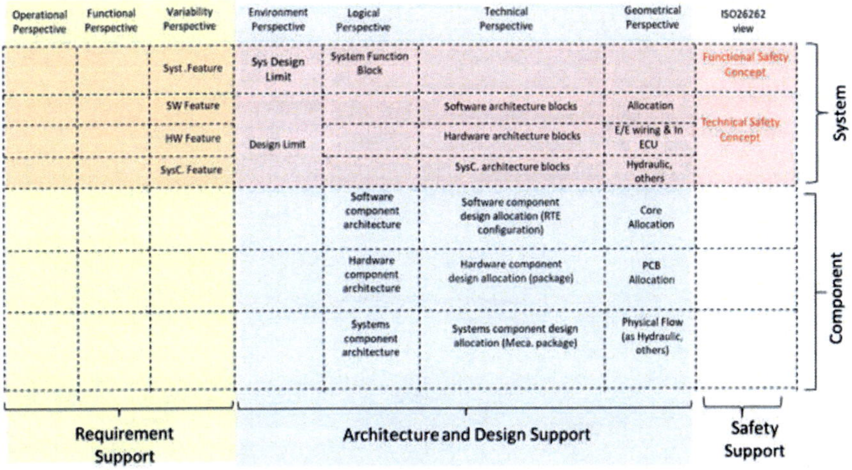

Fig. 4.10 Views and perspectives of an architecture (according to SAFE). *Source* ITEA 2 SAFE, WP 6.1, 2014

- The environmental perspective provides viewpoints on the surrounding context of the considered system and the potential impact or relevance to the considered system.
- The logical perspective provides viewpoints on logical elements or interdependencies of elements and their interfaces.
- The technical perspective provides viewpoints on technical elements to illustrate structures and interfaces or behaviour on such interfaces.
- The geometric perspective provides viewpoints of logical or technical elements in their geometrical environment.
- The safety perspective provides viewpoints on the safety-relevant aspects of an architecture, including different integrity level or risk classes, view on failures or faults or the effectiveness of measures.

The operational, environmental and variability perspective support the requirements development. The logical, technical and geometrical perspective support the architecture, and the safety perspective supports the additional needs for a safety case.

The "safe model" considers for any of those perspectives, system or component horizontal level of abstraction. Which perspectives are needed to specify a system within its context or multiple systems in their operating environment depends on the relevant stakeholder (Fig. 4.11).

ISO 26262 did only consider the risk derived from malfunctions of EE systems. All risk derived from the operational context of the vehicle was not in the scope of the standard. "Safe" did require only relevant architectural views, usually in order to avoid systematic failure the entire architecture and system design need to be

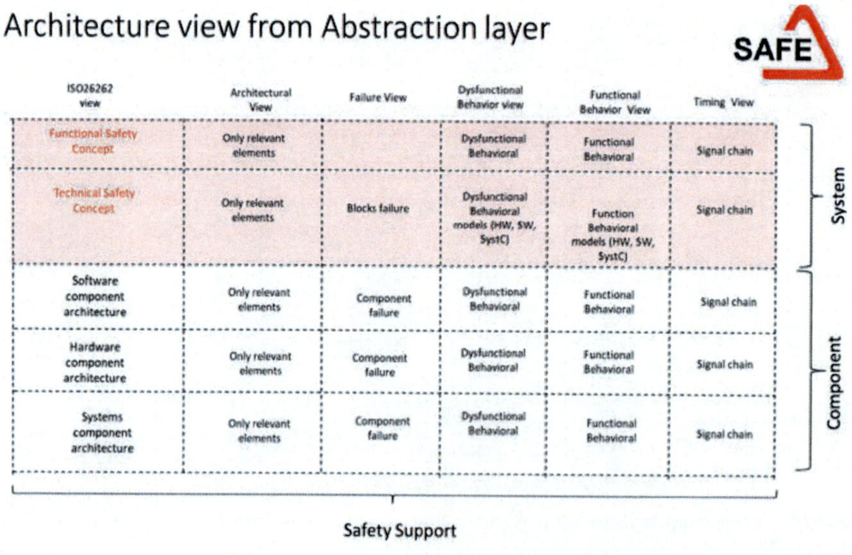

Fig. 4.11 Relevant architecture views supporting ISO 26262. *Source* ITEA 2 SAFE, WP 6.1, 2014

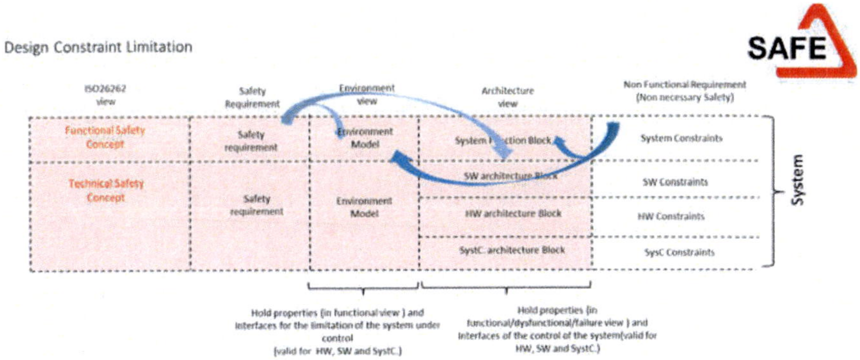

Fig. 4.12 Design constraint limitations according to "safe". *Source* ITEA 2 SAFE, WP 6.1, 2014

considered, unless the sufficient independence or freedom from interference could separate non-relevant parts of the product. For the "failure view" in the system abstraction layer, only block failures are considered, and in the component abstraction layer the component failures are considered. The "functional behaviour view" and the "dysfunctional behaviour view" are in addition since the correct behaviour of the product or the considered system is important for the validation. At least, the "timing view" requires identifying the signal chains within the system so that timing requirements and constraints could be developed.

The ISO 26262 requires maybe just some parts from the system architecture in order to manage systematic or random failure or malfunctions of the considered system. For all other risks, which could lead to accidents even the operational view, variability view and the environmental view need to be considered (Fig. 4.12).

Safety is not just linked to activities, which need to be done, safety requirements provide also many constraints. Constrains are not just limitations to a certain threshold; they could lead also to restriction on material, solutions, production limits or features are simply forbidden by law. The front of a vehicle could not design in a way that in case of accidents pedestrians could get serious injuries.

Many constrains derive from human capability studies or driving tests, etc., because driver could not correctly apply the features. Especially, the so-called safety-in-use aspects lead to many constraints for the integration or the performance of functions and features in a vehicle. Functional performance is quite often limited by the chassis design of the vehicle. If the stability of the vehicle in certain driving situations could not be assured and protection or active safety mechanism like the electronic stability control system (ESC) could not support a certain chassis design, a redesign is unavoidable. The environmental impact is a huge challenge for road vehicles, and the chassis needs to pass a huge summer and winter test program in order to find the right set-up for all the chassis systems before they are released for series production.

4.1.2 ISO 26262 in an Architecture Matrix

ISO 26262 did provide already a matrix in order to explain a hierarchical structure and the information flow within the standard (Fig. 4.13).

The standard made the following basic definitions:

1. A system (a system relates to sensors, controllers and actuators) or more systems which realise one (or more) function(s) on the vehicle level for which ISO 26262 can be used.
2. A system may implement one or more functions, but also one function can be implemented in several systems.
3. An item or system on vehicle level is comprised of one or more systems, where one system is composed of at least one sensor, a processing unit and an actuator. ISO 26262 draws the conclusion that a system should have at least three elements but it could be possible, for example, that an actuator is integrated in the processing unit.
4. A system can be divided into any subsystems but according to ISO 26262 the systems have to be hierarchically structured. In regards to systems, which together should realise functions with a higher ASIL, a clear hierarchical structure of systems has to be defined due to multiple fault control. In the version from 2018 of ISO 26262, also elements of an ASIC have been considered as subsystems.
5. A system (or subsystem) is comprised of one or more components.
6. Components consist of (electrical) hardware components (hardware parts or elements of an ASIC, etc.) or the embedded SW (entire SW in a controller), SW modules or SW units.

The diagram and the definitions provide a view how to read ISO 26262 and make the assignment of the requirements of the standard more understandable (Fig. 4.14).

This figure provides a hierarchical structure of the standard and the information flow considered by the standard between the work products of the standard (Fig. 4.15).

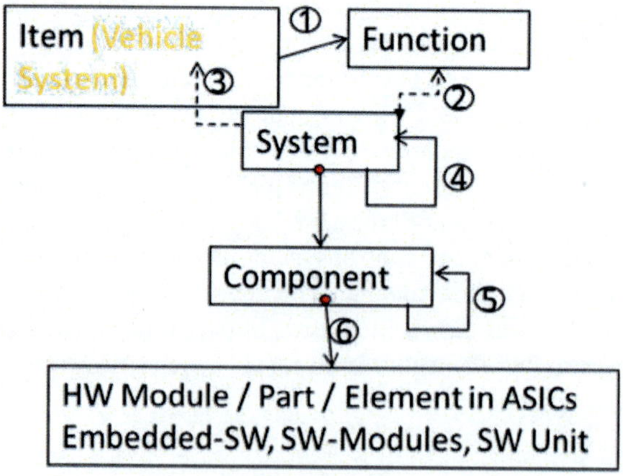

Fig. 4.13 Basic for the hierarchical understanding of ISO 26262

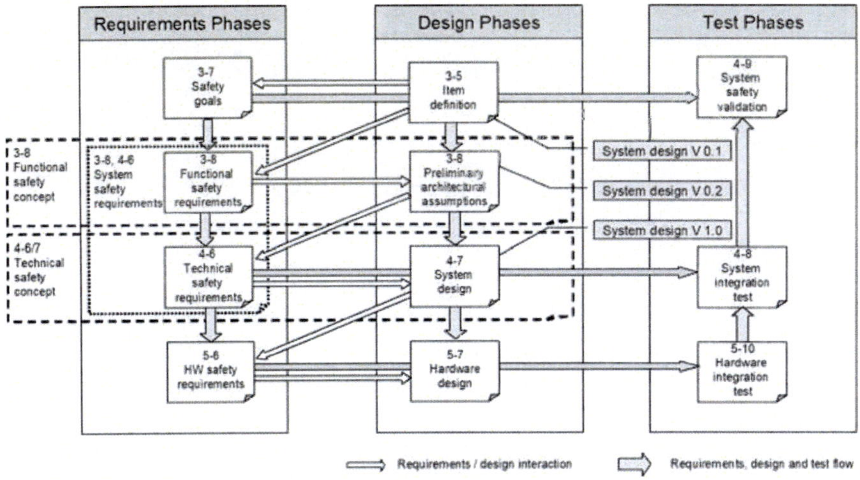

Fig. 4.14 Data flow in system and hardware product development. *Source* ISO DIS 26262

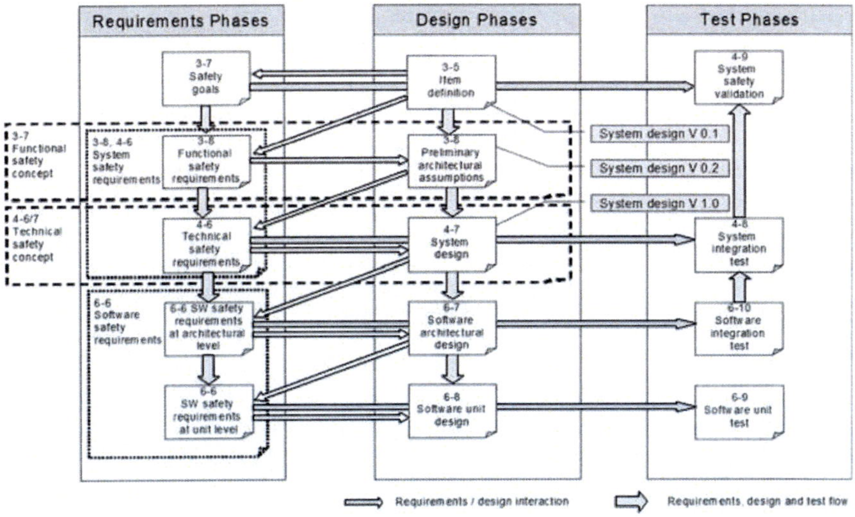

Fig. 4.15 Data flow in system and software product development. *Source* ISO DIS 26262

Comparing the software view with the hardware view shows for hardware one layer for the HW architecture and the design. For the software, an architectural layer and a software design layer had been considered different, and the SW unit had been seen as a representation of the software design (Fig. 4.16).

In the official published standard ISO 26262:2011, the elements of other technology and external measures and for software two integration steps had been separated in software integration testing and software verification. The idea of a maturity

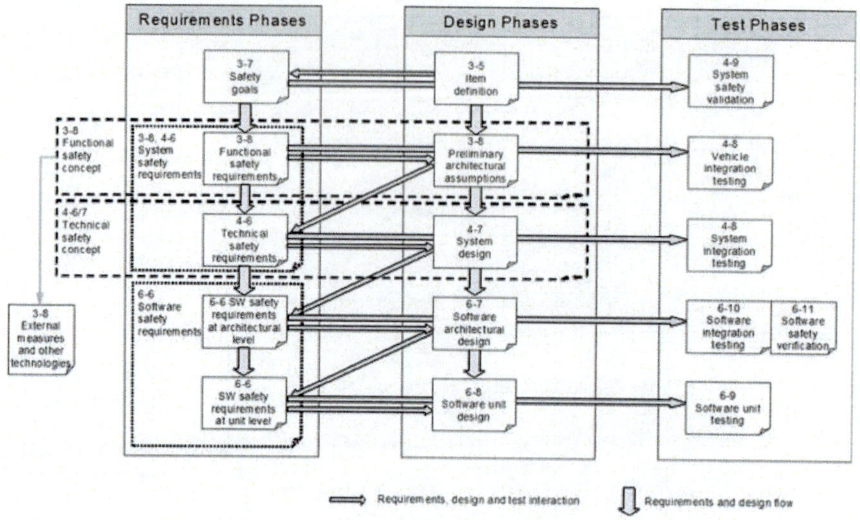

Fig. 4.16 Data flow in system and software product development. *Source* ISO 26262, 2011

model behind the design column had been deleted. The earlier picture did consider that there is an evolutionary grow of the system design from the item definition, via architectural assumptions to the system design. The basic idea was behind that from the intend in the item definition to the entire system design is a continuous architecture and design process. For the application of the standard, the understanding was essential in order to assure the tractability in a hierarchical top-down approach (Fig. 4.17).

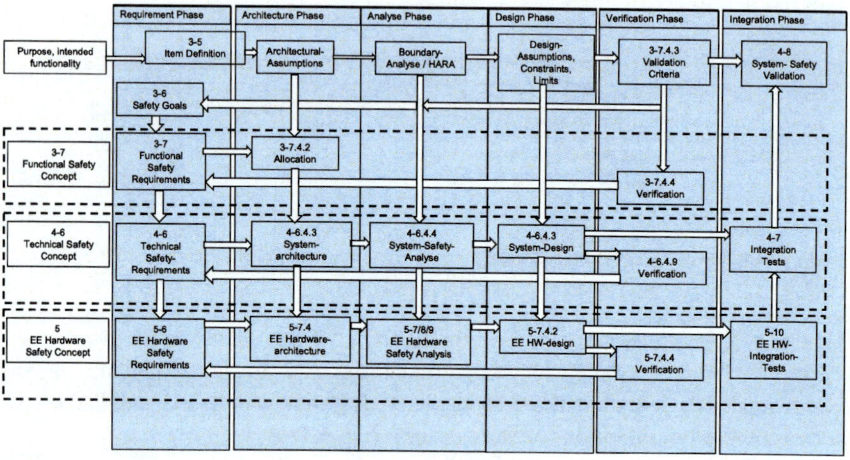

Fig. 4.17 Item development with horizontal hardware layer

The table provides six development phases, which show the safety activities and their work packages as required by ISO 26262. The columns could be also clustered to roles or competency fields.

- The requirement phase addresses all the specification work, and a typical requirement engineer finds in the adequate chapters the requirements about the methodology how to work out the required work products.
- The architecture phase addresses the activities from architectural assumptions down to the realisation or implementation. The architectural assumptions provide the structure for the entire product up to the system design and the basis for the safety analysis.
- The analysis phase addresses all the analysis work in order to assure that the design meets all the requirements. Various artefacts of the architecture provide the necessary input views and perspectives for systematic application of all necessary safety analysis. A traceable top-down approach could start with hazard analysis and risk assessment from the top and could for example provide the activities for the process analysis of the production.
- The design phase provides the technical solutions based on all previous phases. Of course, not only the system safety design should be addressed here, all design work products could contain systematic failure. That design phase addresses also functional design, system design but also the design of the human–machine interface.
- The verification (and validation) phase addresses all activities to assure consistency, correctness and completeness of the output from the previous phases. Some of the activities use also validation methods to provide answers for the necessary assurance. During that phase, the answer to open questions of the analysis should be given if the design fulfils the given requirements and if the given goals are achieved. Also questions: Are the given requirements correct or are the technical solution capable of providing the necessary performance in all relevant use cases should be addressed during that phase?
- The integration phase considers all elements from the steps below that could demonstrate a successful development for the required use cases, required feature and functions in the considered application environment. ISO 26262 considered the achievement and the correctness of the safety goal as objective for the safety validation.

ISO 26262 considers “safety goals” as high-level safety requirements. In order to assure the traceability, all safety requirements should derive from those safety goals. The requirements phase addresses the product development as a triple V-model. The system V uses the specification branch from hardware and software in order to develop the dedicated requirements (Fig. 4.18).

The hardware and the software V-model are liked into the system V-model. The hardware–software integration is in ISO 26262 considered in part 4 and belongs to the product development on system level (Fig. 4.19).

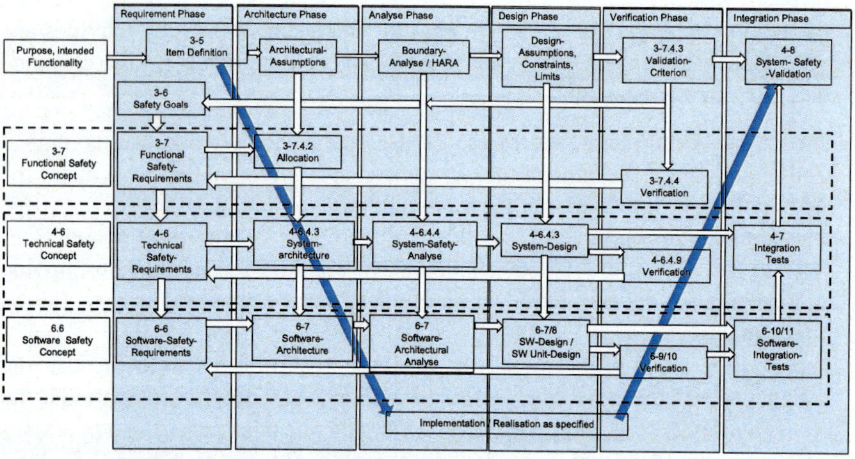

Fig. 4.18 Item development with horizontal hardware layer

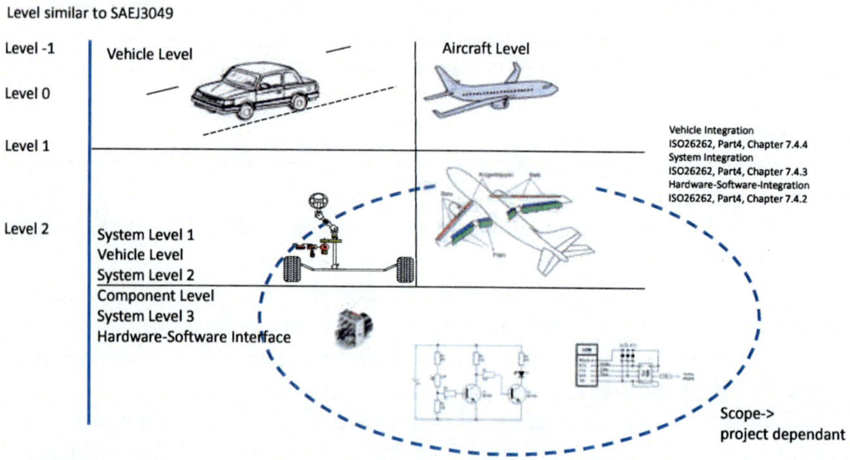

Fig. 4.19 Horizontal integration level according to ISO 26262

From ISO 26262 point of view, there are three system integration levels:

- The hardware–software interface (HSI) is the integration level for the hardware and the software components. Multiple hardware or multiple software component integrations are considered as component integration activity. Hardware–software integration typically addresses the implementation of the basic software into the microcontroller. Usually, the integration of the application software into electronic control unit including the basic software should be considered as a system integration process.
- The system integration considers the integration of the components or subsystems, which are scope in the item under development.

– The highest system integration level of ISO 26262 considers the vehicle integration. In that integration phase, various development items are integrated into the target vehicle. In that phase, all the vehicle interfaces need to be considered, so that the integration environment and the interfaces to the vehicle could be systematically tested.

The scope of ISO 26262 is limited to an "ITEM", which is a system for integration into a vehicle. The "ITEM Definition" is usually derived from similar systems with comparable functionality, so that for the new functionality first off at least the legal requirements, design constraints remain unchanged.

4.2 Road Traffic Perspective

The road traffic should be also considered as a part of a system approach, so that the harmonic coexistence of road traffic participants could be assured.

The "Locomotive Act 1861" did already foresee speed limits; the revision from 1965 did require a person with a flag in front of the vehicle in order to

– slowing down the vehicle, as it was forced to drive at walking speed, and
– warned approaching pedestrians and horse riders of their presence.

In the USA, some states define regulation like,

– 1914, Cleveland installed the world's first electric traffic signal,
– 1996, Buffalo, N.Y. erected the first "No Left Turn" sign in 1916,
– 1901, Connecticut passed the first speed limit law,
– 1908, Detroit was the first city to use stop signs, lane markings, one-way streets and traffic signals,
– 1923, Garrett Morgan patented a traffic signal that stopped vehicles in both directions to change the direction of traffic flow,
– 1935, 39 states did require a driving licence.

Many regulations follow which intends to share the road between different stakeholders. Of course, also the judicial aspects and the executive, e.g. by police had been established.

Sharing of road space has several dimensions based on various principles mainly by road traffic signs or regulations:

– geometrical assignment usually visible by road markings,
– default user (pedestrians, horses, cyclist, road vehicles, lorries, etc.) or category of transportation systems,
– default driving direction, speed limits and expected behaviour,
– priority in junctions and crossings,
– rules for lane change, overtaking, etc.,

- temporal use of lanes or priority by traffic lights, railroad crossing, etc.,
- restriction of users (e.g. driving licence for type of vehicle),
- use of roads or pavements restricted by age (e.g. children below 6 years may not cycle on roads).

The basic assumptions are that any class of users behaves as required by the road traffic regulations. If not, it could be

- errors,
- inadequate use of transportation system,
- inadequate design of transportation system,
- insufficient or wrong instructions,
- kinds of misuse or negligence,
- or any combination of above.

Usually, in case of occurring damages, the road traffic act defines parties to compensate the occurring damages. In case increasing damages without a liable party, the law will be adapted accordingly

In principle, one can read all kind of traffic laws in the world that each group of road users has its preferred range of motion. The roadway is intended for the cars and the sidewalk for the pedestrians. Cyclists are still a problem. It is always a risk for people to walk on the road with wrong speed, in the wrong situation or at the wrong moment in time. It is also a risk if vehicles leave the lane or move at the wrong time, with too high speed or in wrong situations. Roadways are preferred corridors for vehicles similar like railways and air corridors for trains and aeroplanes (Figs. 4.20 and 4.21).

In a system where all elements behave as expected, usually all the rules should be unambiguous. If one element behaves different, all the other elements could react accordingly. The challenge occurs if there are unexpected dependencies, which could

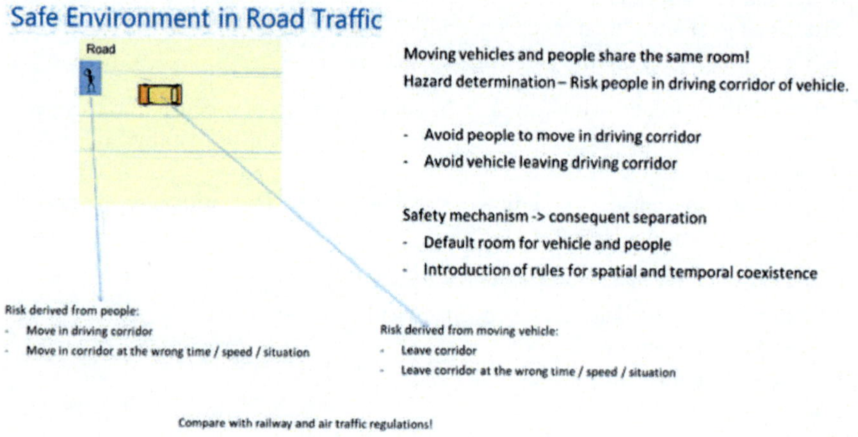

Fig. 4.20 Risk in road traffic environment

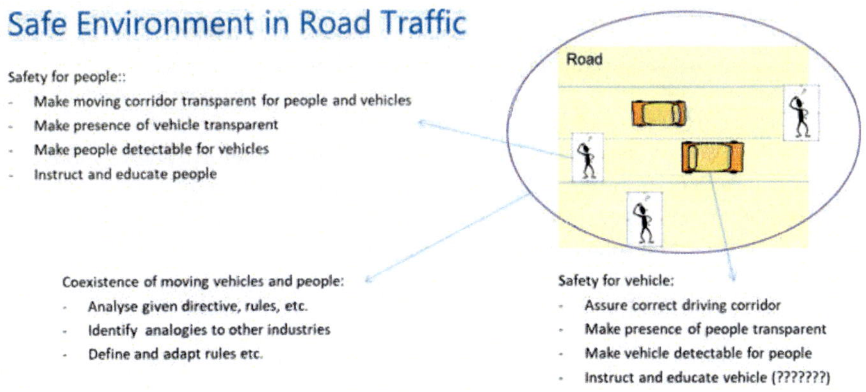

Fig. 4.21 Safety for road traffic participants

lead to complete other behaviour scenarios. Typical chain reactions can be caused by weather changes. It starts to rain, the visibility for pedestrians and vehicle drivers is getting worse, and cyclists are driving more uncertainly. Consequently, an accumulation of spontaneous or reflex reactions can occur, because the persons perceive possible changes rather suddenly. The typical behaviour of drivers, pedestrians or cyclists on a changed situation depends on many impact factors. Where there are few actors, there will be less spontaneous unpredictable responses, and the more the actors, the higher the likelihood of unpredictable reaction cascades. There is a final degradation for vehicles in unmanageable, uncontrollable or even chaotic situations; the vehicle should reduce the speed unless risk-free driving is possible. In case of good engineering of the possible operation scenarios, reactions and degradation scenarios, the traffic performance even in case of high traffic density could be high. The better and more accurately one can predict the traffic behaviour; the better the traffic space could be used within a time interval or traffic could move with higher speed. If pedestrians and cyclists are separated from the driving lane of the road vehicles, there is less risk even in case of higher vehicle speed. Motorways with structurally separate lanes for each driving direction allow higher speeds without significant higher risk. Road design, traffic regulations and restricted use for other road traffic participants could lead to risk reduction, and the level of tolerable risk leads to a question of speed.

The challenge for autonomous vehicles or vehicles that make their own decisions is that they could not share a common viewpoint of a neutral observer.

All road traffic participants have to follow the same rules, independent if their behaviour is dominated by a system or by human intent. Due to the relevant road traffic regulations and social security, foreseeable misuse needs to be considered. Any scenario need to be in its relevant road traffic environment should be analysed for possible misuse scenarios. The possible misuse scenarios lead at least to a more conservative driving strategy. The need of operational degradations or degradations is due to failure in infrastructure, vehicles and so on.

4.2.1 *Road Traffic Environment*

Any country provides their own view on regulations and the tolerance level for various road traffic scenarios. Road condition, characteristics, etc., provide a certain level of awareness of the driver, and various weather conditions require more or less concentration or awareness of the driver in the different driving situations. How much risk the driver takes, if driving slow, with particular care or however a driver behaves depends on his education and experiences.

In order to analyse the road traffic environment, a very systematic approach is necessary. The road surface changes with the type of asphalt, dirt, ice or rain. The visibility of traffic signs and road markings changes with weather. Some of the changes are dynamic or floating like drying a road after a heavy rain in summer. Moreover, the friction of the road is not at any place the same, puddles need to be reckoned as well as wet leaves in autumn (Fig. 4.22).

The funding project, Pegasus, targets to bring road vehicle technology closer to the road traffic environment.

In the Pegasus method, six layers are defined:

- layer 1—road (geometry,...),
- layer 2—road furniture and rules (traffic signs,...),
- layer 3—temporary modifications and events (road construction,...),
- layer 4—moving objects (traffic-relevant objects like vehicles, pedestrian,...moving relative to vehicle under test),
- layer 5—environmental conditions (light situation, road weather,...),
- layer 6—digital information (V2X, digital data/map,...).

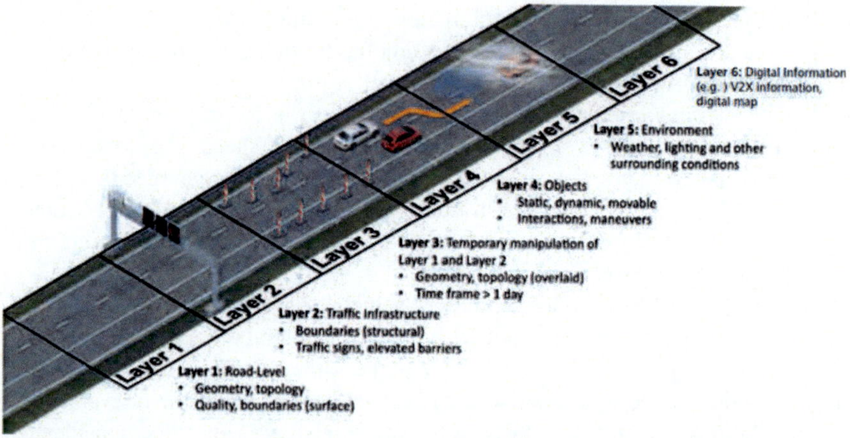

Fig. 4.22 Model for a systematic description of scenarios with six independent layers. *Source* PEGASUS METHOD, https://www.pegasusprojekt.de/files/tmpl/Pegasus-Abschlussveranstaltung/PEGASUS-Gesamtmethode.pdf, download January 2020

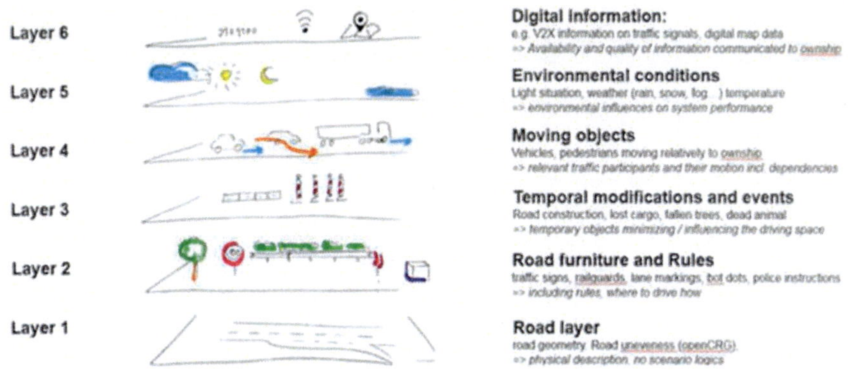

Fig. 4.23 Disjunction layers used to logically derive concrete scenarios. *Source* VMAD-05-03, Proposal for traffic disturbance test scenarios for ALKS, Annex X, Traffic disturbance test scenarios for ALKS

Those layers become in automotive industry more and more a standard. The sixth layer is more a part of normal system engineering, but it provides the interface to an "intelligent road".

Also in the UNECE working group, validation method for AD (VMAD) published a similar approach, which is based on similar layer (Fig. 4.23).

In that scenario for traffic disturbances for automated lane-keeping systems (ALKS), they provide further needs that what needs to be detected and what scenarios they see as the critical elements or corner case scenarios.

In Germany, there are several regulation how to design roads. One example is the "Guidelines for the Design of Motorways English Version of Richtlinien für die Anlage von Autobahnen, Edition 2008 Translation 2011 RAA" which provide principles for safe roads or here in the document-specific "safe motorways" (Fig. 4.24).

RAS-L did define the field of application on roads without cultivation (both country roads and motorways). Since June 2008, the RAS-L has been limited in its validity but is still valid for country roads. For motorways, the guidelines for the construction of motorways (RAA for short) were introduced. In 2012, the RAS-L was replaced by the Guidelines for the Construction of Motorways (RAL). The RAS-L was not valid for so-called city roads. For urban roads, the Guidelines for the Construction of Urban Roads (RASt) were published in June 2007 as a successor to the Recommendations for the Construction of Main Roads (EAHV) and Access Roads (EAE). The RASt applies to the new road categories VS, HS and ES of the Guidelines for Integrated Network Design (RIN), which correspond approximately to the previous RAS-N categories C, D and E.

The RAA considers very specific characteristics, which makes a so-called safe road.

> Motorways have to cope with high traffic volumes and allow vehicles to travel at high speeds. For this reason, safety is of particular importance. Design and operation characteristics are

Table 1: Road categories as defined by RIN and the scope of the RAA

Category group / Link function level		Motorways	Rural roads	Trunk roads in non built-up areas	Trunk roads in built-up areas	Local roads
Link function level		**AS**	**LS**	**VS**	**HS**	**ES**
continental	**0**	**AS 0**		–	–	–
sub-continental	**I**	**AS I**	LS I		–	–
inter-regional	**II**	**AS II**	LS II	VS II		–
regional	**III**	–	LS III	VS III	HS III	
sub-regional	**IV**	–	LS IV	–	HS IV	ES IV
local	**V**	–	LS V	–	–	ES V

Legende:

AS I	Designation of the category as it occurs
	problematic
–	does not occur or is not justifiable

Fig. 4.24 Road categories according to RIN. *Source* the "Guidelines for the Design of Motorways English Version of Richtlinien für die Anlage von Autobahnen, Edition 2008 Translation 2011 RAA"

used to influence road user behaviour and, consequently, road safety. Motorways should be designed and equipped in such a way that

- The characteristics of the road are as homogenous as possible over large sections.
- Road users can adapt their speed in time to suit the course of the road and the traffic situation.
- Junctions can be identified well in advance.
- The occupants of errant vehicles and areas/habitats alongside the road that need protection are protected against the serious consequences of road accidents.
- Road users can make emergency stops alongside the carriageway and
- Road users can call for assistance in emergencies using roadside emergency telephones.

Figure 4.25 contains a selection of important objectives and possible influencing variables. Road safety enhancements are typical aspects such as the following:

- Generous design elements (not minimum elements) are used.
- The sequence of elements in the horizontal alignment is balanced.
- Horizontal and vertical alignments are coordinated and adapted to suit one another.
- Road users can see sufficiently in the distance along the section of road on which they are travelling.
- The cross section (in particular at construction works) is sufficiently wide and has a hard shoulder.
- Signposts are erected sufficiently far ahead and are not ambiguous.
- Surface water is drained away as directly as possible and areas with poor drainage are avoided.
- Lateral obstacles are avoided or
- Action is taken to provide protection against dangerous obstacles (e.g. vehicle restraint systems or safety barriers and crash cushions).

Table 2: Road safety

Objectives	Possible influencing variables
Safe vehicle tracking	- Radii - Three-dimensional alignment - Sight distance - Crossfall and incline - Drainage
Safe travel alongside or behind other vehicles	- Longitudinal gradients - Climbing lanes - Sufficient lane width at roadworks - Hard shoulders - Traffic control systems
Junctions with a low number of conflict points	- Separation of decision points and conflict points - Generous design elements in both the horizontal and vertical plane - Sufficient weaving distances or the avoidance of weaving areas - Sufficient sight distances with regard to oncoming traffic - Sufficient distances between consecutive junctions for directional signing - Adequate sight distance for merging
Safe central reserves and roadside areas	- Density of obstacles alongside the carriageway - Distance between the obstacle and the edge of the carriageway - Quality of the vehicle restraint equipment or safety barrier - Hard shoulders (marginal strips) - Deceleration lanes and coasting areas at the exits - Protected emergency telephones
Safe maintenance procedures	- Width of the hard shoulders - Design of the maintenance access roads - Selection of vehicle restraint equipment

Fig. 4.25 Characteristics and aspects for road safety

- Accidents involving wild animals are minimised by the erection of suitable fences, animal crossings in the form of bridges or underpasses and/or by not planting plants that attract animals.
- Markings and traffic guidance equipment are applied and implemented properly.

The Recommendations for Road Safety Audits (Empfehlungen für das Sicherheitsaudit von Straßen, ESAS) contain other possible ways of improving road safety. In addition, these recommendations require a road safety audit to be conducted before each planning stage is completed.

Quote-Source Source: the "Guidelines for the Design of Motorways English Version of Richtlinien für die Anlage von Autobahnen, Edition 2008 Translation 2011 RAA" (Fig. 4.26)

There is one important example for all the characteristics, which shows the influence on safety.

The EKA classes provide information about the different types of motorways or equivalent and their characteristics.

They propose the following speed limits by considering wet roads:

- 130 km/h for long-distance motorways (EKA 1 A),
- 120 km/h for inter-regional motorways (EKA 1 B),
- 100 km/h for motorway-like roads (EKA 2) and
- 80 km/h for urban motorways (EKA 3).

The German Road Traffic Act allows also higher speed than 130 km/h, unless there is no specific sign the driver is fully responsible to control the vehicle with

Table 10: Design classes and design features

Design class	EKA 1 A	EKA 1 B	EKA 2	EKA 3
Designation	Long-distance motorway	Inter-regional motorway	Motorway-like road	Urban motorway
Signing	Z 330 StVO (motorway)		Z 331 StVO (trunk road)	Z 330 or Z 331 StVO
Directional signing	Blue		Yellow	Blue, yellow
Maximum permissible speed*	None		None	≤ 100 km per hour
Recommended distance between junctions	> 8,000 m	> 5,000 m	> 5,000 m	None
Traffic management around construction works on four-lane roads	4+0 generally necessary**		4+0 not absolutely necessary	

* see explanation given in Section 3.4

** '4+0' indicates the lane configuration during reconstruction. The '4' indicates that four lanes (two in each direction) will be accommodated on the one carriageway and the '0' indicates that there will be no traffic on the other carriageway while it is being reconstructed.

Fig. 4.26 Design classes and design features according to RAA. *Source* RAA, Table 10

higher speed. If roads are not in an adequate condition, the driver is also responsible to drive with a speed which allows to safely control the road traffic situation.

Because of all this, up to know the driver is fully responsible to observe the road traffic and environment and the vehicle condition. Unless both allow the driver to drive with a certain speed, it could be assumed that the vehicle is on a safe journey.

4.2.2 The Context of Automated Road Vehicles

There are many stakeholders, which provide many aspects into the product design (Fig. 4.27).

Some of the influences are important for the entire vehicle and some only for specific systems. Safety will be of minor importance for the car multimedia system, but highly important for the chassis systems like brake and steering systems. The legal requirements are addressing specific devices as the regulations for radio frequencies for the radar, especially the homologation regulations addressing the whole vehicle. If UN ECE R13 is for brake systems, it does not mean that it only addresses the vehicle brake system. It is the job of engineering to analyse the requirements and break them down to the system under development.

The road traffic environment is the major context for road vehicles. The main context for vehicles and especially for automated driving, the legal context provides requirements with a very high priority.

There are the following field of legislation to be considered:

- homologation regulations,
- road traffic regulations,
- road safety regulations.

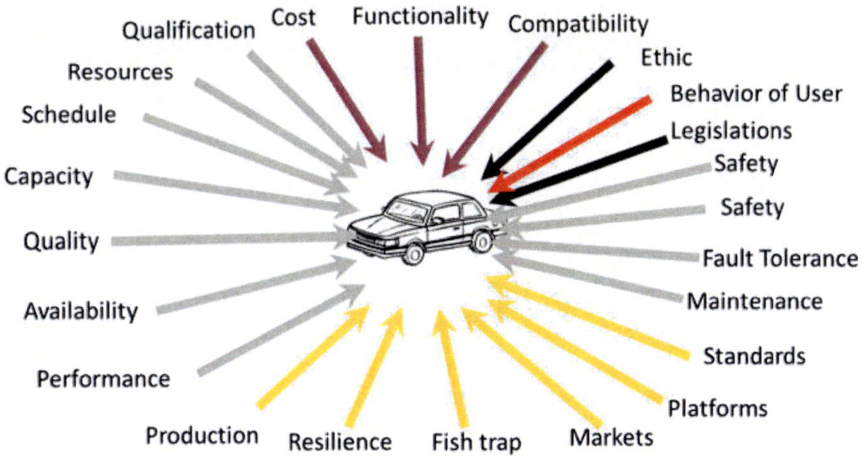

Fig. 4.27 Context aspects for motor vehicle and their systems

Beside that specific legislation of course product safety, product liability and trade law are significant driver of requirements for road vehicles and their associated systems. Trade law is of course also closely related to the target market for the vehicles or associated systems. In our days, several new regulations coming up for pedestrian, cyclist and occupancy safety are coming up. Several requirements derived from consumer protection standards like NCAP are getting more and more apart of legal regulations.

Another major stakeholder group is the users. Users could be in various positions:

- the driver,
- vehicle occupancies (people transported in the vehicle),
- owner (people that own and pay for the vehicle),
- users on a professional basis, like a vans, taxis, transporters, rental cars, dealers, etc.

For these sort of people, the usability is in the focus, and we are close to safety-in-use. The intended use is the basis for the misuse analysis. This analysis is a major focus for development of new type of vehicles. Especially, new handling scenarios like charging of electrical vehicles, new occupancy, pedestrian or cyclist protection systems, or new assistant or automation system require deep usability analysis. The hazard analysis and risk assessment according to ISO 26262 address just the aspect how a driver could control failure and malfunction of a vehicle system. In addition to the ability to control failure by the driver, an analysis needs to provide evidence that the driver, for example, could manage the situation during driving.

Very prominent questions related to usability are as follows:

- Could the driver take over control of the vehicle after automated driving?
- Could the driver manage the driving situation adequately?
- Is the vehicle ergonomic adequate for the safe driving by the driver?

Another big group of stakeholders is people that focus on features, performance, design, etc., which addresses not only the shape of the vehicle. All the driver assistant systems are in the first glance features, which should provide enthusiastic benefits for the users. With all the feature analysis from a safety point of view, those features are

- potential source of failure,
- the driver could be irritated,
- the driver could be distracted, so that he does not pay the necessary attention to driving,
- lead to wrong reaction by the driver,
- it could influence other safety-related functions, etc.

Production, design capability and other limitations of course influencing the design of a vehicle and last and not least the financial aspects have to be taken into account.

4.2.3 Breakdown from Stakeholders View to an Operational View

The basic vehicle functions are

- driving
- braking
- steering.

The basic use case of a standard vehicle is driving. Braking and steering are unavoidable consequences by the design of today´s vehicle (Fig. 4.28).

The feature driving could be decomposed by the following subelements:

- perceive the road traffic environment,
- perceiving vehicle condition,
- evaluating the optimum driving strategy for the mission,
- continuously control the vehicle speed.

The feature braking is basically just negative driving; it could be decomposed by the following subelements:

- perceive the road traffic environment,
- perceiving vehicle condition,
- evaluating the brake demand,
- braking as demanded.

The objective of steering is controlling the vehicle in the optimal lane; it could be decomposed by the following subelements:

- perceive the road traffic environment,
- perceiving vehicle condition,

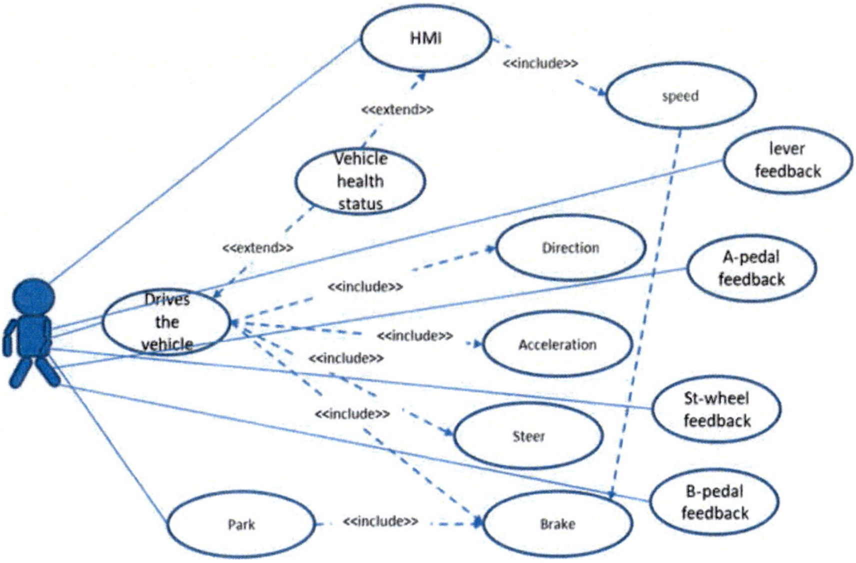

Fig. 4.28 Basic use case during vehicle driving

- evaluating the optimum lane for the mission,
- continuously steering to follow the lane.

Perception of the road traffic environment and the vehicle condition is common for all basic vehicle control function. The key difference is that very different criterion and characteristics need to be perceived which are relevant to the adequate reaction. For a human driver, it is a matter of driving education to control the basic function at the same time. Steering and operating adequately the brake and accelerator pedal at the same, time is a matter of exercises and education. People, who are used to drive automatic gear systems, could give an impression, how difficult it is to drive fast on a curvy road and changing the gear at the same time. If the driver is not capable of changing the gear ratio during driving, an automatic gear vehicle is strongly recommended. An experienced driver could drive faster because of faster perceiving situations or conditions and controlling the dedicated actuators sufficient independently.

4.3 Operational Design Domain as per SAE J3016

It had been realised that the environment in which a specific operation should designed for defines many requirements for the design of the associated systems. SAE J3016 "Taxonomy and Definitions for Terms Related to On-Road Motor Vehicle

Automated Driving Systems" did not only define the level of automation, and it also introduced some common terms and definition for automated driving.

Operational Design Domain (ODD): J3016 defines an Operational Design Domain (ODD) as "operating conditions under which a given driving automation system or feature thereof is specifically designed to function, including, but not limited to, environmental, geographical, and time-of-day restrictions, and/or the requisite presence or absence of certain traffic or roadway characteristics".

Automated Driving System (ADS): ODD is the design domain of an ADS or a feature thereof with respect to its operation. J3016 introduced this concept in order to capture limitations for driving automation at levels 1, 2, 3 and 4. Level 5 ADS (full driving automation) has an unlimited ODD, which offers the same mobility as a human driver. An ODD may put limitations on

- the road environment,
- the behaviour of the ADS-equipped subject vehicle and
- state of the vehicle.

Generic road environment limitations may include types of roads (urban, rural or freeways), specific elements such as roundabouts or tunnels, temporary structures such as construction zones, traffic volumes and weather and visibility conditions. Road environment limitations may also be specific, such as a geographical area, or season, or time of the day. Road environment limitations are the most common elements of an ODD, but an ODD may also place limitations on ADS behaviour and state. Limitations on ADS behaviour may include speed limitations or limitations on manoeuvres, such as reversing. Limitations on the state of the subject vehicle may include the requirement of no trailer being attached, loading limitations or minimum tyre inflation level. The ODD may reflect the requirements of a particular driving automation feature. J3016 defines a driving automation system feature as "a level 1–5 driving automation system's design-specific functionality at a given level of driving automation within a particular ODD, if applicable". Examples of driving modes include low-speed traffic jam assist, high-speed cruise assist, valet parking and urban in-lane driving. J3016 stipulates that an ADS may have one or more features, and each feature has exactly one ODD.

Operational Road Environment Model (OREM): The design and verification and validation of an ADS require modelling the operational environments in which the ADS is to operate the ADS-equipped vehicle. An operational road environment model (OREM) is a representation of the relevant assumptions about the road environment in which an ADS will operate the ADS-equipped vehicle.

Environment models capture the properties of the environment that are relevant to the ADS operation, while abstracting irrelevant details. OREMs can represent generic environments, such as two-lane rural road, or actual roads in a specific geographic area. OREMs can take different forms, including specification documents and executable models. While a closed course used for testing may be designed to model a target deployment environment, within this document, OREMs are meant to be documents or software models. OREMs provide context for specifying driving tasks of an ADS and for verifying and validating the ADS. Executable OREMs can be used in simulation testing. OREMs can also specify the environment for closed course and field tests. An ODD of an ADS implies a set of operational environments in which the ADS can operate the ADS-equipped vehicle. These environments can be specified using a set of OREMs. OREMs may be fully or partially in scope of an ODD, or they may be out of scope of an ODD. For example, an OREM for a highway pilot may include a section of a highway, ramps and parts of roads accessing the ramps. The highway and the ramps may be in scope of the highway pilot's ODD, but the access roads may be outside of the ODD. Including the parts of the roads accessing the ramps in the OREM allows specifying the ODD boundary. Further, out-of-scope OREMs are also useful to specify ODD boundaries and may be used to test the capability of the ADS to detect an out-of-ODD condition and the fallback performance in the event of ODD departure.

The idea of SAE J 3016 became the basis of further legislation like the new approach of UN ECE, with ASCF (ALKS) as the first homologation standard, which covered roughly the Level 3 of SAE J 3016.

4.4 Layered Approach for Automated Driving

Any modelling approach requires an environmental and an implementation model is required. The implementing model represents the intended product, system or functionality, which is the purpose, or intend of the development (Fig. 4.29).

The vehicle is considered as a black box model, which is defined in their degrees of freedom, for example, as defined in ISO 8855 (Vehicle Dynamics and …). The environmental model is the environment for that the vehicle is designed for. The

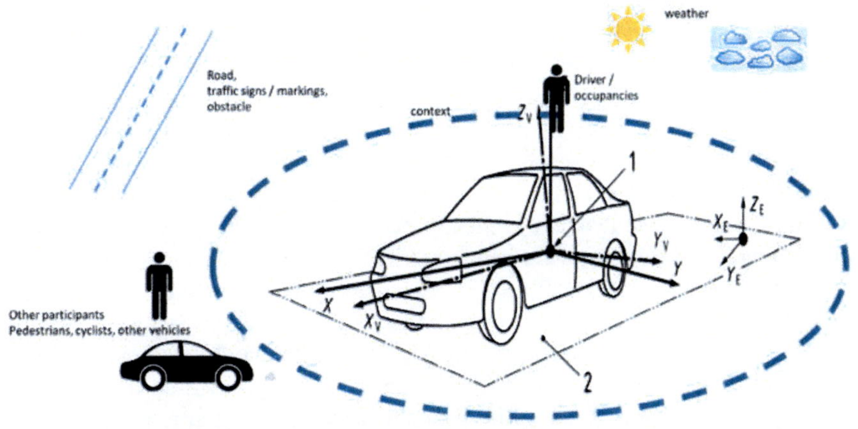

Fig. 4.29 Vehicle in the environmental model

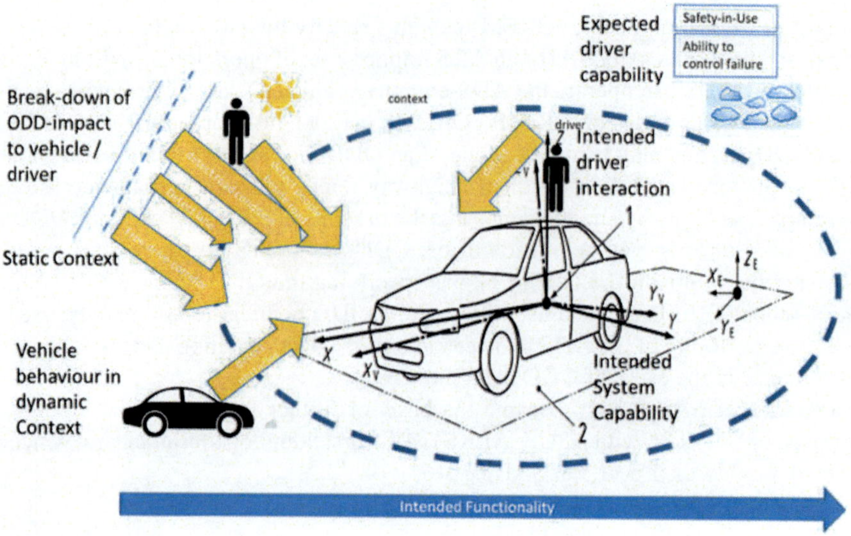

Fig. 4.30 Influence of ODD to the vehicle behaviour and deriving constraints and requirements

approach from SAE J1603 breaking the Operational Road Environment Model (OREM) directly down to the system under consideration or the automated driving system (ADS) leads to negligence of the basic safety feature of the vehicle. All basic driving functions of a roadworthy vehicle require to maintain fully operational during any automation function (Fig. 4.30).

There are many classes of requirements and constraints, which derive from the ODD to the vehicle and a potential driver in the vehicle.

- The road design, the condition like weather, dirt, etc., and the traffic situation restrict the lateral and longitudinal movement of the vehicle. Static elements like the road design, lane markings, buildings, traffic signs, etc., form a static context. They change only in long time intervals and usually the road authorities plan them.
- All dynamic aspects like weather, obstacles, dirt on the road, other road traffic participants or weather conditions form together with the static context the dynamic context. The dynamic context provides immediate changes like obstacles falling on the road or goods from preceding vehicles, which require a sudden reaction by the vehicle. Usually, the driver brakes or modifies the driving lane in order to avoid accidents. In case of automated driving, the driver would not be able to control the vehicle in such sudden situations, so that the system provides an appropriate reaction. A driver approaching a crossing learns during driving school how to behave. Road traffic regulations, other road traffic participants, traffic signs and lights view whether all those factors influence the speed and the lane to pass the crossing.

- In case of not fully automated driving, an adequate behaviour or reaction by the driver is required. Nevertheless, the driver is not always in the position to perform sudden actions to control the vehicle adequately.
- Another aspect of the driver is to monitor the vehicle state or the conditions of the relevant systems. Especially, sudden failure of vehicle systems like the steering or rake system could not be controlled by the driver, so that adequate redundancy needs to be implemented, so that the system could degrade without any drivers intervention.

Based on those analyses, many system performance requirements and constraints are derived as well as significant time constraints (Fig. 4.31).

Beside the requirements and constraints derived from the vehicle level analysis, the vehicle design provides further requirements and constraints, which could influence the "safe" vehicle behaviour for the intended operation environment (Fig. 4.32).

The dependency in the operational layer is very heterogeneous, since the requirements and constraints in various specific design artefacts show significant dependencies.

The typical road traffic regulations provide many constraints for the way of operation of the vehicle, which are weather, road and traffic dependant. Usually, the driver learns during the driving school all necessary elements, but driving capability is a matter of experience.

The homologation regulations are usually born in an evaluative manner from the need for protective measures, which formulated vehicle capability requirements. UN ECE R79 is assigned to the steering system and UN ECE R13 to the braking system, but they consider the braking and steering capability of the entire vehicle. An essential aspect of the approval tests requires the demonstration that the vehicle can ensure braking and steering capability in all relevant situations and conditions.

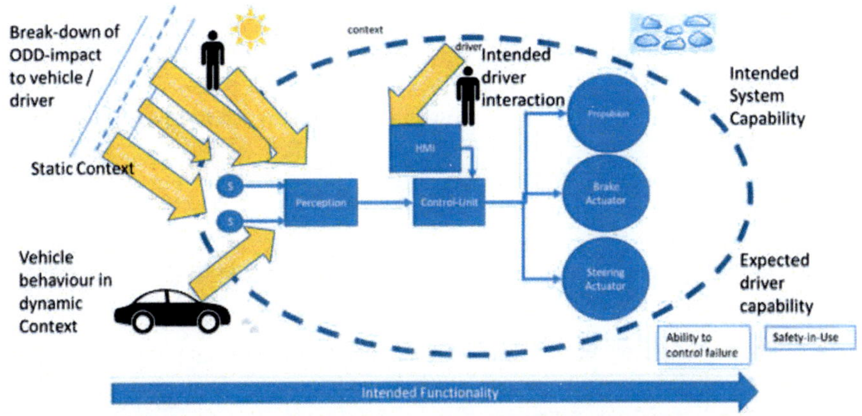

Fig. 4.31 Breakdown of the requirements to the system under consideration

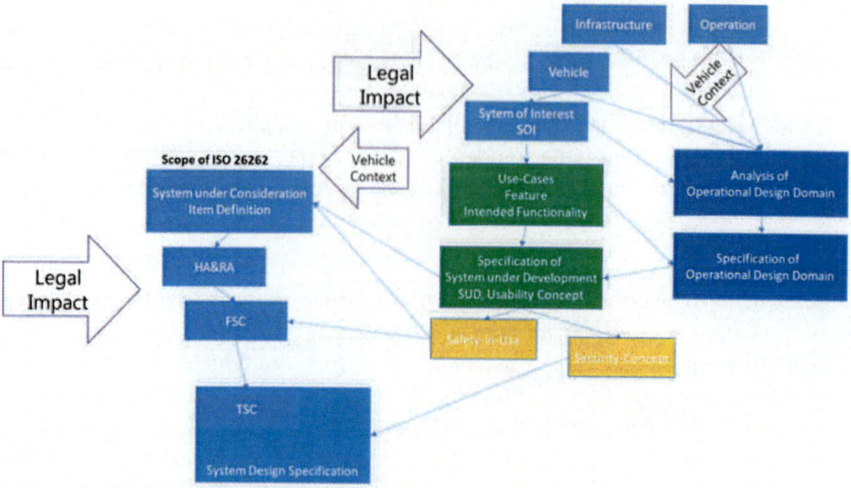

Fig. 4.32 Breakdown of requirements and constraints

There are various classes of requirements and constraints derived from the ODD analysis, which are

- system or functionality capability requirements,
- performance requirements and timing constraints and
- requirements of independence (or non-influence criterion).

The intended use cases, the intended functionality or features, which provide the purpose of the product, are customer requirements in the first glance. Various actors like the driver, occupancies and other traffic participants like pedestrians, cyclists or driver of other vehicles expect traffic typical behaviour of all road traffic participants. Any deviation from the relevant road traffic regulations means a risk for any of the actors. Cybersecurity of course means an active manipulation of data, which lead to damages for any of the road traffic participants, owner of the vehicle, insurances or producer of the vehicle or related systems. Consequently, the usability concept is the basis for the driver capability analysis of the controllability classifier for the ISO 26262 HARA, and as well for safety-in-use, security, authorisation and anti-theft protection (Fig. 4.33).

In order to describe the methodology for the risk identification a typical GSN (GSN—goal structured notation) could be used. A diamond stands for the "strategy", a circle for the corresponding "evidence", a rectangle for the "goals", a rounded rectangle for the "context", and the ellipsis for "assumptions and justifications". The following risk classes are considered:

- fault, errors, failure or malfunctions of functional or technical elements, where EE elements build the typical scope of ISO 26262,
- typical operational impact due to inadequate behaviour of the vehicle, due to system insufficiencies or other errors during development or integration of the vehicle,

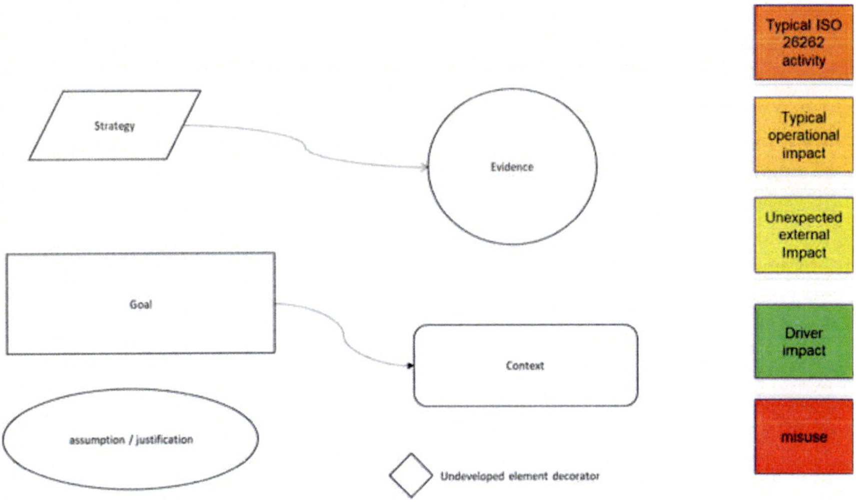

Fig. 4.33 Typical legend for a GSN notification

- unexpected external impact, such as unusual or combination of events,
- drivers influence which lead to erroneous behaviour or any human factor issues,
- all kind of misuse, unintended use, adverse use or cybersecurity attacks by hacker, etc. (Fig. 4.34).

Based on the following risk classes, goals could be defined for the risk management:

- Any risk due to the operation of a vehicle in the given ODD, which leads to unreasonable risk, could be controlled.

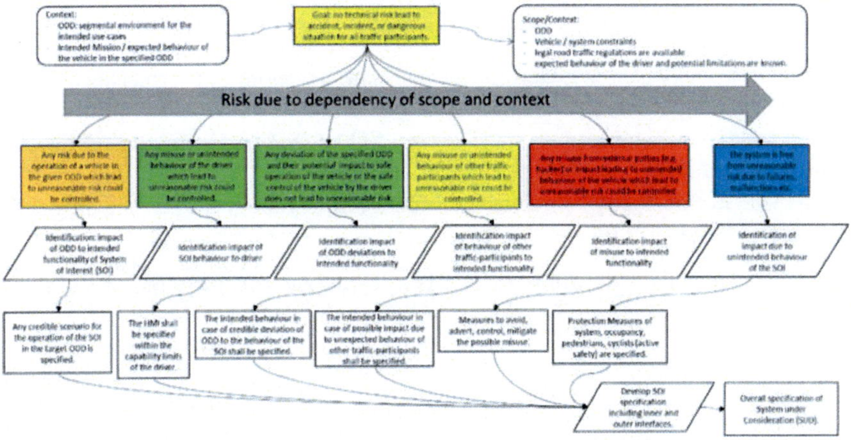

Fig. 4.34 GSN for identification and management of risk

- Any misuse or unintended behaviour of the driver, which leads to unreasonable risk, could be controlled.
- Any deviation of the specified ODD and their potential impact to safe operation of the vehicle or the safe control of the vehicle by the driver does not lead to unreasonable risk.
- Any misuse or unintended behaviour of other traffic participants, which leads to unreasonable risk, could be controlled.
- Any misuse from external parties (e.g. hacker) or impact leading to unintended behaviour of the vehicle, which leads to unreasonable risk, could be controlled.
- The system is free from unreasonable risk due to failures, malfunctions, etc.

Based on those goals, adequate measures for the risk identification could be applied:

- identification impact of ODD to intended functionality of system of interest (SOI),
- identification impact of SOI behaviour to driver,
- identification impact of ODD deviations to intended functionality,
- identification impact of behaviour of other traffic participants to intended functionality,
- identification impact of misuse to intended functionality,
- identification of impact due to unintended behaviour of the SOI.

The adequate measures for risk reduction for those risk classes could be applied by the following:

- Any credible scenario for the operation of the SOI in the target ODD is specified.
- The HMI shall be specified within the capability limits of the driver.
- The intended behaviour in case of credible deviation of ODD to the behaviour of the SOI shall be specified.
- The intended behaviour in case of possible impact due to unexpected behaviour of other traffic participants shall be specified.
- Measures are to avoid, advert, control and mitigate the possible misuse.
- Identification of impact due to unintended behaviour of the SOI.

The main strategy is now to develop a consistent, traceable and correct specification of the "system of interest"—SOI.

The strategy could be formulated: **Develop SOI specification including inner and outer interfaces**.

As a result of this, the system under consideration shall be specified, and the goal could be formulated as follows:

- **System under consideration (SUC) is completely specified based on the context of SOI.**

4.4.1 Phases of an Hierarchical Engineering Approach

Similar to the SPES matrix in any level of abstraction, an engineering process starts with the following phases:

- Requirement phase mainly focuses on elicitation and development of the first requirements. The requirements derive mainly from the stakeholder analysis. One of the major class of stakeholder requirements are instead is the legal requirements, which should determine from the state-of-the-art analysis or for complete new engineering fields from the analysis of relevant state of science and technology. In the lower layer of abstraction, the requirements derive from the previous layer and continuously expand with the activities in the dedicated layer of abstraction.
- The architecture phase considers at least two subphases; the first is the drafting phase to define the structural assumptions. The second phase considers the behaviour and dynamic aspects. The structure should be inherent from the higher level of abstraction down to the implementation or realisation and provide the basis for interface or boundary analysis. Iteratively, the structure also provides the framework for requirement engineering.
- The analysis phase provides a set of methods which starts often with decomposition diagrams like the breakdown of use cases, feature and functions so that the basic elements become evident, which are required to compose the purpose of the development step or the entire product in the dedicated context.
- The design phase is the phase, where the solutions are determined which should be the resulting from the given requirements, the architectural assumptions and the analysis. The design decision should be recorded as part of the requirement management, so that design decision remains traceable.
- The verification phase provides the input for the maturity of the development approach. Typical questions about completeness, correctness and consistency shall be examined.
- The integration phase poses the question if the product or the system fits in the given context and if the decomposed parts are providing still the higher-level purpose in the early phases of the project based on simulations and models of the product or system artefacts. There are more theoretical, mathematical and logical assumptions providing the evidence. After realisation or implementation, the prototypes or real parts of the product could be integrated.
- The validation phase provides the evidence that all the previous activities lead to the desired targets.

The matrix oriented on the activities derived from a SPES matrix and considerations from the function-behaviour-structure framework (Fig. 4.35).

1. Formulation formulates the problem space, by transforming requirements into a function state space (R → F) and transforming functions into a behaviour state space (F → Be).
2. Synthesis generates structure based on expectations of the behaviour state space (Be → S).
3. Analysis derives behaviour from the generated structure (S → Bs).

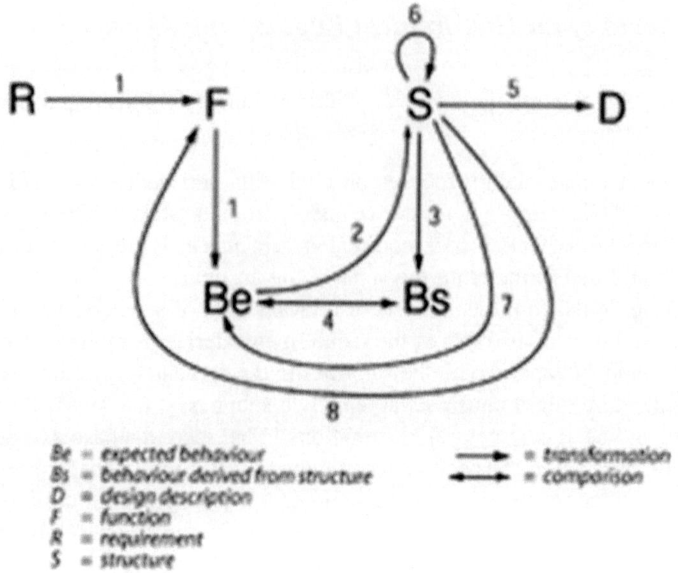

Fig. 4.35 Function-behaviour-structure framework

4. Evaluation compares expected behaviour with the behaviour derived from structure (Be ↔ Bs).
5. Documentation produces descriptions of the design based on structure (S → D).
6. Reformulation type 1 modifies the structure state space, based on a re-interpretation of structure (S → S′).
7. Reformulation type 2 modifies the behaviour state space, based on a re-interpretation of structure (S → Be′).
8. Reformulation type 3 modifies the function state space, based on a re-interpretation of structure and subsequent reformulation of expected behaviour (S → F' via *Be).*

It has been extended with the "Situated Function-Behaviour-Structure Framework" (Situated FBS).

The basic assumption underpinning the situated FBS framework is that designing involves interactions between three worlds: the external world, the interpreted world and the expected world. They are defined as follows:

- External world contains things in the "outside" world (e.g. in the physical environment of the designer.
- Interpreted world contains experiences, percepts and concepts, formed by the designer's interactions with the external world.
- Expected world contains expectations of the results of the designer's actions, driven by goals and hypotheses about the current state of the world.

Source Gero J. S. and Kannengiesser U. "The function-behaviour-structure ontology of design", 2000

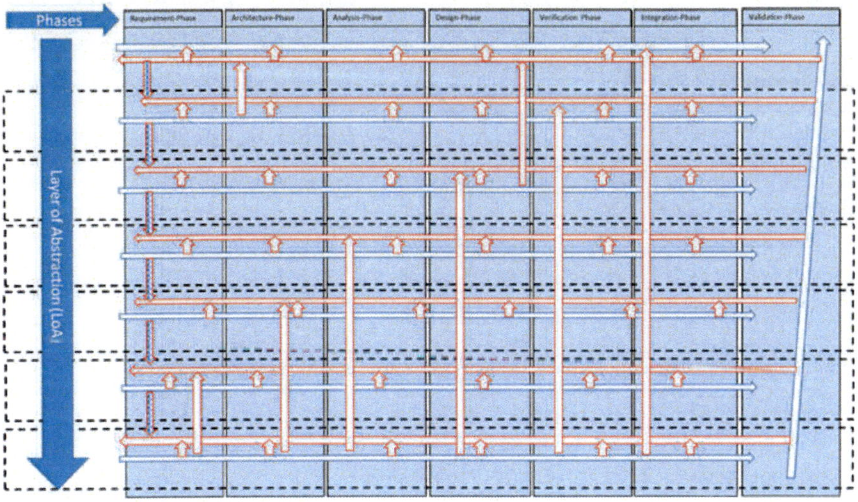

Fig. 4.36 Phases in a layered engineering approach

The herein proposed matrix deals more with the cybernetic approach and considers

- the real world as it is observable in the real world,
- the user view on the real world, how we expect the product or system of interest fits into the considered context,
- the specified word, how the product in the context is specified (Fig. 4.36).

The seven phases are similar within any level of abstraction (LoA) of the development. The main activity flow is always from left to right, with the needed iterations, depending on many factors like the size of the project, the product complexity, number of organisational units involved or number of project members.

The development phase is usually based on a problem space and a solution space. It is important to analyse what are the space for solutions, and what are the constraints and the implications of the considered solutions. On the other hand, the problem areas, which are derived from a stakeholder analysis, are always a compromise between all stakeholders unless the stakeholder tolerance limits are unknown and the solution is not assessable (Fig. 4.37).

The stakeholder requirements or at least their expectations are often driven from various point of views. The most discussed tension is based on the factor "cost". The producer should provide the most cost-effective solution, but the customers expect the highest quality. Usually, it is one of the main skills of a dealer to arrange the best compromise for his customers. The product manager of the producer needs to forecast somehow the expectations from the customers in order to define the solution space for the products. In the supply chain, there are many of the different viewpoints, the supplier should understand not only the direct customer, and the best solution always satisfies the end-user.

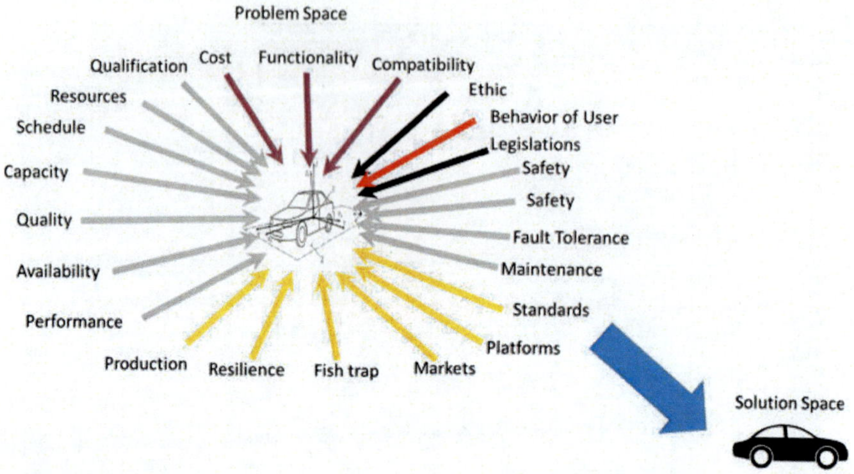

Fig. 4.37 Problem space versus solution space

But what are the tolerance levels for safety? Where is the compromise, between

- safety
- security
- performance
- availability and
- timing?

If you need a car, you buy the car, which is available. Since one dealer could not provide the vehicle right in time, you go to another. In the past especially in Germany, customer accepted to wait up to 1 year for a new vehicle. Since we more and more consider not owning a vehicle, we prefer to use the vehicle only for a limited time. That means there are many new stakeholder preferences for the vehicle. Many different user, driver, etc., will use the vehicle, and the emotion of the driver is difficult to be adapted to the system. A small solution will be "Feature update over the Air", which is maybe just another software configuration, but for the vehicle and the systems in the vehicle, such changes could mean many new design limitations. Just implementing an adaptive cruise control (ACC) increases significantly the demand rate for the ESP pump. The force of the drivers' foot must now be provided by any ACC braking by the ESP pump. The drivers interface becomes with such assistant system completely different. Just changes for the drivers could not be trained in a driving school. A rental car with such new feature will immediately be used by a driver. The introduction of the rental car agent could not teach the new behaviour of the vehicle. Much wider, is the change in case of automation functions, how could the user be trained for the new feature? Usually, the expectation was that I could rent a car for a weekend and test the entire feature on a trip through the public traffic. What happened if the driver is full of emotions and could not perceive a warning message from the dashboard? Today, we expect feature updates by demand; we could rent another 100 horsepower to drive with 250 kph on a German “Autobahn”. In any other country, we order more a comfort package, or better entertainment system on demand. We are asking the question, what it all has to do with safety. We expect, that the dealer, the rental car or operation service provider provides only safe feature for the vehicle. However, who do we know, if any driver, operator or whatever relation the user has to the vehicle, could control the entire feature in any situation? Someone could say, the answer is artificial intelligence. Emotions are not at all controllable by intelligence systems; it is a unique feature of human beings or animals.

4.4.2 Context Layer

The highest level of abstraction is the context development. The stakeholder requirements, the organisation for the project and product architectural framework including the context and environmental impact should be elaborated. Without a basic model of the vehicle, the necessary system breakdown within the entire context derived from the ODD analysis completeness could not be considered. Any detail, which is missing in the specification, means a further risk for the product under development (Fig. 4.38).

A safe road means a road traffic environment that assures road traffic with as low accidents as possible. The typical environmental layer starts with the anthology of the road design. Typical requirements are as follows:

- The road shall be design for passengers, cars and commercial vehicles.
- It shall have separated lanes for each direction.

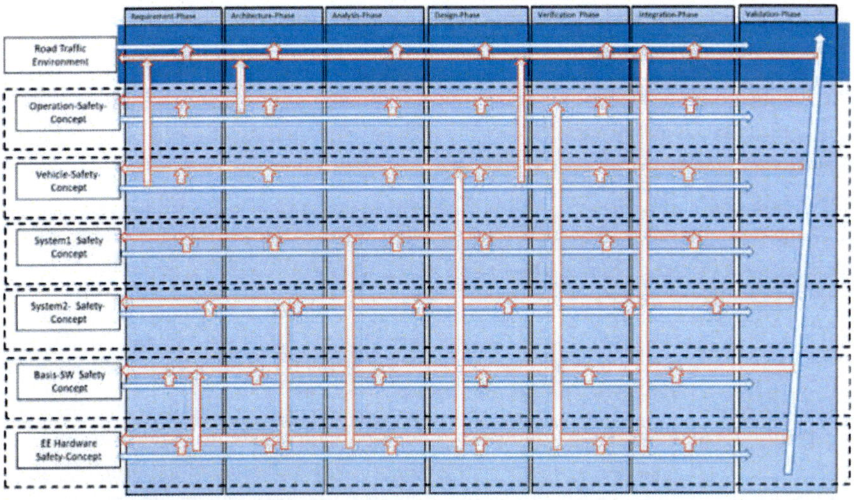

Fig. 4.38 Engineering for a safe road

- It shall have hard shoulders with crash barriers.
- The road surface should be sufficiently slippery.
- Weather conditions shall have minimum influence to the road traffic (water on road, etc., should be avoided).

There are typical constraints like

- The roadway should not have any crossings, and left turns are prohibited.
- Curves shall not have short radius.
- The road radius must not be too narrow.

These are typical requirements and constraints, which derive from the idea, what traffic, and how fast the vehicles should drive on the road.

As more, the road follows consistent design patterns, as more the drivers adapt their driving style for that type of road. The authorities made their experience with accidents and the user profile. They define specific or general speed limits for various types of road. Since the vehicles also continuously improve their design, several requirements derive for the roadworthiness of the vehicle.

The requirements and the architecture for road and vehicles continuously improve, usually it was not a systematic engineering approach, but in case of new planning of roads, today guidelines provide the design principles. Continuous analysis leads to optimisation for the.

- traffic density,
- average traffic speed,
- reduction of accidents, etc.

Verifications of the requirements and constraints are of course a major step before the road construction starts. The integration is more commissioning phase rather than an integration phase, but from the methodology point of view, it could be compared with the typical integration phase in systems engineering. The validations phase is at least a three-step approach.

- the first step is the validation at the end of road design phase,
- after the site acceptance and
- continuously during the road operation.

This three-step approach for validation will repeat also in the validation phase in the lower-level abstractions during system engineering.

4.4.3 Development of the Operational Safety Concept

The phases repeat now but on a more detailed layer of abstraction. All the activities of the higher level are analysed, and their derived requirements are the input for the next layer of abstraction engineering. The operational safety concept is particularly driven by the road traffic regulations. All the performance requirements for "safe operation" about speed for road segments and based on certain weather conditions, driving strategy on the different type of roads (motorway, urban roads, etc.), priority at crossings, coexistence with other road traffic participants are major new input for the requirements phase (Fig. 4.39).

The next layer led to the development of an operational safety concept. The bases were the engineering from the previous layer, so a safe road design is the input for the safe operation.

The engineering sequence is again as follows:

- The requirement phases collect the input from previous layer, and the activity grows with the derived requirements from the current phase. For example, the input from previous layer, as safer the road is as input, as easier is it to develop the requirements for the operation of the dedicated road segment.
- The architecture phase details the structure and the behaviour from the higher LoA and considers the details as relevant for the considered scope.
- The analysis provides additional requirements, constraints and decomposes all the results from previous activities.
- The design proposes measures resulting from the analysis and solutions derived from the relevant requirements.
- The verification phase provides again as in any other LoA evidences or maturity related to completeness, correctness and consistency.
- In the integration phase for the operational concept, evidence should be provided if the vehicle is designed based on given requirements and constraints.
- The validation provides evidence if the targets are fulfilled and the given requirements are correct. Depending on the maturity of the entire development, validations are based more on real-world test rather than models. The models used in the earlier phase could be validated.

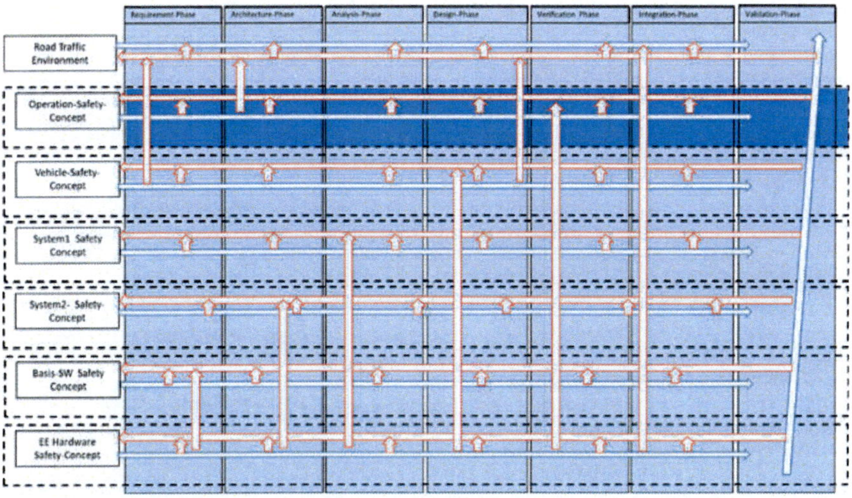

Fig. 4.39 Development of operational safety concept

4.4.4 Development of the Vehicle Safety Concept

The vehicle safety concept is dominated by the homologation regulations. All regulations, which are relevant for the target vehicle type, market and user profile or type of vehicle (sedan, van, SUV, commercial vehicle etc.), need to be considered in order to develop a vehicle safety concept (Fig. 4.40).

The activity phases are the same as in the layer above, but sequences and methodology need to be adapted for the specific LoA.

The engineering sequence is again as follows:

- The requirement phases collect again the input from previous layer. Major input comes now from the homologation regulations and standards and the target vehicle-specific aspects provide new requirements.
- The vehicle EE architecture and the requirements and constraints that derive from the vehicle systems architecture now dominate the architecture phase. Energy management, propulsion, brake, steering and HMI are introduced which need to be adapted to the entire vehicle architecture. Interfaces and boundary conditions need to be defined so that the behaviour in the context could be identified as required functions and functionality.
- The analysis provides additional requirements, constraints and decomposes all the results from previous activities. Since now not only behaviour is considered but also the breakdown of use cases, features, functions and their assignment to elements inside the vehicle, the space in the vehicle and the coexistence of the various systems inside the vehicle need to be evaluated.

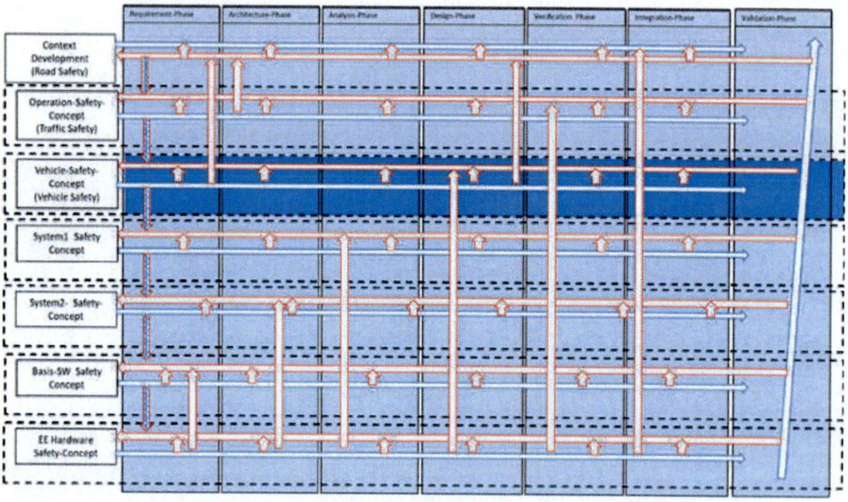

Fig. 4.40 Layer of operation safety concept

- The design activities are now dominated by the vehicle overall design concept. The design proposes measures resulting from the analysis and solutions derived from the relevant requirements.
- The verification phase provides again as in any other LoA evidences or maturity related to completeness, correctness and consistency.
- The integration phase for the vehicle provides evidences if all considered systems fit into the vehicle design. The vehicle test concept needs to be developed so that the defined interfaces could be assessed.
- In the vehicle validation, the two questions are dominating, does the vehicle fit for the intended purpose and do the vehicle architecture and design fit to the given requirements and constraints.

In that phase, the entire homologation needs to be planned. Depending on the homologation strategy and the given architecture and design decisions, the subsystem needs to be specified.

4.4.5 Process Sequence for the Engineering Model

The engineering model follows a typical top-down approach form the highest level of abstraction down to hardware and software, or even into the silicon for semiconductors, controller, ASICs, etc.

The level of abstraction could be defined as follows:

- Context development with the focus on road safety, which is a complete other safety approach as we know from vehicle safety.
- The "operation safety concept" based on the coexistence of all the road traffic participants and the development about the necessary consequences are derived from the operation principles.
- The "vehicle safety concept" based on the requirements for roadworthiness, which means the vehicle needs to fit with the intended purpose into the road traffic environment.
- The various "system safety concepts" derive from the vehicle safety concept. A brake system for a passenger's car would never be able to fit in a 40 tons truck.
- How to cluster the vehicle architecture down to control units requires more than one system layer. However, the activities in any system layer follow the same process activities.
- Inside a control unit, the breakdown to hardware and software provides the requirements and constraints for those LoA engineering (Fig. 4.41).

The LoA is not completely separated process area, but they always follow the same principles. From the right to the left, the requirements and constraints grow with the maturity scale of the individual activities. Like in any process, those activities never happen in proper sequences, but at least during the final assessment for the safety or assurance case, the sequences need to be traceable. Any gap in the higher

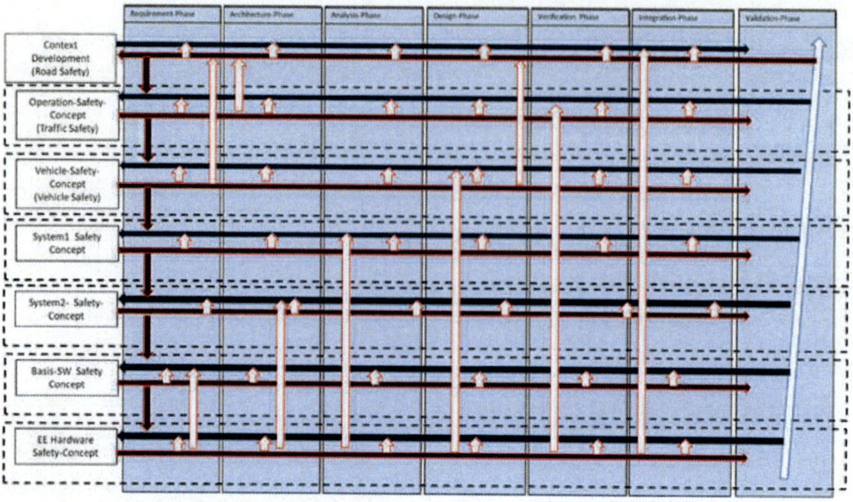

Fig. 4.41 Process flow charts in the engineering model

LoA could lead to new requirements and constraints in lower LoA. Therefore, any incompleteness means a risk for the product and/or the project.

If software development phase is below, the hardware development phase or above is a very high hypothetical question. As a basic rule, the application software derives from the higher-level system requirements, and the base software requirements derive more from the design decisions in the electronic like what microcontroller or microprocessor is considered.

4.5 Software Development

Software development appeared in three basic different categories, which require various different processes.

The following categories of software development need to be considered:

- Software tool development range is from production tools to the typical software factory tools like code generator or compiler, etc., test tools, engineering tools for architecture requirement engineering, etc.
- Basic software development provides the basis for the application software and also major measures to develop a trusted or safe computing environment.
- Application software development lays down the bases for the product purpose. All the typical user expectations on a software-intensive product are features and functions in the application software.

An effective development for software-intensive products requires a complete software lifecycle approach as proposed by ISO 12207 or ISO 15288. The three areas of software development need to be integrated into the product infrastructure process or in the value-added product development process.

4.5.1 *Software Development Based on ISO 26262*

Safety validation according ISO 26262 is just a validation of the safety goals and no other goals in the project. All other evidences during development are defined as verification, according to ISO 26262 the correctness of requirements is a verification measure.

ISO 26262 does not directly address a tool development process; it just considers requirements for the tool qualification in its part 8 as a supporting process (Fig. 4.42).

The typical traceability concept and the structure from ISO 26262 consider the software requirements as a result of the assignment of system requirements from the technical safety concept. From the application software point of view, that would be right.

Unless we ignore tool and basic software, we see a V-model approach with a growing specification model on the falling branch of the V and the product maturity on the ascending branch of the V-model (Fig. 4.43).

The standard considered that the implementation and realisation activities are performed according to the specification. The evidence of correct implementation and realisation derive from the preceding integration tests. A model integration or model-based integration strategy is not addressed in ISO 26262.

For the basic software, the main retirements are addressed at the hardware–software interface (HSI) which is part 4 "product development on system level" (Fig. 4.44).

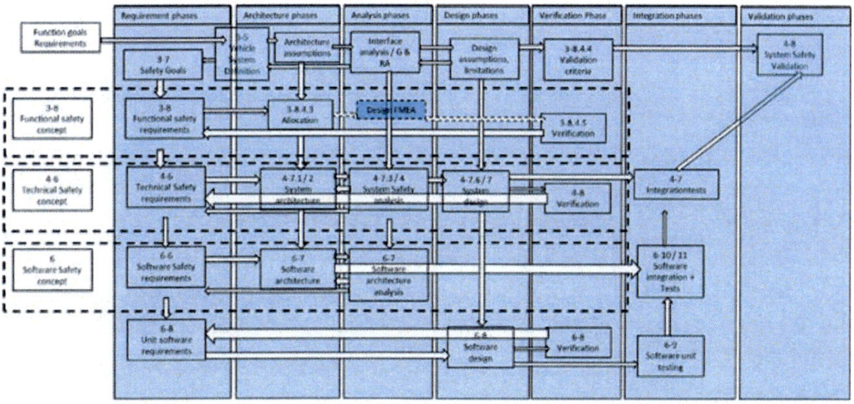

Fig. 4.42 Software safety concept as detailing of the technical safety concept according to ISO 26262

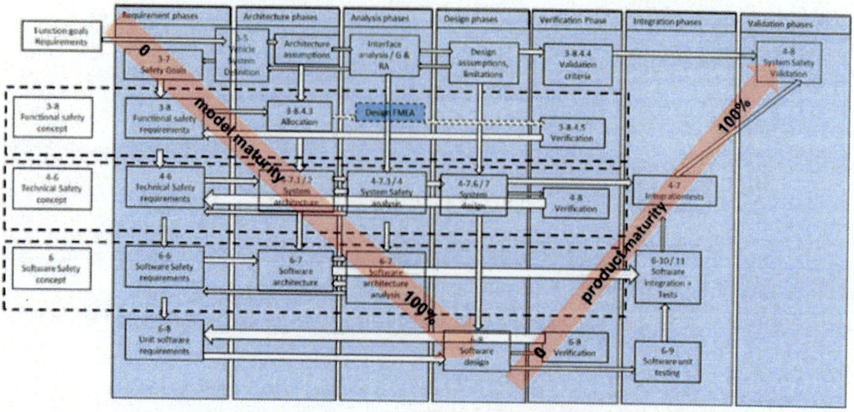

Fig. 4.43 Model maturity versus product maturity according to ISO 26262

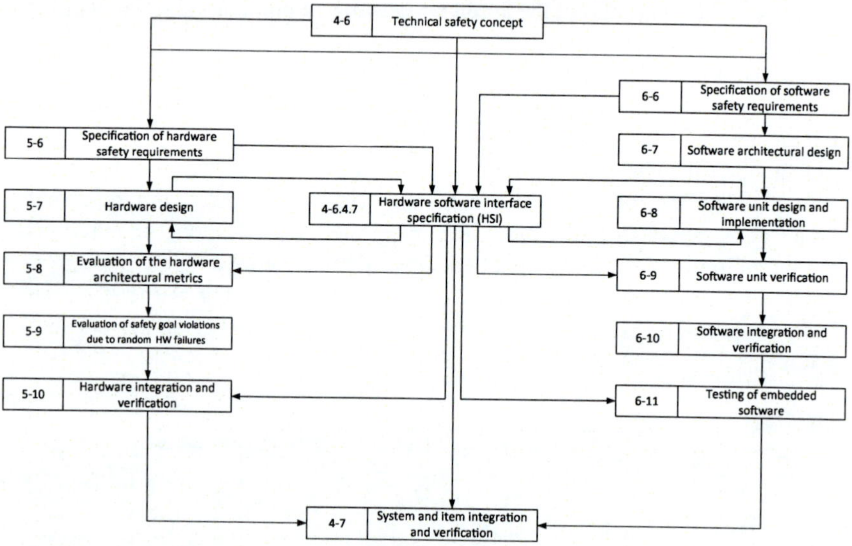

Fig. 4.44 Requirements breakdown according to ISO 26262. *Source* ISO 26262:2018, Part 4, figure B1

The electronic hardware provides requirements to the hardware–software interface (HSI), which are then assigned to the software architecture. From an architectural point of view, the basic software is addressed, which provides an operating system, hardware abstraction layer, several utilities, drivers and further services, etc.

4.5.2 *Safety Mechanism in Basic Software*

ISO 26262 addresses random hardware fault and soft errors, transient or intermediate faults. All sporadic error effects could lead to failure within the data processing. Since a diagnostic could not monitor continuously the fault occurrence, redundancy comparison is the only solution. Redundant processing, like the lockstep principle, considers that sporadic hardware faults do not occur at the same time in a redundant path. For any systematic fault, common cause effects redundancy is not an adequate measure, adequate fault tolerance or avoidance measures are required (Fig. 4.45).

Today's microcontrollers do not assign hardware ports and peripheral channels in a direct link to the interface to the application software. The hardware provides in the port register data, which are multiplexed and provide different input depending on their address configuration. Handler could be designed very efficient, so that handler cycle and hardware cycle operate in the same time slot. For robustness reasons, usually the handlers are clocked in a way that they could read the data from the hardware and get via redundant calls or checksums the evidence of correct reading of the data. Due to optimised timing and redundancy mechanism in the handler, especially random hardware faults could be detected or event corrected. If those data correction could be realised with sufficient safety integrity, random hardware faults could be controlled by the safety mechanism in the safe software handler (Fig. 4.46).

There are two basic data flows in a typical microcontroller:

- from peripheral elements to the control core (including ALU, arithmetic logical unit) and memory,
- from application software to control core and to memory.

Except the instruction set of the ALU, the communication ways are hardware dominated. For example, AUTOSAR proposes an E2E-protection (E2E = End2End) from the source of information to the consumer of information (Fig. 4.47).

For fault-tolerant systems, the only way in case of failure is shutting down of the system. Typical so-called safety controller supports via built-in self-tests (BIST) the diagnostics and the interface to an internal and/or external watchdog.

In order to avoid continuous resets by the watchdog, the diagnostics of potential faults and errors need to be separated from the shutdown device. In case of physical

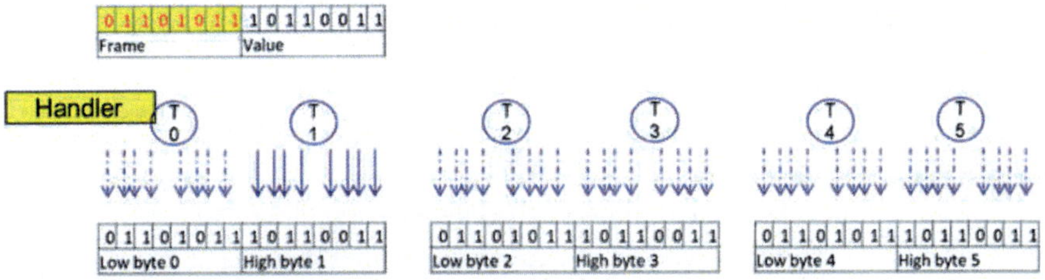

Fig. 4.45 Simplified data handler in the basic software

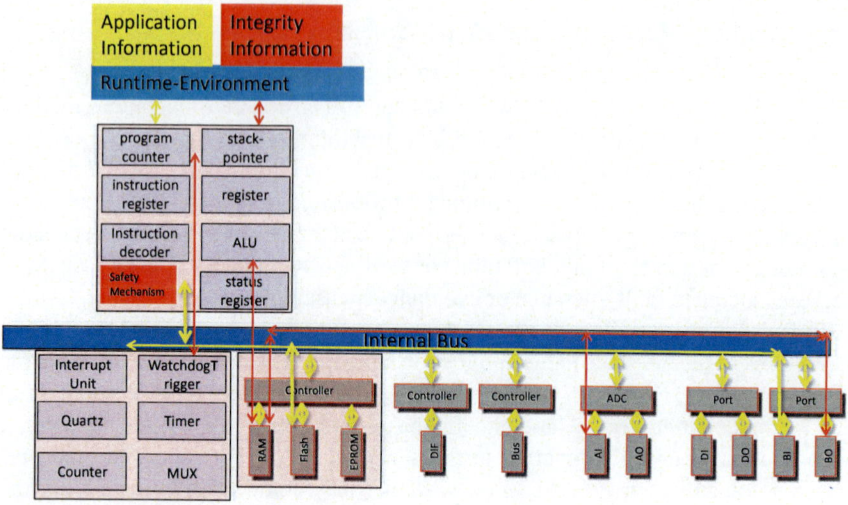

Fig. 4.46 Typical data flow in an embedded microcontroller

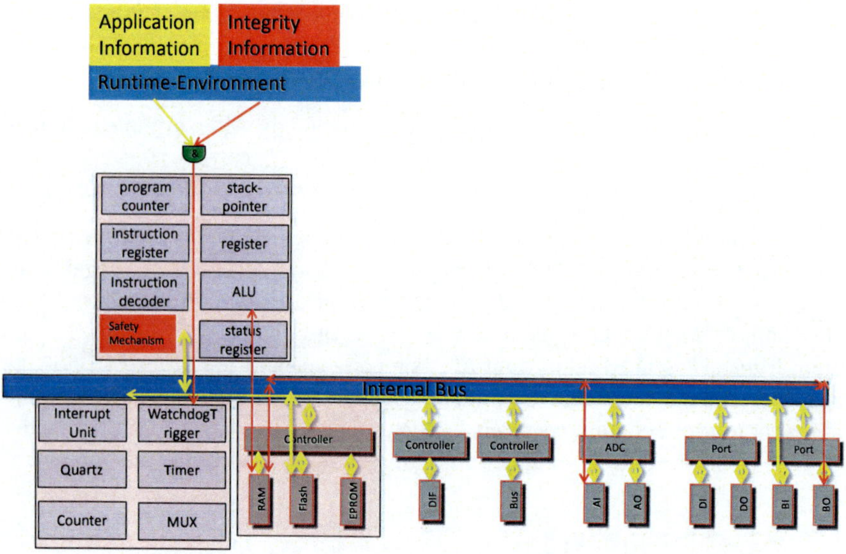

Fig. 4.47 Integrity flag for application software

faults in the hardware that could cause significant thermal incidents or other significant damages of the electric, an immediate shutdown is required. For all systematic faults, an immediate shutdown is not necessary but the fault impact needs to be communicated. AUTOSAR offers for that case on qualifier, which could be used as an "integrity qualifier". If the application software is designed for the use of "safety

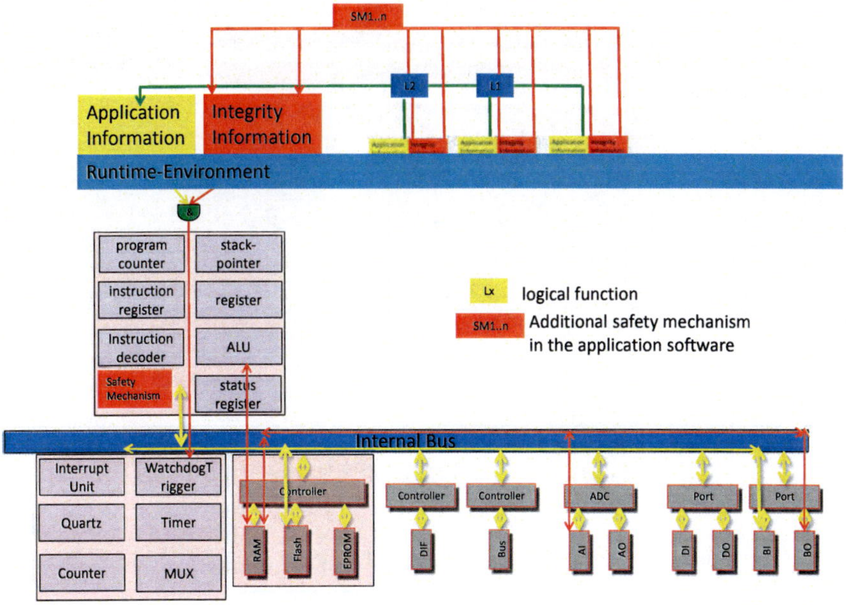

Fig. 4.48 Safety mechanism in the application software

qualifier", the application software may not use disqualified data for safety-relevant functions, tasks or processes (Fig. 4.48).

A typical EGAS concept (E-Throttle according VDA safety concept) which is based on multi-layer concept with different safety integrity levels of the data flow could be protected by E2E-protection. Additional safety mechanism needs to be implemented in the application software, for example, in EGAS-Level 2. The entire basic software needs to be implemented with the highest safety integrity level. An adequate partitioning or fault containment concept is necessary (Fig. 4.49).

The structure offers also a diverse computation approach. In the application software, the intended function (e.g. increase speed) with the maximal performance could be processed and the safety mechanism calculates with the inverted function (accelerate until limit). In order to avoid systematic failure in the intended software function, a logical dissimilar function should be defined. By validation during development and by logical monitoring during operation, systematic failure could be identified. This approach is very interesting in case of using machine learning algorithm (ML). By using two or more independent machine learning approaches based on dissimilar logical principles, the algorithm needs to be compared or voted and only the positive result provides an valid control output. The challenge is to identify which channel provides the right result; therefore, graceful degradation concepts based on diagnostic are required.

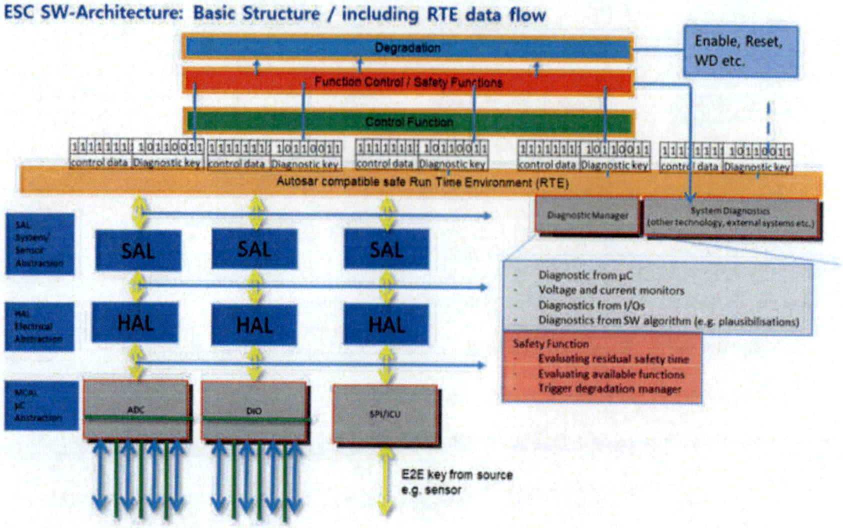

Fig. 4.49 Multi-layer software architecture similar to VDA EGAS concept

4.5.3 *Safety for POSIX-Based Architecture*

Today, more and more also embedded Linux applications are considered. Linux operating system is derived from UNIX and comprises two main blocks known as the user space and kernel space.

In Linux, since, there are multiple vendors contributing to the development of drivers for different hardware components in the kernel space. This resulted in the creation of POSIX standard.

Portable operating system interface (POSIX) is a set of standard operating system interfaces based on Unix operating systems. UNIX was chosen as a basis for a standard system interface because it is considered "vendor neutral". The applications or code running on the user space executes on a restricted underprivileged level with access to limited portion of the memory, while the kernel space (code running on Linux Kernel) is the block that runs on a privilege mode with no restrictions and limitation to the type of operations it could perform.

The kernel space is the region that runs the operating system modules such as invoking and allocating memory in RAM for applications, scheduling tasks to be executed by the processor, sending and reading data back and forth to the hardware.

Process management: This component comprises code that is responsible on how individual processes are managed by a processor. The scheduler forms the major part of this component.

Each process has its own virtual memory and has no knowledge about other processes (Fig. 4.50).

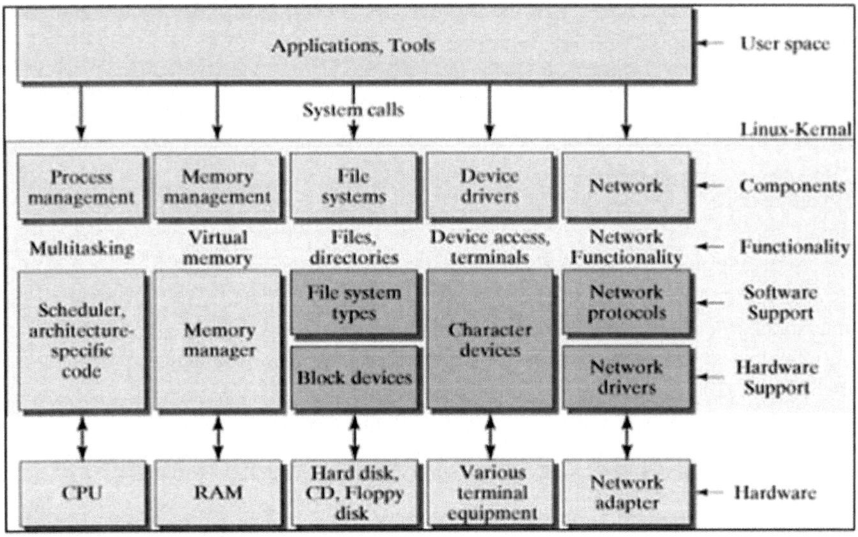

Fig. 4.50 Linux components in a modular view

- Memory management: It deals with storage and operation of data on a persistent and non-persistent storage such as a hard drive and RAM. Linux uses virtual memory to provide a wider memory region for concurrent execution of multiple processes. Virtual memory uses page tables for addressing different physical storage addresses. This component also uses a technique called demand paging that is not ideal for conditions under system load and faster memory access and should be avoided for real-time requirements for a system.
- File systems: Linux treats character and block devices, directories and few IPC mechanisms such as pipes and FIFO as files. Linux has multiple layers of abstraction unlike the miniature embedded solutions that write and read data directly from memory blocks.
- Linux uses device drivers that performs operations based on the storage file system type. Linux allows the root user or administrator to define rules for permissions to files.
- Device drivers: In Linux device, drivers provide a layer of abstraction over the hardware. Devices such as block devices and character devices are available as files to the user space. Upon reading or writing data into these device files, the corresponding kernel driver code is invoked and the operation is performed. For a specific operation on a peripheral device when an event happens, an interrupt is triggered to indicate the system for the operation to be executed by processor.

POSIX standard is the basis for many variants of Linux operating systems such as Ubuntu, QNX and Solaris. POSIX achieves compatibility for applications by providing APIs that operates on different segments like memory management, process management, file system operations, socket programming API for network communication, command line interface, etc.

The pre-empt-RT patch on Linux is known to achieve real-time behaviour by providing a deterministic window under which the operations could be completed.

POSIX stands for portable operating system interface. Initially, when many variants of UNIX started emerging, there was a need to make a standard that could make the source code of the applications more compatible among the family of UNIX like operating systems. The source code developed in one operating system should be able to recompile in the other system without any compatibility issues. This lays the initial development of POSIX.1 that covered basic UNIX calls, followed by many subsequent releases. Adding to this, a real-time extension was made in POSIX.4 approved in 1993. POSIX.1-2008 is the recent release, which is also known as IEEE Std 1003.1™-2008.

POSIX offers a wide range of APIs in the standard C Library, and these functionalities could be categorised as follows:

- file operations
- process management
- memory management
- inter process communication (IPC) mechanisms
- network management.

The various blocks from POSIX perform the following function per default:

IPC Mechanism: In this block, the operations involving in creation of FIFO or so-called pipes functions trying to access the named pipes using API and performing read/write operations are specified. It also includes the type of inputs and the expected outputs from those functions. Following are functions specified in this block.

- mkfifo(…): Interface responsible for creating a FIFO.
- open(….): Opening a specified device file or special file(FIFO in this case) with given access parameter and provides a descriptor for read/write operations.
- read(….): Performing a read operation on the special file.
- write(…): Perform a write operation on the special file.
- close(…): Closing the given descriptor.

Process Management: This block represents the operations involved in creating POSIX timers including inputs and expected outputs. The number of timers created per process is restricted to a specific limit. The operations involved in POSIX threads are also included. In Linux, threads run similar to a process with a shared address space. Following are some of the functions specified in this block.

- pthread_create(…): Operation used to create a thread in an application with the provided priority and thread handler.
- timer_create(…): This operation creates a timer with the given properties and provides a ID of the timer running.

- Timer_settime(…): This operation defines the properties such as periodicity and system clock variant used for the timer.
- fork(…): This function is invoked in Linux when an application is run in user space. It creates a child process when invoked inside a process.

Network Operations: Sockets are used in Linux to achieve network operations with either TCP/UDP protocols. In the application, UDP protocol was used to transmit/receive messages. The operations involved in creation of socket, binding socket to a port and listening through socket are listed out in this portion. Following are the function described in this block.

- socket(…): This function creates a socket based on the given parameters that describe protocol (TCP/UDP) and domain.
- bind(…): The operation results in binding the socket to the port, and this operation is also responsible for creation of a server in network that listens in the network with specified port.
- recvfrom(…): The operation receives messages from the provided socket descriptor to a buffer.
- sendto(…): The operation sends messages to the destined address and port in the network.

File Operations: In Linux, the devices and directories are treated as files, most of the device files are found in/dev/ directory of Linux system. This block specifies basic operations involved on normal files and device files. This also includes inputs and expected outputs for the interfaces. Following are some operations specified in this block.

- fopen(…): The operation opens the file given in the parameter and then provides a file descriptor that could be used to perform read/write operations.
- fread(…): The function performs a read operation on the file descriptor and stores it on the given buffer.
- fwrite(…): The function performs a write operation with the given buffer on the file descriptor.
- fclose(…): It closes the given file descriptor.

Memory Management: POSIX provides interfaces to create mapping of the file descriptor with virtual address space of the calling process and dynamic memory allocation interfaces even though they are not recommended unless necessary. This block specifies operations of these interfaces along with inputs and outputs.

- mmap(…): This operation creates a new mapping in the virtual address space with the provided file descriptor.
- mlock(…): This operation locks part of the process virtual address space in a way that prevents the memory to be paged.
- munlock(…): This operation unlocks the memory to be allowed for paging.

– malloc(…), calloc(…): These operations allocate dynamic contiguous memory and provide the buffer to application (Fig. 4.51).

The most important aspect of a POSIX architecture in comparison with conventional embedded control system is the virtual memory management.

Each process is assigned its own private memory in RAM. This memory may be located anywhere in physical RAM. The addresses used within a process are virtual addresses, which are translated at runtime to their physical memory locations by the MMU. These virtual addresses compose the process' virtual address space. A process' virtual-to-physical address space translations are private to that process. The memory addresses owned by the process can only be read or written by that process and cannot be accessed by other processes. This prevents processes from corrupting or snooping on each other's memory (Fig. 4.52).

Technically, a thread is defined as an independent stream of instructions that can be scheduled to run as such by the operating system. Threads are scheduled in the way that they run simultaneously and/or independently by the operating system, which is called a "multi-threaded" program. They call it "virtual threading" because the threads are not physically separated (Fig. 4.53).

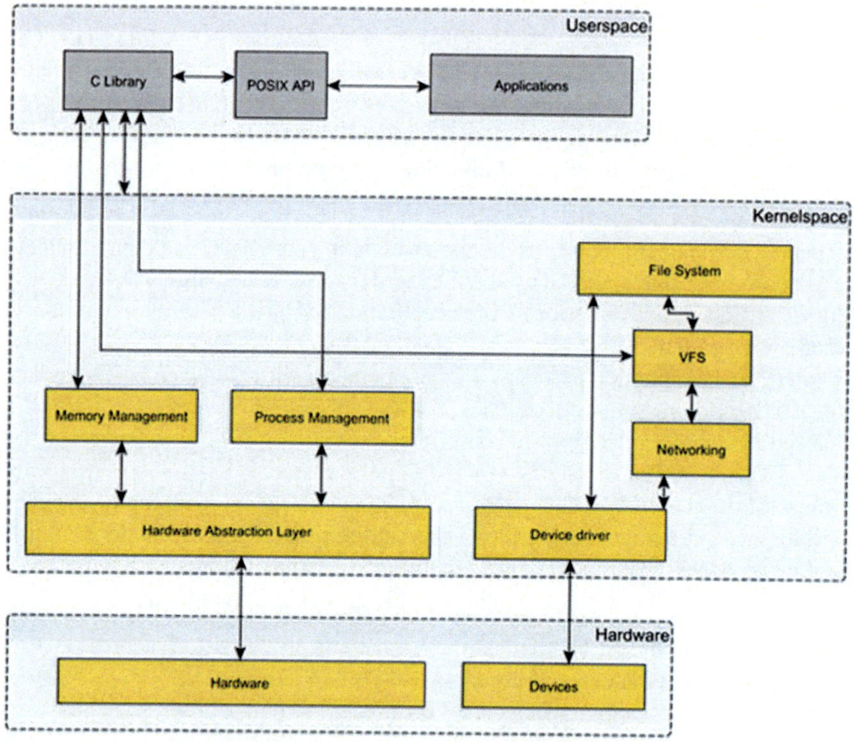

Fig. 4.51 Standard POSIX architecture

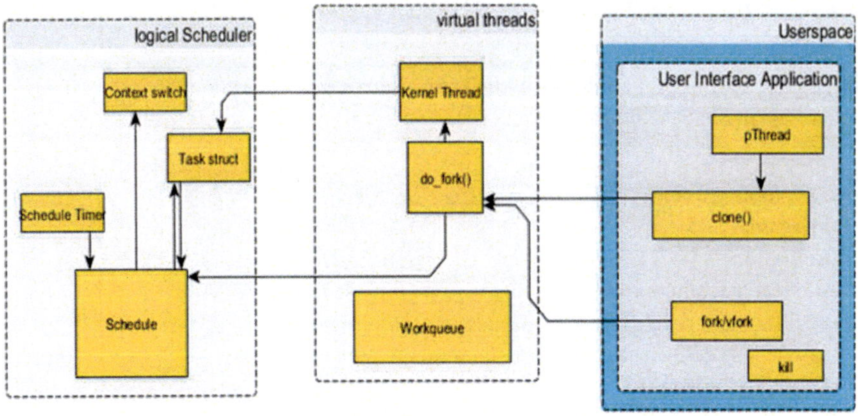

Fig. 4.52 Model for virtual threads

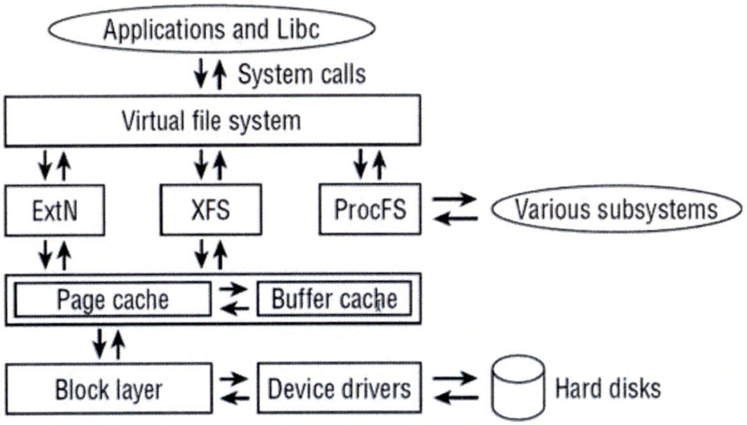

Fig. 4.53 Virtual file system in POSIX. *Source* Speicherübersicht (Maurer 2008, S. 19)

A virtual file system (VFS) switch is an abstract layer on top of a more concrete file system. The purpose of a VFS is to allow client applications to access different types of concrete file systems in a uniform way. The virtual file system (also known as the virtual file system switch) is the software layer in the kernel that provides the file system interface to user space programs. It also provides an abstraction within the kernel, which allows different file system implementations to coexist.

To manage the system's resources, the kernel must be able to control the running process. Therefore, it needs to be able to suspend and resume the execution of a process, which requires so-called process, task or context switches. Consequently, the entire context or processing environment of a process, thread, task, etc., is based on a virtual environment.

4.5.4 Failure and Errors in POSIX-Systems

Such a POSIX system could show the typical error modes at the interfaces.

Failure modes at POSIX API:

- stuck-at,
- D.C. fault model (including signal lines outside),
- drift and
- oscillation,
- delay of information,
- blocking access to system resources,
- incorrect synchronisation between elements.

Failure modes at file system.

- stuck-at
- incorrect addressing information,
- corruption of information,
- incorrect sequence of data,
- insertion of information.

Failure modes at networking.

- stuck-at
- asymmetric information sent from a sender to multiple receivers,
- blocking access to a communication channel,
- incorrect sequence of data packets,
- traffic congestion and message loss.

Failure modes at communication (process management).

- message corruption,
- message repetition,
- message delay,
- message loss,
- unintended message repetition,
- resequencing,
- insertion of message,
- masquerading.

Failure modes at memory management.

- stuck-at and D.C. fault model for
- data,
- addresses and control interface,
- lines and logic (includes address lines within same block and inability to write to cell),
- corruption in data or information,
- soft error model for bit cells.

The causes of those errors are not traceable, so that their cause could be whether controlled nor avoided in a single core application.

4.5.5 *Hypervisor Approach*

In order to provide safe execution on multi-core system, a Linux hypervisor approach is often applied. In a single processor, multiple tasks can be divided and performed on different cores. Hypervisor approach allows multiple operating systems to be run on dedicated cores. In this scenario using hypervisor, applications on Linux could be run to meet the needs of a designer. Alongside with it, a real-time operating system can be run that has specific applications to monitor the flow of events. This achieves monitoring of application at different abstraction of layers. However, this approach has increased effort in the abstraction layer present on top of hardware. There are possible effects on performance of the individual systems due to multiplexing of activities on processor that could result in running out of memory. Docker could possibly be used that is similar and simplified variant to hypervisor but more analysis needs to be done for the dependencies to be met with hardware (Fig. 4.54).

The typical hypervisor approach does not allow processing of safety integrity task in the Linux partition. Since the cause of failure in a POSIX architecture is not traceable, for random hardware failure, soft errors, etc., hardware redundancy is unavoidable. For systematic failure, the possible failure that has the potential to influence the application software needs to be controlled by adequate measures. By pure software qualification, Linux or any other POSIX-based system could not be used for safety integrity tasks. By implementing on a same core or on secondary cores, safety integrity functions on Linux systems are applied also for automotive solutions.

Alternatively, to a hypervisor approach, an "asymmetric multi-processing approach" runs two operating systems in a completely different processors or controllers. The high performance function could be performed on a POSIX-based

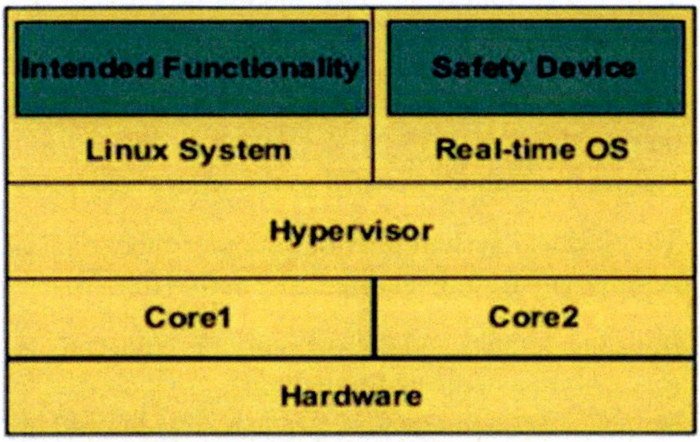

Fig. 4.54 Hypervisor component layout view

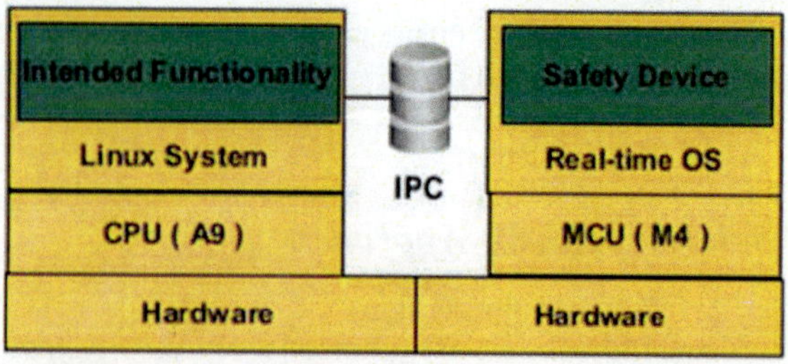

Fig. 4.55 Example for an asymmetric multiprocessor approach

microprocessor system; here even deep learning or other machine learning algorithms could be implemented on GPU or even dedicated hardware for AI systems. A secondary system completely independent could work as a safety integrity controller. This controller could directly influence the microprocessor systems by timer and periodisation functions or just work as a monitoring system. Usually, the safety integrity system runs in a time disc in a hard real-time mode and provides or controls the deterministic behaviour of the entire asymmetric multiprocessor system (Fig. 4.55).

The safety integrity system should be developed according a dedicated safety integrity standard like ISO 26262 or IEC 61508. The microprocessor systems operate like a basic control system as it is for limited risk reduction measures considered in IEC 61508. According to IEC 61511, the safety integrity system (or SIS—safety instrument system) does fault control of the application and the basic control system.

There are microprocessor systems considered with a separated safety operation system, but most of the safety operation system require at least also safe hardware support in order to assure sufficient diagnostic coverage to meet the basic safety requirements derived from the given safety integrity standards. The same is recommended in order to provide a trusted operation environment for security mechanism.

4.6 Real-Time Embedded Systems

A real-time system is any information processing system which has to respond to externally generated input stimuli within a finite and specified period.

- The correctness depends not only on the logical result but also the time it was delivered.
- Failure to respond is as bad as the wrong response.

Time-critical system provides another functional or virtual dependency. Timing constraints and timing requirements derive from many influence factors for example

- from the context and the operational requirements (e.g. detect obstacles on the road within a dedicated time limit),
- from a closed loop control (feedback is required in a time interval),

- resources are only available in a specific time interval (scheduler provides a specific time slot),
- faults need to be control within a specific time interval (typical FTTi, fault time tolerance interval),
- sequential processes require the result from other processes,
- parallel processes require data within a specific time slot for comparing or synchronisation purpose, etc.

Delays could lead to very different consequence depending on the consequence, there are

- hard real-time or
- soft real-time systems (Fig. 4.56).

There are common definitions available for hard and soft real-time systems.

One approach defines hard real-time systems by the distance or robustness margin to the critical time threshold. It is based on the capability of the system.

- In case if the threshold is so conservative that the worst case execution time could not violate the critical threshold, the system is called a hard real-time system.
- If the average process time is significantly below the critical threshold and only in extraordinary worst case scenarios, the threshold could be violated, the system is called soft real-time system.

Another definition is based on the characteristic of the threshold (Fig. 4.57).

For a hard real-time system, an immediate violation of the threshold leads to a serious hazardous event. A deadline is a given time after a triggering event, by which a response has to be completed (Fig. 4.58).

A soft real-time system means that the critical threshold has a soft course over time, i.e. it rises slowly over time. Missing a deadline does not directly lead to

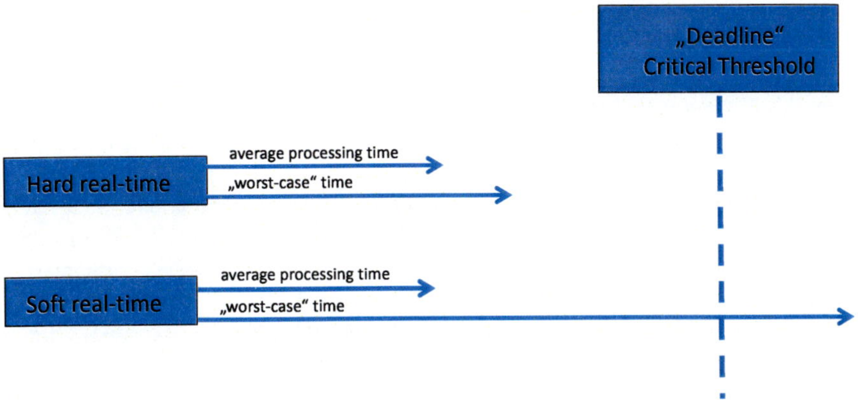

Fig. 4.56 Hard and soft real-time based on distance to critical deadline

Fig. 4.57 Hard real-time for an intermediate critical threshold

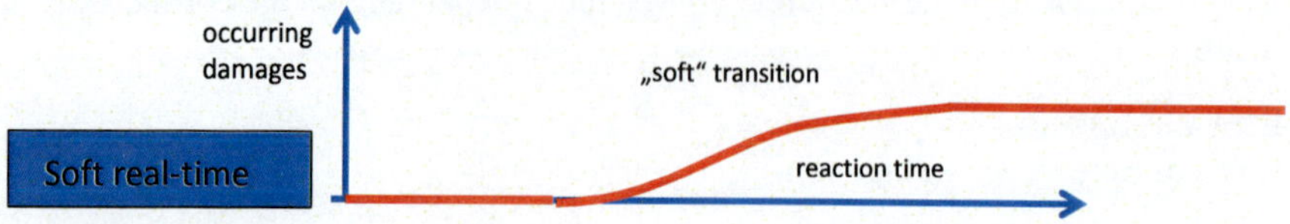

Fig. 4.58 Soft real-time for a soft threshold

hazardous event. The hard real-time system, on the other hand, ultimately leads to a hazard when the critical threshold is exceeded.

4.6.1 Timing and Determinism

One of the key characteristics of real-time systems is determinisms. An embedded system or a communication system provides a call determinism if the response is predictable and could be executed before a critical deadline. Consequently, there is no significant impact to the execution or response time by

- systematic faults,
- random faults,
- common resource usage,
- or external influences.

The execution or response time does not necessarily mean a short time; the timing constraints derive from the higher level of abstraction and with proper architectural planning, bottlenecks could be avoided. Similar to the worst case execution time for hard real-time systems, the time prediction is the key characteristic. A serial data communication allows deterministic behaviour like time-triggered solutions or FlexRay, but in a dynamic network, responses are difficult to predict. Very common is carrier sense multiple access/collision detection (CSMA/CD) principle. It is a media access control (MAC) method used most notably in early Ethernet technology for local area networking. It uses carrier sensing to defer transmissions until no other stations are transmitting. This is used in combination with collision detection in which a transmitting station detects collisions by sensing transmissions from other stations while it is transmitting a frame. When this collision condition is detected, the station stops transmitting that frame, transmits a jam signal and then waits for a random time interval before trying to resend the frame. CSMA/CD is a modification of pure carrier sense multiple access (CSMA). CSMA/CD is used to improve CSMA performance by terminating transmission as soon as a collision is detected, thus shortening the time required before a retry can be attempted. IEEE802.3 does not use anymore CSMA/CD. But all improvements up to time-sensitive networking (TSN), which is a set of standards under development by the time-sensitive networking task group of the IEEE 802.1 working group, real time could not be assured just by the network architecture. The dependency to network load has always the impact to the response time.

The system requirement for determinism derives from many aspects for example

- A sensor must provide all 5 ms a new record.
- A simulation model in an embedded system must provide all 50 ms of updated values.
- A position sensor in a brushless engine must provide the engine position continuously in a fixed time framework, etc.

In order to reduce bus or network load, today's applications transmit only changes. Due to safety reasons, it became necessary to send frequently an inverted information in a fixed interval.

Examples are frequently sending inversions:

- no change flag, to indicate missing change information,
- remain torque, to indicate continuous torque command,
- confirm system status, to maintain current status,
- validity flag, to confirm actuality of system parameter,
- sending empty messages, to indicate no changes.

The principle provides evidence that the communication system is operating and the receiver could degrade in case of missing data. The data frame should have a time-stamp and counter, so that the receiver could monitor sequence and age of the message.

Isochrones data communication systems like FlexRay work precisely periodically, for example, in a time interval of 1 ms. The bus cycle jitter is usually nominal under one millisecond. Bit transmission jitter, etc., could also influence the determinism negatively.

Synchronised communication and control algorithms are coordinated in the way that control mechanism prioritises the sequence of transmission based on the current relevant data traffic.

There are two basic synchronisation principles for synchronous communication systems:

- Time slot method: The synchronisation is based on sending a synchronisation signal, which is received by all, network participants cyclically and processed accordingly. In order to achieve minimum timing variances from sender to receiver, the synchronisation signal should be sent in a fixed time interval.

- Based on the principle of distributed clocks, IEC 61588 defines "Precision Time Protocol (PTP)". The distributed clocks offer a precise time base, independent from runtime or system variances. In order to synchronise the data at the receiver, the messages need to be early enough at the receiver, because the data transmission itself is not deterministic.

4.6.2 Scheduling in Real-Time Systems

Real-time embedded system should be designed in a way that the real-time aspects are covered as precise as necessary, and performance and timely responses should be achieved as good as possible. This principles lead to a conflict by using microcontroller resources. Usage of available resources from the microcontroller has to be planned, and adequate principles like periodisation, etc., have to be implemented.

In real-time embedded systems, the key challenge is how to share the resources of the microcontroller, libraries, functions, etc., within the given time. If there is only a single functionality or task in a controller, any shortage of resources could be forecasted during development, but in multitasking system, which are usually also realised by different development teams, need to be controlled and managed during runtime. A scheduling analysis for real-time computing system includes the analysis and testing of the scheduler system and the algorithms used in real-time applications. Real-time scheduling analysis is the evaluation, testing and verification of the scheduling system and the algorithms used in real-time operations. For safety-related tasks or operations, a real-time system must be tested and verified for performance. The performance verification and execution of a real-time scheduling algorithm should be examined by the analysis of the algorithm execution times. Verification for the performance of a real-time scheduler will require testing the scheduling algorithm under different test scenarios including the worst case execution time. These testing scenarios include worst case and unfavourable cases to assess the algorithm performance. The time calculations required for the analysis of scheduling systems require evaluating the algorithm at the code level.

Pre-emption is a scheduling strategy. Processes that are in the middle of execution can be interrupted and suspended while other processes are scheduled.

Processes that are designed to work with pre-emptive scheduling must be explicitly designed to give up the processor at any point. Any data that they share with alternate threads must be protected by locking mechanisms.

The scheduler of a real-time operating system manages the usage and accessibility of resources in microprocessor- or microcontroller-based systems. Real-time operating system (RTOS) offers multitasking functions, which allows handling more than one task in parallel or time slots on shared resources within a specific real-time window. A computer program is divided into several tasks that can be executed sequentially one after the other.

A basic operating system has three main functions:

- manage the computer's resources, such as the central processing unit, memory, disc drives and printers,
- establish a user interface and
- execute and provide services for applications software.

It provides the following features:

- tasks and scheduling policies,
- events, interrupts and traps,
- counters, alarms and schedule tables,
- resources and error management.

A RTOS serves real-time applications that process data as it comes in, typically without buffer delays by scheduling. The basic states of a scheduler are as follows:

- running (executing on the CPU)
- ready (ready to be executed)
- blocked (waiting for an event, I/O for example) (Fig. 4.59).

The scheduler could be time or event controlled, a time-triggered scheduler changes the state by time sequence. An event-triggered scheduler changes the states by a triggering event (Fig. 4.60).

A task is usually in a buffer and waits for execution or is running on a central processing unit (CPU). In a single CPU system, there is only one task active, so the time and the resources need to be shared adequately (Fig. 4.60).

Periodisation tables could provide information for the execution sequence of tasks.

Fig. 4.59 States of a scheduler

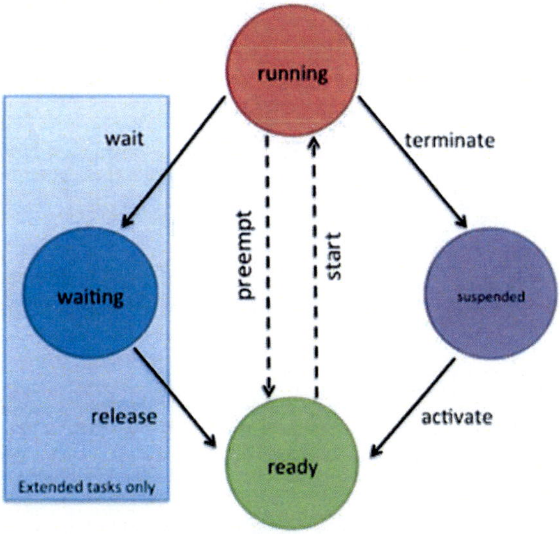

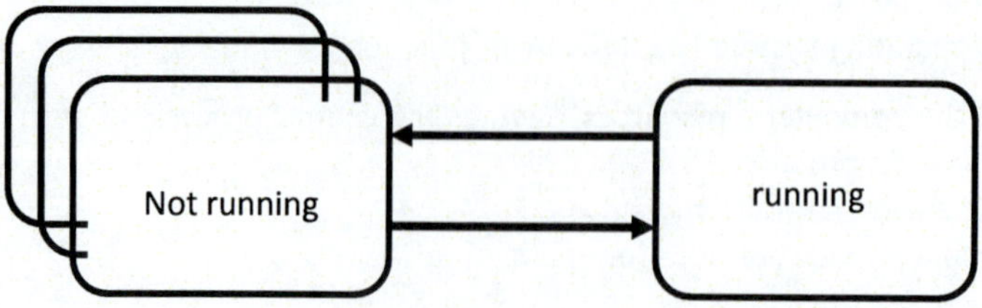

Fig. 4.60 States of a task in a real-time scheduler

In new version from FreeRTOS, there is the difference between "suspend" and "blocked". The difference is that tasks in the blocked state always have a timeout. Tasks in the suspended state are suspended indefinitely, and there is no timeout. From the users point of view, tasks that block on an event are in the blocked state. It is possible that tasks are actually moved into a suspended queue rather than a blocked queue so that a task is still blocked rather than suspended. The functions and the states in RTOS are very different in detail, so that a "safe" scheduling requires detailed knowledge about the operating system (Fig. 4.61).

Events, Interrupts and Traps

An ISR (interrupt service routines also called an interrupt handler) is a software process invoked by an interrupt request from a hardware device. It handles the request and sends it to the CPU, interrupting the active process. When the ISR is complete, the process is resumed.

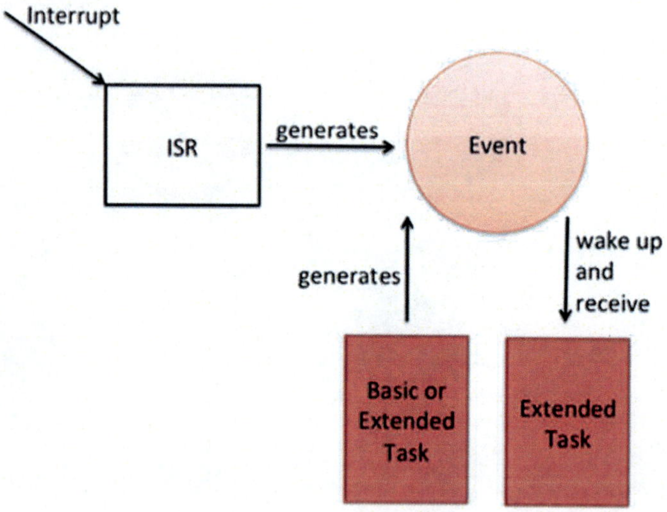

Fig. 4.61 Interrupt handling

Traps are programmer-initiated interrupts and usually not automatically generated "exceptions".

Events can be generated from ISRs or tasks, they can be extended tasks, can wait for the generation of an event, but a task cannot determine the source of an event.

Interrupt service routines (ISRs):

- directly triggered by hardware-invoked interrupts,
- highest priority than all tasks,
- activate a task, pre-empt a task or trigger an event.

There are typically two different categories of ISR:

- ISR category 1—system services cannot be called.
- ISR category 2—system services can be called.

ISR category 1 has a highest priority than ISR category 2.

In computing and operating systems, a trap, also known as an exception or a fault, is typically a type of synchronous interrupt caused by an exceptional condition (e.g. break point, division by zero, invalid memory access). A trap usually results in a switch to kernel mode, wherein the operating system performs some action before returning control to the originating process. A trap in a kernel process is more serious than a trap in a user process and in some systems is fatal. In some usages, the term trap refers specifically to an interrupt intended to initiate a context switch to a monitor program or debugger. In some OS, interrupt service routines (ISR) and threads follow different policies. Threads can be dispatched as efficiently as interrupt handlers, and interrupt handlers can be scheduled as flexibly as threads. Deriving from this original usage, trap is sometimes used for the mechanism of intercepting normal control flow in some domains.

Traps occur because of an event such as non-maskable interrupt (NMI), an instruction exception or an illegal access. They are always active, and they cannot be disabled by software action (Fig. 4.62).

Interrupts	Traps
Caused by external hardware events	Software invoked interrupts
They get disabled by hardware action	Cannot be disabled by software action
Interrupt flag affects hardware	Clearing the flag does not prevent traps
Interrupts restore the previous state of CPU	It does not restore the previous state of CPU

Fig. 4.62 Difference between interrupts and traps in automotive OS

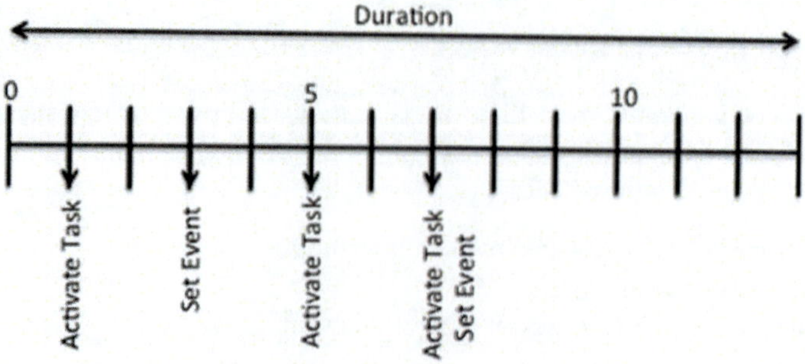

Fig. 4.63 Events and task activation

Counters, Alarms and Schedule tables

Counters: Used to count events, e.g. timer ticks.

Alarms: Bound to a counter and expire when a certain counter value is reached.

Expiration of counter will generate an event (or) activate a task. A counter can hold more than one alarm, and they are cyclic (or) linear.

Schedule tables: They are pre-defined sequence of expiry points. On each expiry points, a task is activated or an event is set. It is attached to a counter, which provides the underlying interval measurement (Fig. 4.63).

Resources Management

Resources are data structures for the protection of critical regions, for example, exclusive access to a port or a memory location. Resources can be used by tasks and optionally by ISRs. A deadlock prevention occurs when two tasks wait for resources that are held each by the others. Priority inversion arises when a high priority task is forced to wait for an indefinite amount of time for a lower priority task to be completed (Fig. 4.64).

Scheduling

There are two basic CPU scheduling principles for real-time systems.

Pre-emptive scheduling is used when a process switches from running state to ready state or from waiting state to ready state. The resources (mainly CPU cycles) are allocated to the process for the limited amount of time and then is taken away, and the process is again placed back in the ready queue if that process still has CPU burst time remaining. That process stays in ready queue until it gets next chance to execute.

Algorithms based on pre-emptive scheduling are as follows:

- round robin (RR)
- shortest remaining time first (SRTF)
- priority (pre-emptive version), etc.

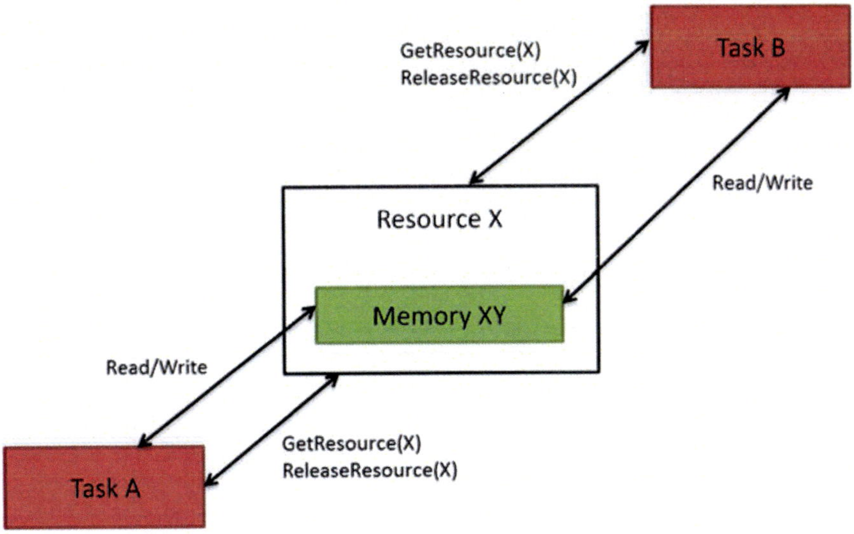

Fig. 4.64 Resource management

Non-pre-emptive scheduling is used when a process terminates, or a process switches from running to waiting state. In this scheduling, once the resources (CPU cycles) are allocated to a process, the process holds the CPU till it gets terminated or it reaches a waiting state. In case of non-pre-emptive scheduling, it does not interrupt a process running CPU in middle of the execution. Instead, it waits till the process completes its CPU burst time and then it can allocate the CPU to another process.

Typical algorithms based on non-pre-emptive scheduling are:

- shortest job first (SJF basically non-pre-emptive) and
- priority (non-pre-emptive version), etc.

The main differences between pre-emptive and non-pre-emptive scheduling are as follows:

- In pre-emptive scheduling, the CPU is allocated to the processes for the limited time, whereas in non-pre-emptive scheduling, the CPU is allocated to the process until it terminates or switches to waiting state.
- The executing process in pre-emptive scheduling is interrupted in the middle of execution when higher priority one comes; whereas the executing process in non-pre-emptive scheduling is not interrupted in the middle of execution and waits till its execution.
- In pre-emptive scheduling, there is the overhead of switching the process from ready state to running state, vice versa, and maintaining the ready queue. Whereas in case of non-pre-emptive scheduling, it has no overhead of switching the process from running state to ready state.

- In pre-emptive scheduling, if a high priority process frequently arrives in the ready queue, then the process with low priority has to wait for long, and it may have to starve. On the other hand, in the non-pre-emptive scheduling, if CPU is allocated to the process having larger burst time, then the processes with small burst time may have to starve.
- Pre-emptive scheduling attains flexibility by allowing the critical processes to access CPU as they arrive into the ready queue, no matter what process is executing currently. Non-pre-emptive scheduling is called rigid as even if a critical process enters the ready queue, the process running CPU is not disturbed.
- The pre-emptive scheduling has to maintain the integrity of shared data that is why it is cost associative as it is which is not the case with non-pre-emptive scheduling.

The term pre-emptive multitasking is used to distinguish a multitasking operating system, which permits pre-emption of tasks, from a cooperative multitasking system wherein processes or tasks must be explicitly programmed to yield when they do not need system resources. Pre-emptive multitasking involves the use of an interrupt mechanism, which suspends the currently executing process and invokes a scheduler to determine which process should execute next. Therefore, all processes will get some amount of CPU time at any given time. In pre-emptive multitasking, the operating system kernel can also initiate a context switch to satisfy the scheduling policy's priority constraint, thus pre-empting the active task. Pre-emptive scheduling means the high priority task at that instance seizes the currently running task. Pre-emptive multitasking allows the computer system to provide each process a regular "slice" of operating time. It also allows the system to rapidly deal with important external events like incoming data, which might require the immediate attention of one or another process.

At any specific time, processes can be grouped into two categories: those that are waiting for input or output (I/O bound) and those that are fully utilising the CPU (CPU bound). In early systems, processes would often "poll", or "busy wait" while waiting for requested input (such as disc, keyboard or network input). During this time, the process maintains complete control of the CPU. With the advent of interrupts and pre-emptive multitasking, these I/O bound processes could be "blocked", or put on hold, pending the arrival of the necessary data, allowing other processes to utilise the CPU. As the arrival of the requested data would generate an interrupt, blocked processes could timely return to execution.

Time Slicing

The period of time for which a process is allowed to run in a pre-emptive multitasking system is generally called the time slice or quantum. The scheduler is run once in every time slice to choose the next process to run. The length of each time slice can be critical to balancing system performance vs process responsiveness—if the time slice is too short, then the scheduler will consume too much processing time, but if the time slice is too long, processes will take longer to respond to input.

An interrupt is scheduled to allow the operating system kernel to switch between processes when their time slices expire, effectively allowing the processor's time to be shared between a number of tasks, giving the illusion that it is dealing with these tasks in parallel (simultaneously). The operating system, which controls such a design, is called a multitasking system.

4.6.3 Mix-Criticality Application in Hard Real-Time Systems

New vehicle control units need solutions for different ASIL within a single control unit. If one of the safety-related function even requires time constraints, the assurance of sufficient independence is difficult to demonstrate. There are two principles, which allow different ASILs in one microcontroller:

- Multitasking.
- Multi-core.

Multitasking generally is based on single core or single core lockstep controllers. Most approaches with single core lockstep have only the advantage that random hardware faults of the core functions need no additional software safety mechanism, even for ASIL D. However, single core lockstep could not avoid systematic faults or errors in the embedded software. On both cores of the single core lockstep operates the same identical software. Due to the need of separation of, or at least sufficient independence of software safety mechanism and the intended function of the system, a separation on task level becomes necessary.

Running different tasks on a single core requires sharing of resources. For higher-level ASIL, the dependent failure analysis becomes endless complex; even what resources are used depends on compiler settings and programming style. Even very good software coding guidelines have to be questioned in case of new instruction sets of the microcontroller core. Changing of compiler or even microcontroller could have a tremendous safety impact.

The following example considers a safety task, which needs to be repeated each 1 millisecond. The millisecond is necessary to synchronise I/O data, especially data with no time-stamp and to trigger the watchdog, or any other degradation mechanism in the system. Three basic functions are considered which run on a microcontroller with only these tasks for 3, 5 and 7 ms (net task time). The pure addition of the runtime leads to 15 ms. By having each millisecond, an interrupt for the higher-level safety task is the time double to 30 ms (Fig. 4.65).

Since the task itself does not have any time constraints, such solutions are acceptable.

Typical active safety functions such as for chassis control, where for example typical control loops for the application software are required, higher scheduling applications are required. If the safety task requires a higher-level ASIL and the application a lower-level ASIL or even only QM (e.g. legacy code), timing constraints violate with context switch between the tasks with different ASIL. Depending on the complexity of the application function and the ability of the microcontroller and operating system, such context switches lead enormous time exposure for the time.

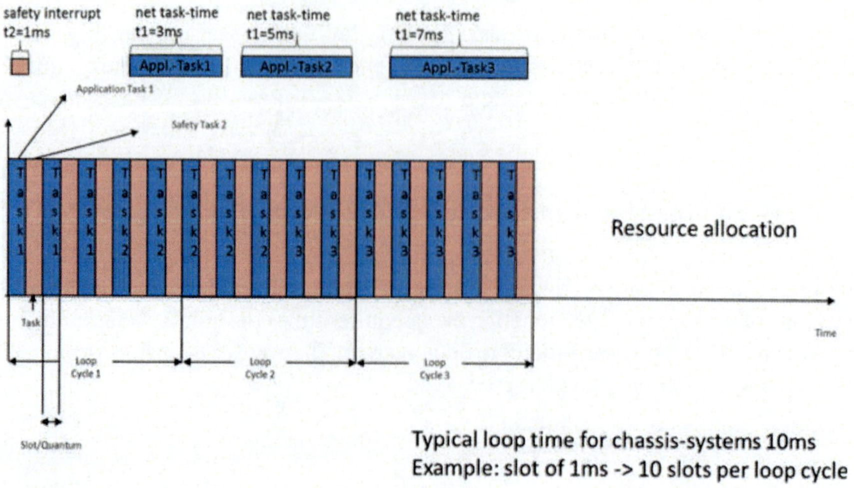

Fig. 4.65 Mixed criticality based on multitasking solutions

There are interrupt routines, which run in an order of magnitude of nanoseconds, but the lowest time for the context switch is in an order of microseconds. That leads to very limited time slots for the application tasks.

In today's multi-core solution, the microcontroller offers an assignment of the safety function to one core and using the second core for the lower ASIL application function. Since even here the lockstep solution offers no measures versus systematic faults or errors, a sufficient independent monitoring layer is needed also on the safety core (Fig. 4.66).

By using multi-cores, impacts due to dependent faults or errors are a challenge, since the kind of microcontroller, the compiler and the operating system could decide

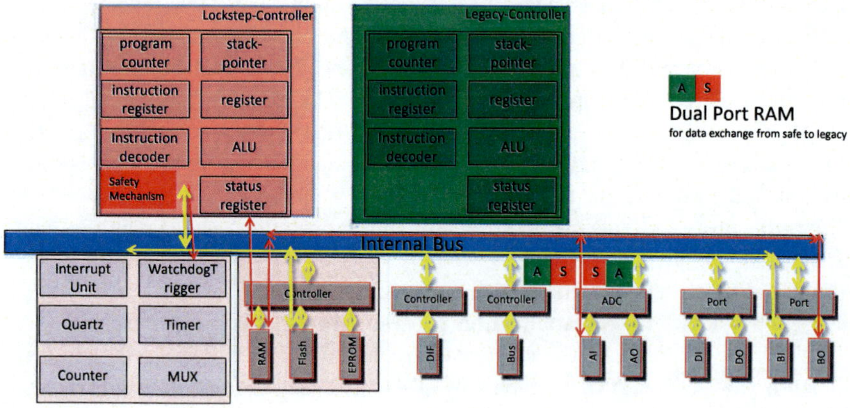

Fig. 4.66 Mix-criticality solution on multi-core controller

on which common resources are used within the microcontroller (core functions use different resources). Changing of microcontroller, operating system and/or the compiler (or just settings) are nearly impossible and if necessary highly safety critical.

In order to avoid interferences between the data exchange of peripheral elements and with common memory resources, a dual-port RAM could be used, which would be monitored by the higher-level ASIL core. Such a RAM interface would be handled like a communication interface. The higher-level ASIL core shall control any other external peripheral elements.

4.6.4 Control and Data Flow Monitoring

Control flow and data flow analysis are basic safety activities during development of a safety-related software system in any industry. The analysis methods are forecasting the ability of the system to perform the intended software, when it comes to the real execution. There are methods with and without intrusive functions that offer the possibility to prove data- and control flows by tests, so systematic checks can be performed to prove the functional completeness of the requirements.

Data flow analysis is a technique for gathering information about the possible set of values calculated at various points in a computer program. Control flow charts represent the parts of a program to which a particular value assigned to a variable might propagate (Fig. 4.67).

Control flow analysis is analysis of the order in which individual statements, instructions or function calls of an imperative program are executed or evaluated.

Challenges of data and control flow analysis.

In software programs, there is very often a close coupling between a data flow and a control flow; however, software should have an infinite number of data and control flows. A complete proof of the correctness of all data and control flows is impossible in practice.

Non-intrusive system observation allows the monitoring of software parameters without influencing the behaviour of the system. Such tests can be fully automated compared to the debugger tests. This supports the systematic measurements of data and control flows in integration and system tests. The integration and system tests prove the functional requirements. Important data and control flows are always reflected in the functionalities of the system. Therefore, the non-intrusive system observation supports the checking of the requirement completeness.

For time-sensitive hard real-time safety systems during development, not all conditions could be forecasted and predicted. Control flow and data flow monitoring should be implemented during runtime in order to change prioritisations, degradations, scheduling or any other control measure during runtime. Activity diagrams specify data/control flows and make the expected dynamic behaviour transparent, and it could be communicated to users, assessors or other stakeholders. Non-intrusive

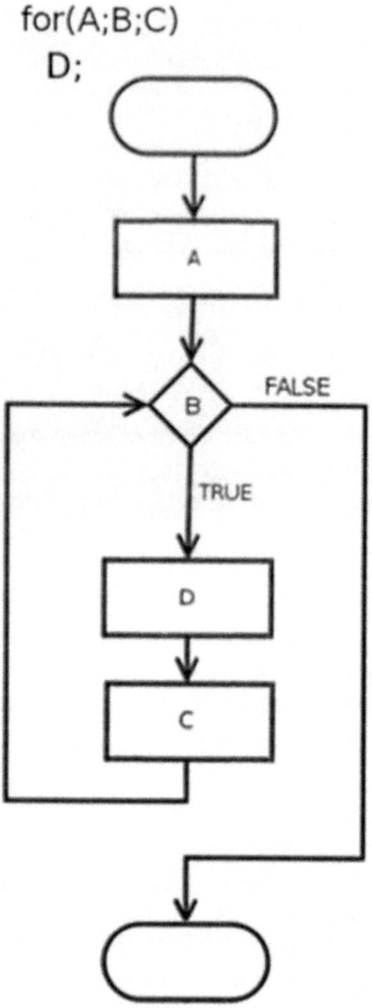

Fig. 4.67 Typical flow chart showing control flow

system observation also draws attention to a further weakness of the existing development approach, so that validation during runtime provides forecasts to the use in series development applications. Both the systematic proof of the data and control flows by test, as well as the completion of existing functional specification gaps are only successful if at least the most important data and control flows are defined in the software architecture. Describing the software architecture in UML, activity diagrams are a good way to do this. Such an approach is required in ISO 26262 (Part 6), DO178C and EN 50128/EN 50657 require data and control flow analysis as well as they consider control flow monitoring during runtime. During development, static analyses or unit and SW integration tests are often used to prove the data and control flow. Non-intrusive system monitoring will make it possible to systematically verify data and control flows at integration and overall system level tests. This approach offers not only the possibility to test complex error scenarios. It also allows verifying the completeness of functional requirements. A software architecture systematically specifying the data and control flows is unambiguous.

Implemented monitoring and diagnostics are considered as

- control mechanisms and
- supervision mechanisms (monitoring mechanism).

During the development, one will look at different levels of the requirements and the possible solutions for the design and must analyse with regard to the implementation and the later implementing in the controller. This data flow and control flow analyses lead to the fact that also some imponderabilities (stream for systematic error) after the design remains which leads to inevitable errors during runtime in the controller. Therefore, one also implements in the programmes control mechanisms, which requires monitoring during runtime (Fig. 4.68).

The sketch shows a typical EGAS or safety bag approach. The nominal or the intended function runs in a dedicated program flow. Specific implemented control mechanism like counters, sequence or checkpoints provides information about the specified data flow. For safety and security aspects, the implemented control mechanism should be provided with adequate keys, depending on the highest safety integrity of the entire system.

Beside the architectural aspects of data flow and control flow, the correctness of the data contents against specifications could be monitored during runtime by such implementations.

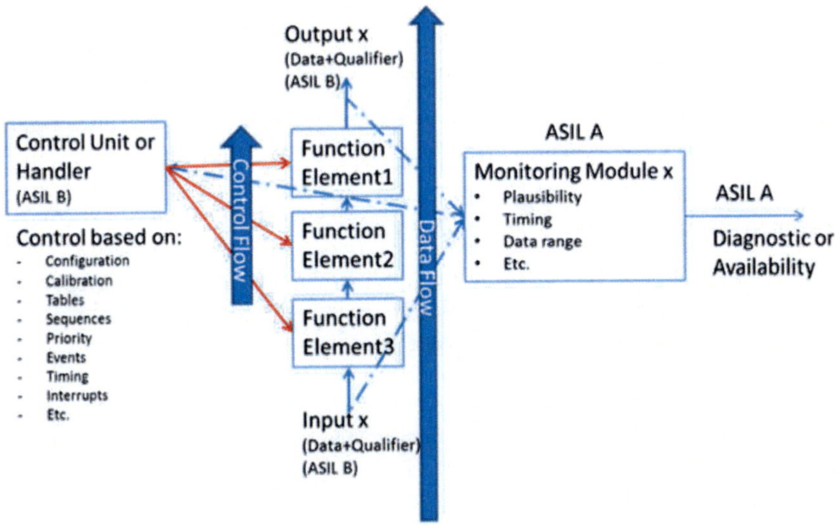

Fig. 4.68 Data flow and control flow in a mix-criticality approach

- configuration check, if various software elements are called with different parameter depending on the context of the call function,
- calibration check (software elements provide evidence of correct calibration data),
- data and sequence checks look-up, periodisation tables, etc.,
- availability of resources and their correctness or correct functional ability,
- temporal aspects,
- interrupts, hooks or other control mechanisms and
- authentication, security checks, etc.

The monitors usually could not distinguish between the cause of faults or errors, so that a systematic failure analysis needs to support the monitor design during development.

4.7 Operating Systems in Vehicles

Most of the embedded control systems in automotive industry are using an OSEK derivate, since they claim AUTOSAR compatibility with their control devices. In many actuator control systems for chassis systems, there are also other operating systems considered or there are dedicated solutions implemented that mange the entire software by timer and tables.

OSEK (*Open Systems and their Interfaces for the Electronics in Motor Vehicles*).

The abbreviation OSEK stands for "Offene Systeme und deren Schnittstellen für die Elektronik in Kraftfahrzeugen" in English: "Open Systems and their Interfaces for the Electronics in Motor Vehicles"). It is a standardised framework, which provides specifications for an embedded operating system, a communications stack and a network management protocol for automotive embedded systems.

A German automobile company consortium, with companies like BMW, Bosch, Daimler, Opel, Siemens, Volkswagen group and the University of Karlsruhe, founded OSEK in 1993. In 1994, the French car manufacturers Renault and PSA, who had a similar project vehicle distributed executive (VDX) have joined to the syndicate. They merged to OSEK/VDX (Fig. 4.69).

The main functional units of OSEK are as follows:

- OSEK/VDX-OS: OSEK/VDX is a pre-emptive multitasking operating system. It is defined as a static operating system to reach an optimised usage of CPU time and hardware resources. The detailed functions are described in the osCAN product description.
- OSEK/VDX-NM: the network management is mainly made for applications in the automobile. The main task is the coordinated and synchronised transfer of all bus nodes from sleep mode to normal mode and vice versa. Thus, it is not necessary to set the CAN bus to sleep mode and do the wake up manually. You can find further information at the CANbedded pages.
- OSEK/VDX-COM: this standard describes the data exchange between tasks and ISR. These can run either on a target platform or they can exchange data over a

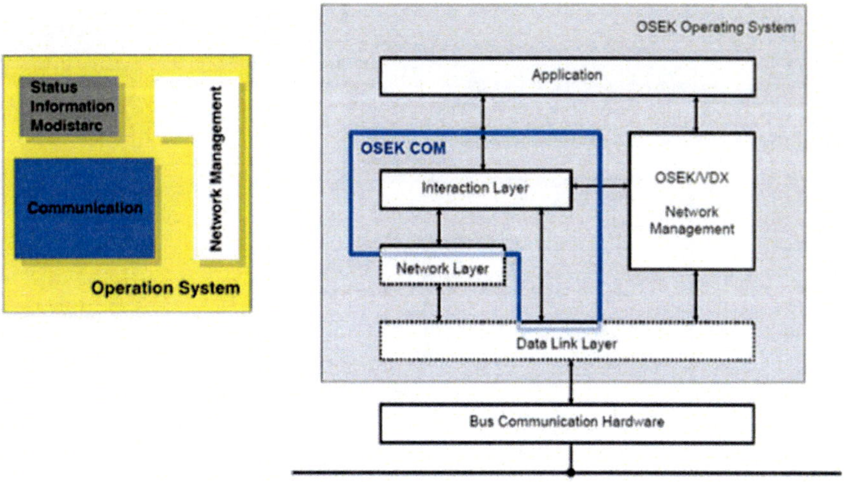

Fig. 4.69 Parts of OSEK

network. The data exchange within the target hardware is always included in the osCAN delivery because this is a basic mechanism also for stand-alone applications. You can find further information about the data exchange with CAN at the CANbedded pages.

The OSEK standard specifies interfaces to multitasking functions with generic I/O and peripheral access, which is architecture dependent. OSEK systems are expected to run on chips without memory protection. Features of an OSEK implementation can be usually configured at compile time. The number of application tasks, stacks, mutexes, etc., is statically configured; it is not possible to create more at runtime. OSEK recognises two types of tasks/threads/compliance levels: basic tasks and enhanced tasks. Basic tasks never block; they "run to completion" (coroutine). Enhanced tasks can sleep and block on event objects. The events can be triggered by other tasks (basic and enhanced) or interrupt routines. Only static priorities are allowed for tasks. First in first out (FIFO) scheduling is used for tasks with equal priority. Deadlocks and priority inversion are prevented by priority ceiling (i.e. no priority inheritance). The specification uses ISO/ANSI-C-like syntax; however, the implementation language of the system services is not specified as well as an application binary interface (ABI) (Fig. 4.70).

The OSEK COM scope covers partly or entirely the following layers:

Interaction Layer (IL)

The interaction layer (IL) provides the OSEK COM API which contains services for the transfer (send and receive operations) of messages. For external communication, it uses services provided by the lower layers, whereas internal communication is handled entirely by the IL.

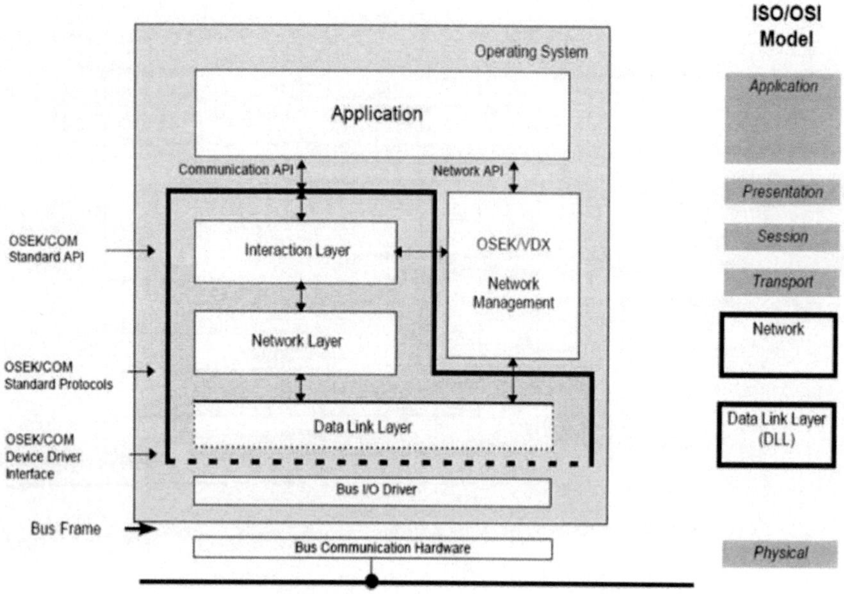

Fig. 4.70 Layer model of OSEK COM

Network Layer (NL)

The network layer handles—depending on the communication protocol used—message segmentation/recombination and acknowledgement. It provides flow control mechanisms to enable the interfacing of communication peers featuring different levels of performance and capabilities. The network layer uses services provided by the data link layer. OSEK COM does not specify the network layer; it merely defines minimum requirements for the network layer to support all features of the IL.

Data Link Layer (DLL)

The data link layer provides the upper layers with services for the unacknowledged transfer of individual data packets (frames) over a network. Additionally, it provides services for the NM. OSEK COM does not specify the data link layer; it merely defines minimum requirements for the data link layer to support all features of the IL.

Source OSEK Communication, Specification 3.0.1

The OSEK is specified in ISO 17356:2005, road vehicles—open interface for embedded automotive applications—Part 3: OSEK/VDX operating system (OS).

ISO 17356-3:2005 describes the concept of a real-time operating system, capable of multitasking, which can be used for motor vehicles. It is not a product description, which relates to a specific implementation. It also specifies the operating system application program interface (API). General conventions, explanations of terms and abbreviations have been compiled in ISO 17356-1. ISO 17356-6 discusses implementation and system generation aspects.

The specification of the OS represents a uniform environment, which supports efficient utilisation of resources for application software in automotive control unit. The OS is a single processor operating system meant for distributed embedded control units.

OSEK VDX considers looks at the following main components:

- OSEK OS: The OSEK operating system is responsible for the program flow within the control unit. It provides the interfaces for the user by which he can develop his applications for the suitable control unit and implement.
- Network management OSEK (network management): It provides standardised functions to administer a network system inside and with other control units and to monitor.
- Communication system OSEK: The communication system OSEK provides the functions required for the data exchange and protocols.
- OSEK-OIL (OSEK implementation language): The OSEK standard and especially the specification of the operating system should be independent of hardware. OESK-OIL makes available system functions and developing interfaces (APIs).

The OSEK OS defines four conformity assemblies for different system requirements, which allow to the user to configure the operating system for different target hardware systems. OSEK OS does not define modifications during the project duration.

- OSEK/VDX-COM provides several levels (orientate themselves by the ISO/OSI layer model) for the communication:
- OSEK/COM device driver interface: data link layer
- OSEK/COM standard protocol: protocol base for the network layer
- OSEK/COM standard API: application interface.

Source ISO 17356-3:2005

The OSEK standard provides no specific safety requirements as defined by ISO 26262, which leads to various heterogeneous safety solutions, but no systematic structured solutions.

4.7.1 AUTOSAR

AUTomotive Open System ARchitecture (AUTOSAR) is a worldwide development partnership of vehicle manufacturers, suppliers, service providers and companies from the automotive electronics, semiconductor and software industry (Fig. 4.71).

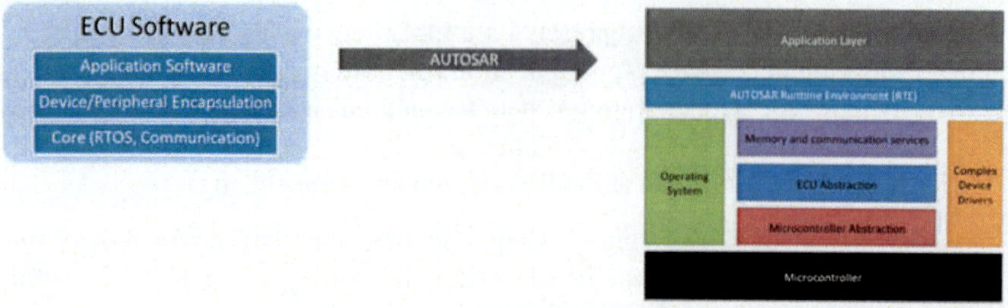

Fig. 4.71 AUTOSAR basic overview

The AUTOSAR classic platform architecture distinguishes on the highest abstraction level between three software layers, which run on a microcontroller: application, runtime environment (RTE) and basic software (BSW).

– The application software layer is mostly hardware independent.
– Communication is between software components and access to BSW via RTE.
– The RTE represents the full interface for applications.
– The BSW is divided into three major layers and complex drivers as follows:
  • services, electronic control unit (ECU) abstraction and microcontroller abstraction.
– Services are divided furthermore into functional groups representing the infrastructure for system, memory and communication services.

AUTOSAR considers the following elements:

– application layer,
– runtime environment,
– system services,
– memory services,
– crypto services,
– off boards communication services,
– communication services,
– onboard device abstraction,
– memory hardware abstraction,
– crypto hardware abstraction,
– wireless communication HW abstraction,
– communication HW abstraction,
– microcontroller drivers,
– memory drivers,
– crypto drivers,
– wireless communication drivers,
– communication drivers,
– I/O hardware abstraction,

- I/O drivers,
- complex drivers,
- microcontroller.

AUTOSAR supports also safety requirements and provides a safety specification (Overview of Functional Safety Measures in AUTOSAR CP Release 4.3.0).

The content of this document is structured into separate chapters as follows:

Functional safety mechanisms: This chapter contains AUTOSAR functional safety mechanisms related to freedom from interference between AUTOSAR SW-Cs.

- Memory: partitioning mechanisms of AUTOSAR with the context of application software development and deployment,
- Timing: temporal program flow monitoring mechanisms using the watchdog manager and timing protection mechanisms using the operating system,
- Execution: logical supervision mechanisms using the watchdog manager,
- Exchange of information: communication fault detection mechanisms using the end-2-end library and extensions.

Functional safety measures: This chapter contains topics related to the development of safety-relevant systems. The following items are covered:

- functional safety measures of AUTOSAR, such as traceability, development measures and the evolution of the standard,
- functional safety measures not delivered by AUTOSAR,
- safety use case: an exemplary safety-related system using AUTOSAR based on the guided tour example Front Light Management,
- safety extensions: how safety requirements can be expressed within the AUTOSAR models and documents by means of the AUTOSAR meta-model.

Hardware diagnostics: This chapter contains topics related to the premise that the provided functionality of the microcontroller can be trusted. The following items are covered:

- core test
- RAM test

The AUTOSAR safety specification considers particularly the handling of AUTOSAR and the scope, which is the basic software and the related architecture. Safety-in-use refers to handling of AUTOSAR software and not the typical safety-in-use aspects derived from the system user of the final product. The timing aspects are closely linked to the watchdog function, so that high available and real-time systems mean a challenge on AUTOSAR-based systems.

4.7.2 ARINC 653

ARINC 653 is the avionics standard Avionics Application Software Standard Interface created by ARINC. Aeronautical Radio, Incorporated (ARINC), established in 1929, was a major provider of transport communications and systems engineering solutions for eight industries: aviation, airports, defence, government, health care, networks, security and transportation. ARINC had installed computer data networks in police cars and railroad cars and maintains the standards for line replaceable units.

ARINC 653 deals with the virtualisation in computer systems with regard to the special requirements in the aerospace industry. One such requirement is the predictability of the behaviour of a computer system. An important feature of ARINC 653 is the allocation of resources (computing time, memory, etc.) for individual applications running on the same processor or memory. This ensures that the applications do not influence each other in an undesirable way. This feature makes it possible to run safety-critical software with different certification requirements on one CPU. The definition of an application programming interface (API) also reflects the trend towards modular avionics (IMA).

ARINC 653 proposes an improvement of the federated architecture. The federated architecture was specified for digital fight control functions such as autopilot, flight management, displays and yaw damping. Each function has its own computer system (separate resource), and the functions can interact with each other through limited shared resource. The advantages are as follows:

- Fault in one computer system does not propagate easily into other functions due to the limitation in sharing data.

The disadvantages are as follows:

- Hardware that is been required individually with own system.
- Each function has a different and unique computer platform.
- The cost employed in developing and certifying the software to run on these distinctive platforms is high.

Integrated Modular Avionics:

Integrated modular avionics (IMA) offers the following:

- single common computer system is used for handling all functions,
- potential to diminish the fault containment between the functions
- OS of IMA architecture, which is designed in a way of partitioning in space, time, input/output and communications.

Effective partitioning methods in IMA standard have been the reason for ensuring the relevant set of safety and real-time guarantees in aviation industry (Fig. 4.72).

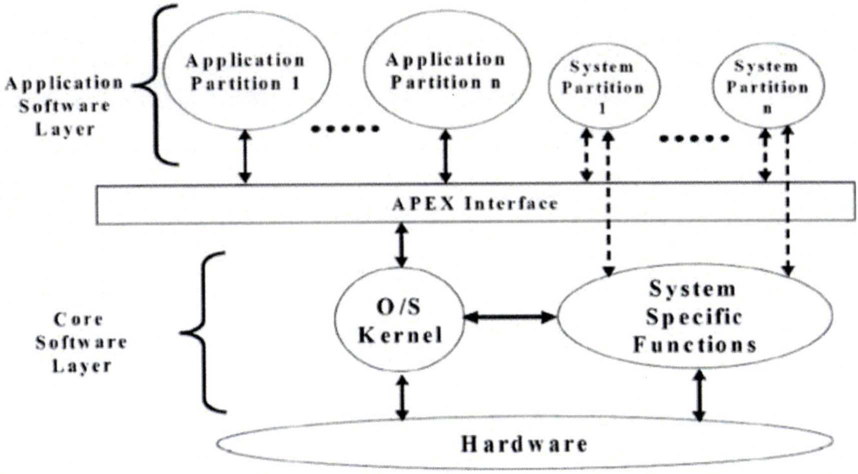

Fig. 4.72 Logical real-time operating system structure created according to ARINC specification 653. *Source* ARINC specification 653 ([17], p. 4)

ARINC 653 defines generic structure of the software architecture. The logical structure is similar to RTOS. The basic elements are

- hardware,
- core software layer including O/S kernel, system-specific functions,
- APEX interface (the runtime environment),
- application software layer including application partition and system partition.

The key concept introduced in the specification is the partition. It creates a kind of container for an application and guarantees that execution of the application is both spatially and temporally isolated. The partitions are divided into two categories: application partition and system partition. The application partitions execute avionics applications. They exchange data with the environment by means of specific interface—APEX (**AP**plication/**EX**ecutive) which is their runtime environment. The system partitions are optional, and their main role is to provide services not available in APEX, such as device drivers or fault management. The system partitions could deal with other resources or with the controller hardware independent from kernel functions, which allows a dissimilar I/O interface including independent core functions (Fig. 4.73).

The ARINC 653 framework provides

- based on IMA and strictly supports robust partitioning,
- freedom of hardware and software development irrelevant to each other,
- **APEX (APplication EXecutive)**: for achieving good time and memory partitioning,
- **Health Monitor**: isolates faults and prevents the errors from propagating from one layer to the other (Fig. 4.74).

The basic software in ARINC 653 is called the "core software".

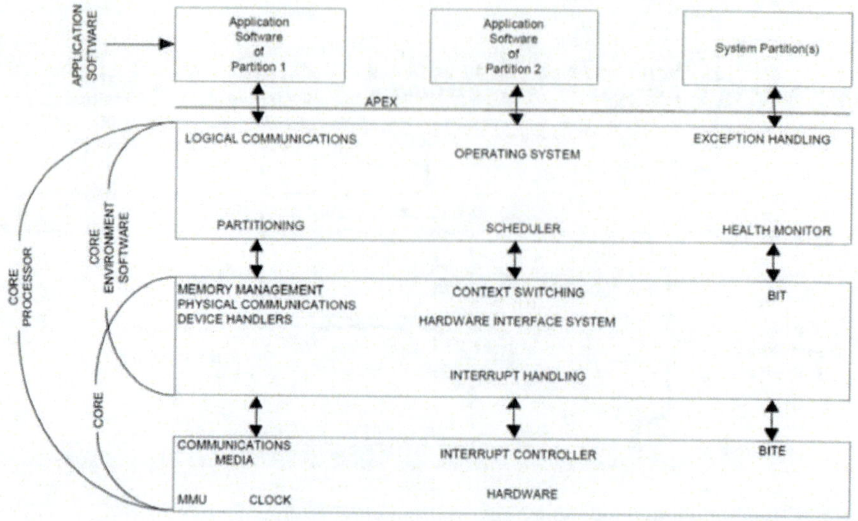

Fig. 4.73 Hardware–software architecture of typical module according to ARINC 653 specification. *Source* ARINC 653 specification ([17], p. 11)

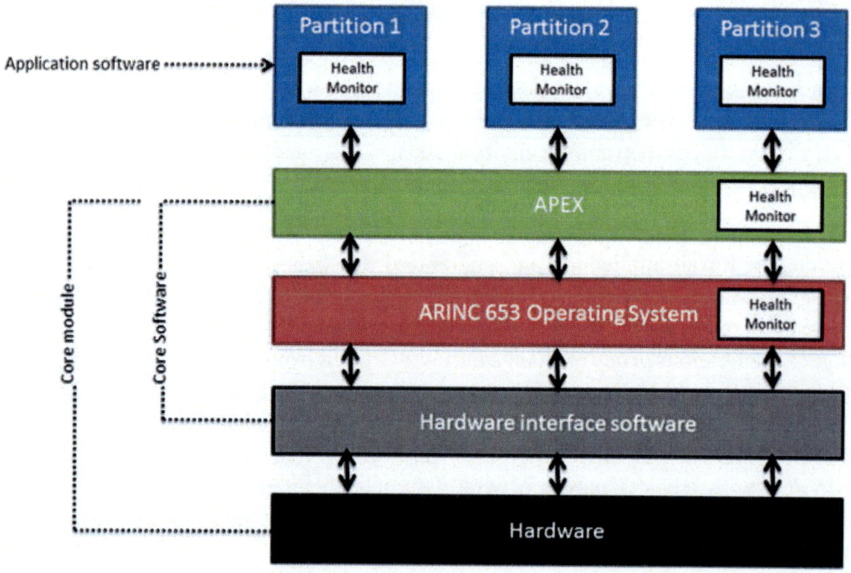

Fig. 4.74 Layered structure of IMA

The APEX manages both aperiodic and regularly periodic activated processes. Additionally, a separate parameter called "time capacity" is attached to each of them. It defines time frame (deadline) within which single instance of task must finish computations. When a process is started, the deadline is set to current time plus time capacity. The operating system periodically checks whether the process completes processing within the allotted time. Each process has a priority. During any rescheduling event, the O/S always selects for the execution the highest priority process in "ready" state. Inter-partition or inter-application communication is based on queuing port and sampling port communication units. The queuing port provides inter-partition message queues, whereas the sampling port shares variables between the ports. During system integration, the ports are connected by means of channels defined in system configuration tables. The ports communicate with other partitions or device drivers within the module, or exchange data between modules.

The ARINC 653 health monitor is an advanced exception handing engine. Three error levels are defined:

- process level which affects one or more processes in the partition,
- partition level with only one partition affected,
- module level which affects all partitions within the module.

Both partition level and module level errors are handled by a set of procedures, which need to be installed by the system integrator. During development, the process level errors could be handled. Separate sporadic task called "error handler" can be registered for each of the partitions. When the health monitor detects an error at the process level, it calls the error handler.

Many of those mechanisms entered also AUTOSAR standards, but especially the partitioning concept and health monitoring are still a challenge in automotive solutions.

4.7.3 Comparison of AUTOSAR with ARINC 653 Framework

There are various features in ARINC 653, which makes the IMA more suitable for safety-related real-time systems with multitasking and multifunction modes in mix-criticality applications (Fig. 4.75).

A typical automotive operating system provides usually resource partitioning with help of the hardware of the microcontroller. The memory protection system works only with adequate microcontroller hardware; virtual memory management, etc., could not be used in safety-critical partitions.

A partitioning OS based on ARINC 653 offers the following:

- a single common computer system is used for handling all functions,
- applications from different vendors share a common hardware resource,
- multiple encapsulated environments within the ECU,

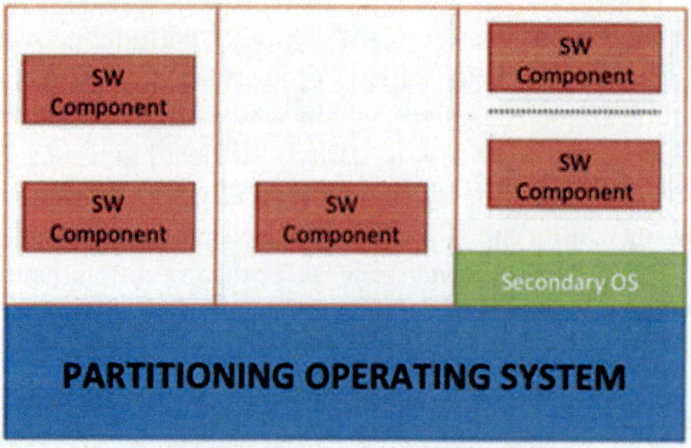

Fig. 4.75 Typical automotive architecture based on a portioning OS

- spatial partitioning: It ensures that the software component in one partition does not change or interfere with the software component in other partition
- temporal partitioning: It ensures that one software component does not affect the ability of receiving services from a shared resource (Fig. 4.76).

Even our brain maintains maps of body parts which reflect an orderly organisation of the somatosensory system with adjacent areas on the skin represented by adjacent neurons in the cortex. Similar to this, a control device needs such separations to safely execute independent function in the same control device.

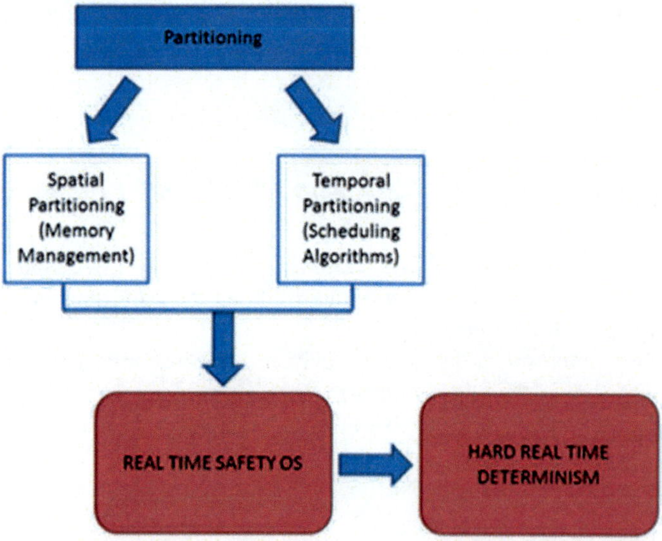

Fig. 4.76 IMA with spatial and temporal partitioning

There are many established ways of handling the effect of scheduling overrun and thereby avoiding extreme scenarios, these include

- minimising the processing that occurs within interrupt handlers,
- analysing the worst case interrupt processing within periodic frames and if necessary taking steps to limit the maximum interrupt processing,
- analysing the worst case processing within tasks and using this to determine average and peak CPU load,
- wherever possible, making effective use of priority based scheduling so that essential tasks are performed in as deterministic manner as possible.

Even when all potential scheduling problems have been addressed and the system is working correctly, we could not prove temporal separation. For some systems, this may not be a problem; however, when dealing with a mixed criticality system, it is necessary to be able to prove that software operating at a lower criticality cannot interfere with the operation of software designed to a higher integrity level partition.

4.7.4 Safe Processing Environment

A **safe runtime environment** is for safety-related applications that need safety qualifier to show if the relevant data provide the sufficient safety integrity for a safe user application.

In case of any safety-related absolute or frequent time constraints, at least a safety-related time monitor needs to be implemented. This time monitor could be used to trigger a watchdog so that the system could be degraded in a safe state by a controller shutdown. If a shutdown of the controller is not the safe state, for example in fail-operational systems such degradations do not provide any safe reactions.

In case of soft real-time safety requirements, such monitors could change periodisation for the scheduler, so that safety-critical tasks are processed. In case of deterministic update of, e.g. sensor data at the runtime environment information could be used to identify delayed information. In this case, the application could provide adequate safety-related function, so that the given safety requirements would not be violated.

In hard real-time safety systems, timing must be monitored. In many applications, it is a matter of the safety architecture, if the lower time interval always requires hard reactions of the controller. Even if the controlled deterministic timing for fail-operational functions leads in case of errors to violation of safety goals, a shutdown of the microcontroller does not provide any safe reaction to the system. In this case, the operation of the microcontroller is the only possible solution. In aeroplane applications, (TMR) triple modular redundancy systems are compulsory. Those systems mainly do not react on diagnostics and comparisons; they are mainly based on a two out of three(2oo3) majority voting principles. If two of the three results are equal, they provide the input for the actuator or any other final element. To assure continuous operations, the systems do continue their operation, even in case of detected

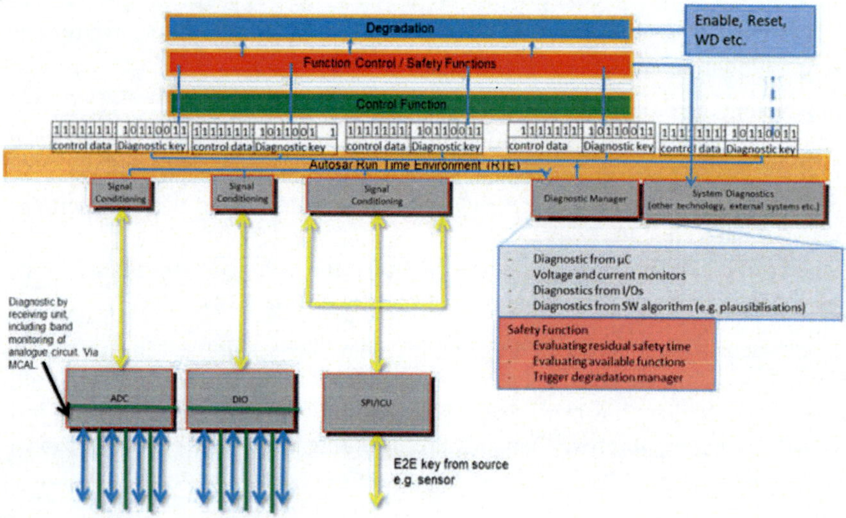

Fig. 4.77 BSW for safe computing environment

faults. So that the defect identified element could run in a recovery mode (reset of a controller) and after recovery all three elements could continue to control the safety critical actuator. The challenge is that such a 2oo3 system is well synchronised. In case of time delays, the voter gets too much unequal results, which usually does not lead to a shutdown but in decreased performance of the system (Fig. 4.77).

Such a typical AUTOSAR compatible architecture requires various safety mechanism in the BSW such as

- E2E protection,
- diagnostic and monitoring functions,
- control and data flow monitoring,
- error handling (Fig. 4.78).

In order to avoid any hardware-related impact to the computing environment, AUTOSAR considers a hardware abstraction layer (HAL) and a microcontroller abstraction unit, called microcontroller abstraction layer (MCAL). The MCAL is usually a driver that provides the access to the hardware elements of the microcontroller. The HAL provides

- filters,
- error correction mechanism and
- diagnostic information.

In addition to the two abstraction layers, a system (including sensor) abstraction layer (SAL) is proposed in order to:

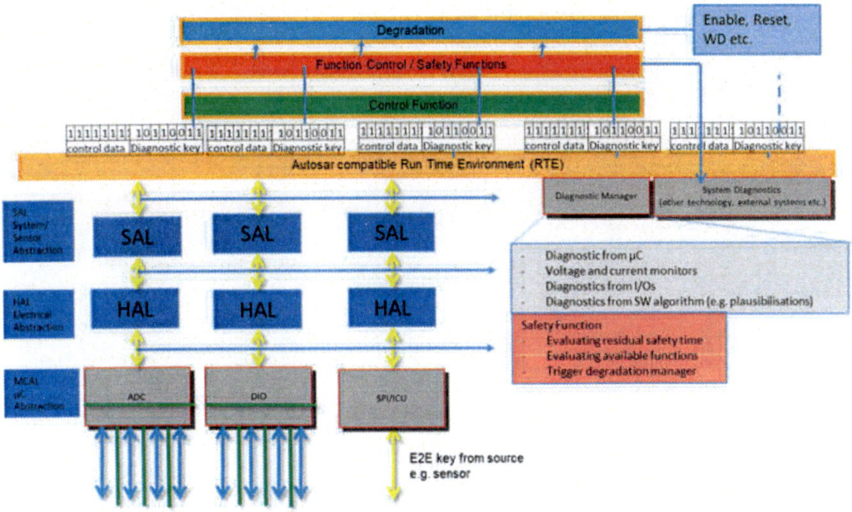

Fig. 4.78 Hardware and system abstraction

- filter, error correction and diagnostic from any system element within the relevant signal chain from the peripheral, input and output signals, etc.,
- physical signal adaption and scaling of data range,
- logical plausibility to provide correct information for the safe computing environment,
- time synchronisation, so that only data are processed with a common time context.

Adequate qualifiers are foreseen in the AUTOSAR specification for the signals at the runtime environment, so that the diagnostics and monitoring results could be assigned to any information for the "safe processing environment".

Within the safe computing environment, at least due to systematic failure, the following layers are proposed for the application layer, which build the safe computing environment:

- control function layer, the intended function,
- function control layer, the monitor and the systematic faults in the control function,
- degradation layer, which degrades based on function control layer (Fig. 4.79).

Consequently, a safe hard real-time environment needs the nominal input information (e.g. from sensors) and a qualifier providing integrity information about the input signal and in addition a second qualifier that provides the timing information. Such a qualifier could provide information that the signal is within the required time interval, but also about their synchronisation such as delayed or too early. In case of safety-related control of vehicle direction, even the time-stamp from the detecting sensor (the origin of the electrical signal) could be provided (Fig. 4.80).

Especially, a perception could not just be monitored by EE monitors, also physical plausibility checks or voting comes necessary. In FPGA, with safe computing

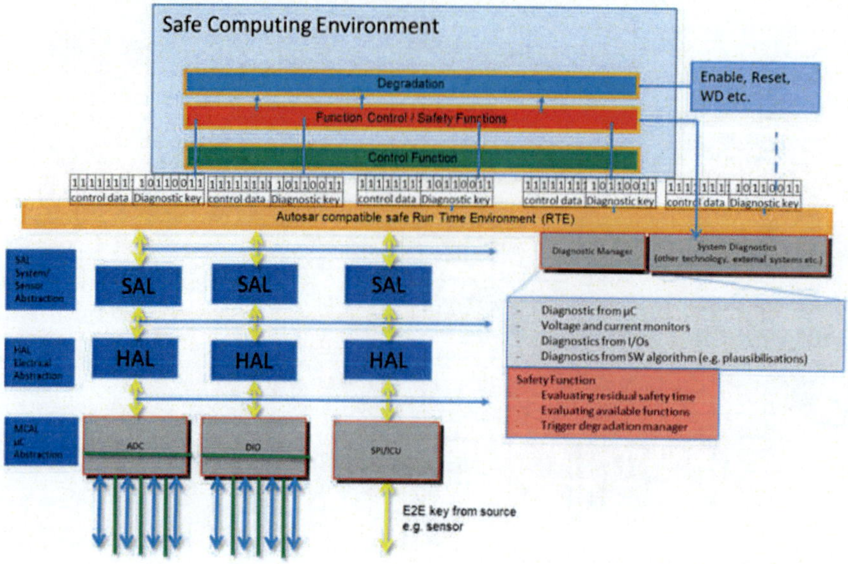

Fig. 4.79 Elements for safe computing environment

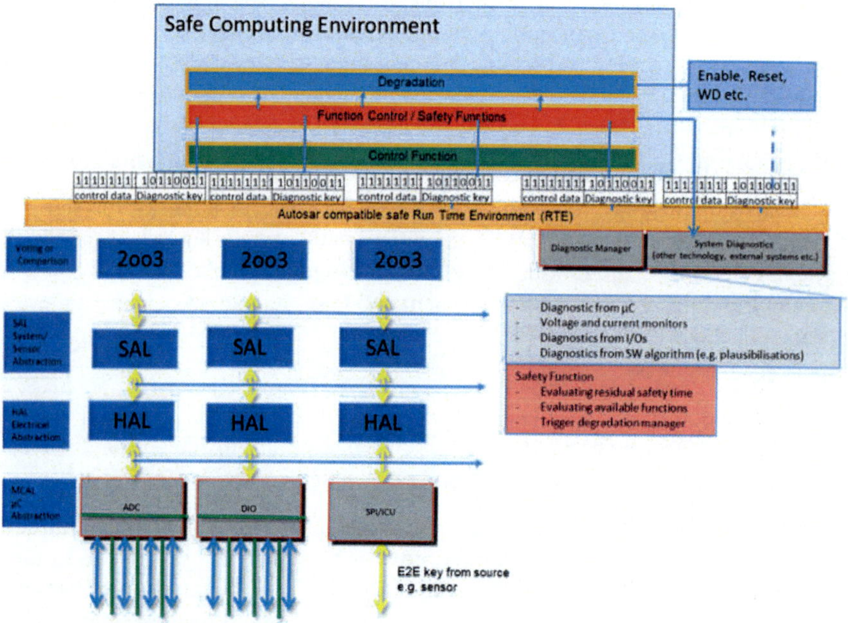

Fig. 4.80 Additional voting layer for the runtime environment

environment, also voting architectures are used for error diagnostics in the peripheral elements. For that, a voting or comparison layer could be introduced in order to avoid failure impacts to the "safe computing environment". Votings are very fast and could be often designed by passive electronics, so that such highly time-sensitive information is provided with a low deterministic delay. During development of a sensor, it is impossible to identify all relevant application use-cases. Consequently, it is possible to specify and implement the sensor capability including its robustness and ability to control possible failure, but it is not possible to specify what the sensor could not perceive or measure.

Such principles could also run on task level, so that three dissimilar tasks provide adequate safety-related control information and by 2oo3 voting the actuator (or any output) provides the control information. In a single microcontroller application, such software-based voting does not provide a solution against entire controller impacts such as lightning flashes, which immediately kills the system. However, what level of hardware redundancy necessary for a system is a matter of the system analysis of the entire item (the vehicle system within its road traffic environment), rather than a question of the software or microcontroller architecture.

4.7.5 Predictive Health Monitoring

In other industries, in particular of the aviation and with other transport systems, how of the astronautics one looks at alternatives to this 1:1 allocation of error mechanism, cause and effect. One borrows himself from the medicine and places on the principles of the forecast and recognising symptoms. Among experts, one finds the abbreviation PHM, often as

- predictive health management and
- prognostic health management.

In many publications, also the letter "M" stands for "monitoring".

In plant automation or machinery industry, similar principles are used for

- condition monitoring,
- predictive maintenance,
- predictive machine health monitoring.

Their motivation is based on early fault detection, identification of ageing effects due to stress, etc. In addition, advanced manufacturing depends on the timely acquisition, distribution and utilisation of information from machines and processes across spatial boundaries. These activities can improve accuracy and reliability in predicting resource needs and allocation, maintenance scheduling and remaining service life of equipment. As an emerging infrastructure, cloud computing provides new opportunities to achieve the goals of advanced manufacturing. The idea is using well-tried principles like prognosis theories and techniques and projects and their future growth enabled by the emerging cloud infrastructure. Techniques for cloud computing are applied as well as the influence of these techniques on the paradigm of cloud-enabled prognosis for manufacturing.

Condition based and predictive maintenance are two maintenance strategies, which aim at optimising equipment efficiency and reducing service timing and costs during equipment lifecycle. Condition monitoring (CM) is the monitoring of several parameters such as equipment vibration and temperature to identify potential issues such as misalignments or bearing failures or any other design-related dynamic effects, which could lead to failure. Condition monitoring tools can for instance map equipment degradation when a vibration analysis shows a change in the harmonic frequency of rotating equipment components. Based on both vibration measurement and noise data by microphones, etc., frequencies could indicate system weaknesses.

Continuous condition monitoring techniques are in machinery applied on several pieces of equipment such as compressors, pumps, spindles and motors and can be implemented to identify partial discharge on machine or vacuum leaks. Predictive maintenance is based on condition monitoring, abnormality detection and classification algorithms and integrates predictive models which can estimate the remaining machine runtime left, according to detected abnormalities. This approach uses a wide range of tools, such as statistical analyses and machine learning (ML) to predict the state of the equipment.

In high available systems such as aviation systems, PHM is often used also for diagnostics as input for

- preventive resource usage,
- degradation for system and performance or
- initiation of limp home scenarios.

A PHM consists often of various feature and functions such as

- health channels (detection and physical monitoring),
- modes (thresholds, limits to be relevant),
- condition which triggers valid event,
- expression for the relevance (if–then and other consequences based on the input),
- rules and consequences for actuator or further processing activities,
- actions at the actuator or for following data processing actions, etc.

Such a sequence offers a structured implementation and handling of a huge amount of health channels with multichannel heterogeneous actions with consequences for the behaviour of the system.

Typical effects to be monitored by the health channels could be like the following:

- detection of fault, errors failure,
- ability to provide nominal performance by the system, or communication equipment, etc.,
- detection of deviation from design,
- environmental data (temperature, EMC, pressure, etc.),
- system stress (noise, etc.),
- monitoring of sequence, conditions, modes and states in dynamic systems,

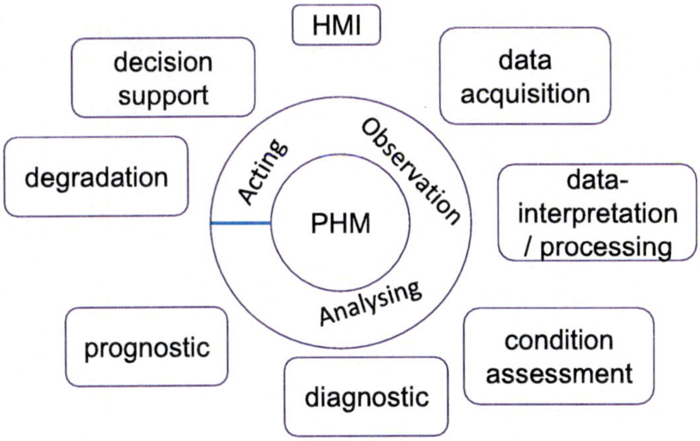

Fig. 4.81 Typical predictive health management (PHM) cycle

- pulse form monitoring,
- sequence and synchronicity monitoring,
- measurement of frequencies, accumulations, point of inflection, etc.,
- measurements of robustness margins,
- monitoring of security, privacy, authorisation policies,
- usability or adequate use monitoring,
- monitoring of safety barriers, integrity monitoring and other separation mechanisms,
- deadline monitors, timing constraints or time-related policies (also duration in states, conditions or operation modes, etc.),
- quality of service monitoring for characteristics in complex software algorithms, AI, ML or software with unknown contents like COTS (Commercial-off-the-Shelf), etc.

The PHM uses similar principles like data and control flow monitoring; it could be integrated as well as the enable, reset, error handling and watchdog management (Fig. 4.81).

In medical industry, the following cycle is often used:

- monitoring
- analysing
- action.

which includes a wider cycle with the double activity at the human–machine interface (HMI):

- monitoring of the human (patient),
- acquisition of relevant data,
- data processing,
- condition assessment,

- diagnostic determination,
- determination of health prognostic,
- degradation,
- decision support,
- monitoring of the effect at the human (patient).

This follows a very similar sequence as we consider it for technical systems.

Of one of the most important tasks with highly available systems, the evaluation to the system is the information in which resources or elements for a "sure function" are still available. The "sure state" is not with complicated systems mostly any more the weak state; hence, this information will correspond to the requirements of the highest security integrity step of the system.

4.7.6 Safety and Security Error Propagation

Safety and Reliability follow similar principles for the failure analysis. Especially, dependent failure and their analysis show that the typical sequences of fault, error, failure or failure cause, failure mode and failure effect are not always applicable. Similar challenges are affecting security analysis. Measures to control the different security threads like integrity, confidentiality and availability show different relations to their possible effects and effectiveness. Hackers use the intended functionality in order to introduce Trojans or masqueraded messages. In dynamic systems manipulation of state-machines, their entry conditions are well-known threats or possible failure scenarios in safety and security. In closed loop control systems, the manipulation of the feedback could lead also to significant dangerous behaviour at the actuator.

Deep Medhi provided a keynote where he presents a common "dependability and security model".

The possible threats (mainly security) had been defined on a level of faults, so that further propagations lead to errors, failure and accidents. Unauthorised access to data and unavailability had to be considered also as an accident in this case.

Safety is defined as "attributes" as well as other quality characteristics like security, availability, confidentiality, integrity, privacy, authenticity, performance, reliability, robustness, resilience, survivability and maintenance. Similar considerations are in railway standards with "RAMS" approach (reliability, availability, maintainability, safety), the typical error propagation in a FMEA follows a similar approach

- failure cause
- failure
- failure effect.

Typical measures follow proposed measure in the FMEA standards like

- fault/intrusion prevention,
- fault/intrusion tolerance,

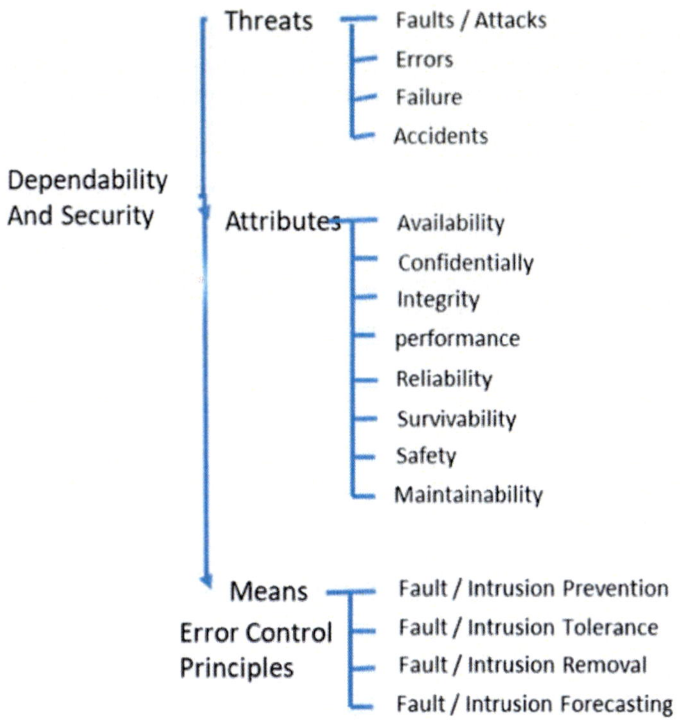

Fig. 4.82 Dependability and security models. *Source* Deep Medhi, Proceedings of 7th International Workshop on the Design of Reliable Communication Networks (DRCN 2009), Washington, DC, October 2009

- fault/intrusion removal,
- fault/intrusion forecasting (Fig. 4.82).

This shows that similar analysis approaches for safety and security could be applied, but that there no one2one relation to be expected.

Security analyses are as similar to safety analysis as reliability analysis. By considering, such well-tried analysis principles, all kind of threats, critical impacts or any other unwanted event to systems or products could be examined.

4.8 Development of Software-Intensive Tools

The typical waterfall process was always the basic process for development of tools. Tools are used along the entire development cycle for safety-related product development. Tools are used during development from the first idea until the decommissioning of the products, even due to liability reasons, the entire product lifecycle history needs to be recorded in an adequate data storage system. Tools are used during the development for the following activities:

- architecture and design tools, computer-aided engineering (CAE),
- requirement engineering/management,
- reliability or safety analysis,
- verification and test during development and production,
- modelling, simulation and validation,
- tools for document handling, archiving and management,
- change, configuration, version, problem resolution management, etc.,
- software factory and software test.

The tools show a significant source for failure and for hackers' attacks. Hackers could implement Trojans into the product and the product documentations give perfect information, how to attack a product. Nevertheless, systematic documentation is needed during entire product lifecycle, up to maintenance and adequate disposal or decommissioning of the product.

ISO 26262 addresses the failure impact from but not the impact due to wrong tool handling. The aviation industry addresses the tool impact as a major activity in the development cycle. The reworked avionics software safety standard DO-178C, along with its supplemental software tool qualification considerations (DO-330), has clarified and expanded the tool qualification guidance provided in DO-178B.

DO-330 addresses these issues various use cases for the different tool application. Specific guidance for previously qualified tools allows reuse of the qualification artefacts as long as nothing has changed that would affect qualification. It considers three scenarios as follows:

- Reuse of a previously qualified tool without change,
- An example is when a tool is used for related projects or on multiple phases of an existing project. The developer needs to identify the approach and rationale in the plans.
- Changes to the tool operational environment,
- The developer needs to update one or more of the plans, but the bulk of the original qualification artefacts may be reused as it is. Only the updated artefacts related to the operational environment need to be reviewed by the certification authority.
- Changes to the tool itself,
- A change impact analysis has to be provided, but tool requalification still has a reduced cost, essentially only requiring activities associated with aspects that have changed or are affected by the change. The key is to be able to exactly determine and specify what has changed and what these changes impact, or perhaps more importantly, what they do not influence.

The new standards show a significant change in the tool development cycle. The tool could not be considered as a stand-alone development step, and it needs to be integrated into the target product development cycle. The tool application context changes significantly the tool safety requirements in terms of

- tool handling,
- error handling,
- usability and
- integration.

Verification and validation of the tool need the same context as the product under development, which view on the entire lifecycle of the system-of-interest (SOI) (Fig. 4.83).

Many tools, especially the tools for modelling, simulation and validation, need to follow a dedicated process. The tool needs to guide and fit to the typical development process. The dissertation from Nicolas Denz provides an interesting waterfall representation, where various aspects are shown. The picture shows various maturity activities from the system down to the "simulation results" and verification and validation steps from bottom-up.

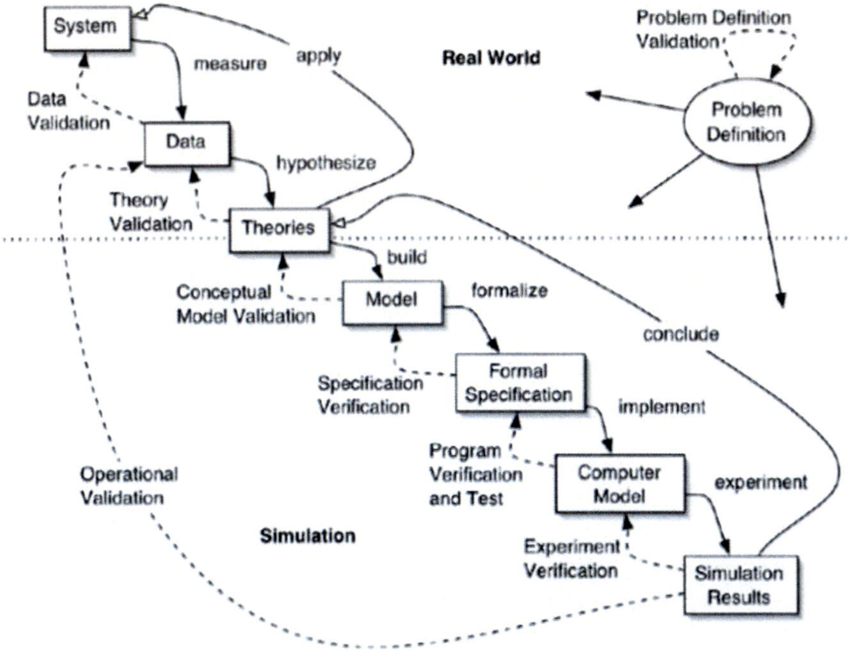

Fig. 4.83 Refined validation process. *Source* Process-Oriented Analysis and Validation of Multi-Agent-Based Simulations, Nicolas Denz, Dissertation, 2013

4.9 Verification and Validation

During the development of the product, ISO 26262 asks frequently for verifications, most likely always, if a development activity relies on the input of a former development step. In the descending branch of the V-model, verifications are always required at interfaces of horizontal abstractions.

In this context, the verification is seen as the completion of a higher-level activity, and the lower-level activities usually begin with a requirement analysis. ISO 26262 considers also tests especially in the lower horizontal abstraction levels, particularly during component design as verifications. However, methodical, the method for the correct derivation from a higher level would need to be a validation. What is important though is that this verification regarding correctness, completeness and consistency is based on the same consistent verification method. Most likely allocations are done before the verification, which means that requirements are allocated to the underlying levels elements. In this case, the relationship between those two levels needs to be analysed. The verification does not initiate process iteration only if the results of verifications are flawlessly positive. Depending on how deviations are assessed during a verification and what measures are initiated for the repetition of the verification, we need to go back to the corresponding previous activity. These could be the requirements, architecture, design or also the test case specifications within a horizontal level as well as a jump back into another horizontal level (e.g. from a component into a system, or even on vehicle level, so that safety goals could be affected by changes). The first activity during verification of requirements should be a requirement analysis.

Therefore, the question is:

Are the requirements of the lower level of abstraction derived from the requirements of the upper level or from constraints, architecture or design of the higher level of abstraction including their context?

Verification of requirements could not sufficiently address the needs of verification. Further semi-formal or even formal descriptions are needed to answer the typical verification question about,

- correctness
- completeness and
- consistency.

by assuring the traceability for later assurance or safety cases.

However, those should be tested or verified before they are passed on to a further user of this information. This shows that there are different ways to develop work results. Basically, we distinguish between a requirement specification and a design specification. However, there are different manifestations and definitions as to how both specification types can be structured. At this point, the requirement specifications should provide the general conditions, which are necessary as foundation for the design. The design specification describes the implemented characteristics, which can be measured by the product. A requirement specification defines “How it should be" and the design specification defines the "how it is designed". We now also reach the performance limits of a process models.

Does the verification really only happen during the development of the requirements? Is the development of requirements completed before the realisation? Obviously not! When the result is validated and all requirements are correctly implemented at the product, there are always new aspects that can occur in the usage phase, which have not been sufficiently considered.

In addition, in this case, constant iteration loops occur and because of today's short innovation cycle, products are often only mature after years of their usage, and each change becomes a risk for other characteristics. This of course is not acceptable when it comes to the safety characteristics of a product. It is true that an inexperienced development team often does not know the influence factors, but an experienced team can also make incorrect assumption. Unfortunately, there are certain amounts of risk even in the approach itself. If requirements are systematically developed and properly derived according the given processes, the known influence factors will also be incorporated. If experienced people perform these analyses, some aspects will also be included in the analysis, which go beyond the requirements and the experience of the designer. At the verification, certain levels of experience can also be incorporated through the test planer. Also through the individual analysis or verification methods systematically, complementary influence factors are considered. However, it is hard to say or maybe even impossible to assume that all influence factors will be considered or even all application scenarios and relevant conditions. If we now have a test case for each requirement (according to the process model derived, e.g. from SPICE or Automotive SPICE, CMMi, etc.), which shows that the requirement is implemented correctly, there will certainly be doubts regarding the significance of the tests. How many tests are necessary will depend on a variety of factors, and even on the diligence, with which the requirement analysis has been performed at the beginning of the processing of the abstraction level. Requirement and design specifications should formally be stored in a way that there is actually only one parameter in the requirement but through the further derivation of the design information, essentially more parameters are developed. Otherwise, the lower levels cannot be sufficiently supplied with information. The most concise example occurs at the hardware–software interface. The design of the microcontroller provides the essential software requirements, not the requirements from the upper levels. Therefore, ISO 26262 requires the verification of all SW requirements but it is not possible to directly derive them from the system requirements, which are allocated to the software. All basic structures of the microcontroller have to be included in the requirements of the hardware–software interface (HSI), which is often unable to fulfil the relevant system requirements for the software components. However, this is a very concise example, but definitely not an unusual exception.

Besides the safety analyses and tests, more and more verifications are necessary for the determinations of the safety maturity for the product under development. At each organisational interfaces and all horizontal interfaces as well as the in between elements, all characteristics should be verified at the end of the development. In general, verification could show the fulfilment of requirements, from a methodology point of view you only get answer if the targets or goals fulfilled could be only shown by validation. The activity to validate the correctness of requirements, by evaluating higher-level requirements or constraints to their correct derivation to lower-level requirements called ISO 26262 “verification”.

4.9.1 *Verification of Diverse Objectives Such as Safety and Security*

Design, architecture of a product or even the requirements should be consistent for any characteristics or features during different steps of the development cycle. Completeness and correctness are in general basics to assess consistency. If the design, the architecture or requirements are incorrect or incomplete, any corrected or added characteristics could violate the consistency requirements.

Questionable is, if completeness or correctness verification could be done in one step for diverse objective like safety and security. Consequently, both have to be complete and correct before their common consistency could be verified. If safety and security mechanism block each other, the implementations of the mechanism makes neither safe nor secure maybe with some compromises but that violates very often objectives like performance, availability or other characteristics.

The following table could be an example for a joined concept development process (Fig. 4.84).

Process step	Activities for Safety	Activities for Security	Common Activity
tool qualification	according to ISO 26262	analysis security impacts	agreement of measures
Security in development environment	respect and planning of safety activities based on the security constraints for development and production	development of a security concept for development and production	check the adequacy of the measures
Item Definition	analysis of system boundaries	authorization concept	
Hazard Analysis & Risk Assessment	risk identification and ASIL	threats identification	
Functional Safety Concept	safety measures and allocation	measures definition	
Verification of FSC	verify of measures	verify of measures	evaluating the ability of coexistence of both sets of measures
System specification / Architecture	define requirements, architecture, behaviour and internal interfaces	requirements, implementation concept	allocating security mechanisms on product architecture
analysis	inductive / deductive	effectiveness analysis	Analysis - error behaviour of all functions and mechanisms
System - Design	define system, parameters, requirements for components	derivation and allocation of security mechanisms	Verify the coexistence capability of both set of measures
Verification of components requirements	verify feasibility, completeness, correctness, consistency of standards for component	verify feasibility, completeness, correctness, consistency of requirements for components	analyse consistency of both sets of requirements.
Component specification / Architecture	define requirements, architecture, behaviour and internal interfaces	requirements, implementation concept	allocating of security mechanisms to component architecture
Analysis of component	inductive / deductive	effectiveness analysis	analysis - error behaviour of all functions and mechanisms
Components - Design	component design, parameters, requirements for components	derivation and allocation of security mechanisms	verify the coexistence capability of both set of measures
Verification of the components prior to implementation / deployment / realisation	verify feasibility, completeness, correctness, consistency of requirements before implementation	verify feasibility, completeness, correctness, consistency of requirements for implementation	analyse consistency of both sets of requirements
implementation / deployment / realisation	implement as specified	implement as specified	verify the specification-compliant implementation
Integration / Test	tests according to specification	tests according to specification	tests of ability of coexistence and effectiveness of both sets of mechanisms.

Fig. 4.84 Verification of safety and security

The common activities are in the focus of verification, so that the activities and results are

- consistent
- correct and
- complete.

The usages of common resources as well as the coexistence of

- intended functionality
- security and
- safety measures.

This coexistence is a major planning aspect of the integration activities and the related test plan.

4.9.2 Validation

Validation means in other standards a much wider view as it is considered only as "safety validation" in ISO 26262. Validation needs to be considered different in the context of safety validation or any other validation activity. Furthermore, safety validation in ISO 26262 is a specific activity and the discussed topic for verification to see validation as a methodology. Validation in general is often described as the confirmation of targets. A certain general validity is required, and man could say: "Validation is the proof that targets are reached reproducibly". ISO 26262 sees safety validation on vehicle level, but safety validation is not considered as a single activity. Safety validation should be accompanied the whole safety process until the end of the development phase. An interesting aspect from a functional safety concept of ISO 26262, part3 is the development of "validation criterion". Other standards and particularly the homologation regulation consider validation in a wider view, which questions all development of requirements during the entire development process. Validation becomes an iterative activity during any development step; such steps are an inherent part of a spiral process or automotive typical APQP activities. However, in contrast to verification, validation still has kind of a blurry character since targets are often not as precisely formulated as requirements, which are generally verified according to ISO 26262. In the automobile industry, we can often find the following definition:

"The customer requirements are validated, but the requirements for example in the product and technical specifications are often verified."

The term "validation" could be used very differently in various industries and the different application following aspects are associated with the term validation:

- Latin: validus; strong, effective, healthy.
- Validity: weight of a statement, investigation, theory or premise.
- Validation is a method of communication with dementia patients.

- Validation: proof that a process, a system and/or the production of an active substance reproducibly fulfils the requirements in practical use.
- Validation for a semiconductor says that it can be produced according to the specifications.
- Validation: external proof of large-scale projects and theirs sustainability reports.
- Validation in computer science is the proof that a system meets the requirements in practice.
- Validator: A method or program, which should confirm the verification with respect to a standard.
- Validation or testing of the validity of statistical values or their plausibility.
- Method validation proves that an analytical method is suitable for its purpose.
- Validation often describes a statistical proof.
- Validation of educational attainment.
- Model validation should show that the system developed through the implementation of a model reflects reality sufficiently precise.

4.9.3 Validation According VMAD

Validation method for automated driving (VMAD) is a working group in the UN ECE WP.29.

Based on the new regulation for the "automated lane-keeping system—ALKS", they follow the new approach for the homologation.

They define in a paper with their targets:

> This paper was drafted by the EC with subgroup 2a to describe the concept of the "audit/virtual testing/in use data reporting" pillar in accordance with the layout proposed by VMAD at the last meeting (Why, What and How). It also includes as an annex the main building blocks identified at this stage for possible requirements under this pillar. Another paper proposes a draft Annex to the new UN Regulation on automated lane-keeping systems (ALKS).
>
> **–Why–**
>
> With automated driving, the amount of possible (traffic) scenarios to be compared to current driver assistant functions such as anti-lock braking system (ABS) or advanced emergency braking systems (AEBS). The complexity of the situations to be managed by automated driving systems will also increase, as the systems will take care of the entire dynamic driving tasks, in particular environment monitoring and interaction with other road users, which are managed by the driver in case of driver assistant systems (Fig. 4.85).
>
> Consequently, the "classical approach" based mostly on a set of pre-defined physical tests is no longer sufficient. It will be impossible to demonstrate the performance of the system solely on the basis of a limited "representative" set of tests on a test track.
>
> **–What–**
>
> As a result, it proposed that manufacturers will be required, in addition to pre-defined physical tests, to demonstrate that an acceptable thorough consideration of functional safety (safety issues due to possible system failures) and operational safety (safety issues due

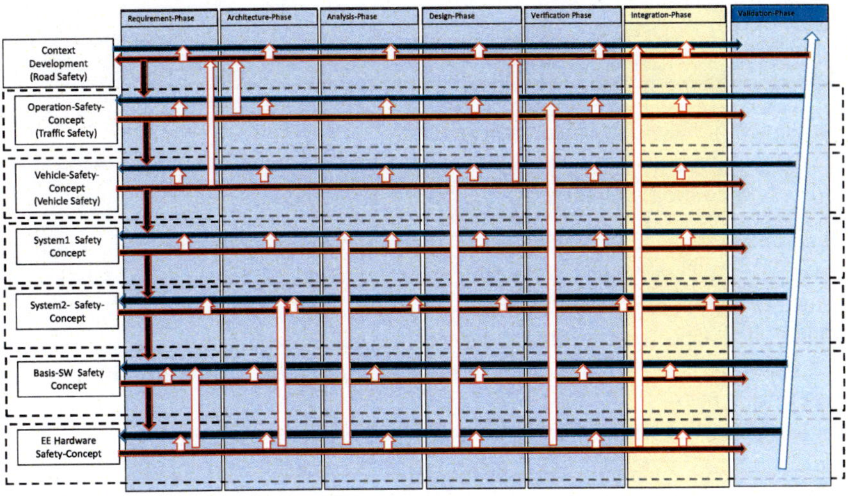

Fig. 4.85 Validation phase

to other causes than failure, e.g. perception problems) have been implemented during the design and development processes of the automated driving system and will continue to be done throughout the vehicle lifecycle (development phase, production, postproduction phase, operation, decommissioning).

Industry standards already exist for functional safety (ISO 26262) or are under development for operational safety (ISO PAS ISO/PAS 21448 and UL 4600). Similar process-based standards are being developed for cybersecurity (ISO/SAE DIS 21434).

–How–

The manufacturer will be required to demonstrate through documentation, physical tests (on track and on road) and/or virtual testing that:

1. Hazard and risks relevant for the system have been identified, and a consistent safety-by-design concept has been put in place to mitigate these risks.
2. A robust validation process of the design (virtual testing, track tests, real driving tests) confirmed that the system meets the legal performance requirements, traffic rules and is free of unreasonable safety risks to the driver, passengers and other road users before the vehicle is placed on the road.
3. Robust processes/mechanisms/strategies (safety management system) are in place to ensure continuous safety through the vehicle lifecycle including development phase, production, but also operation on the road and decommissioning (e.g. the manufacturer shall have processes to monitor incident/accidents with their vehicles and to react appropriately in case of incident).

On the basis of the evidence provided by the manufacturer, authorities will be able to audit/assess whether the design, validation and the safety culture of the manufacturer are robust enough with regard to functional and operational safety. Authorities will also be able to check through targeted physical tests (on track and on road) and virtual testing whether the safety by design concept is implemented on the vehicle. ff.

Source VMAD-05-05, UN ECE link: https://wiki.unece.org/pages/viewpage.action?pageId=60361611

The working group VMAD builds the umbrella for various activities such as

- validation of hazard and risk management,
- design validation,
- process validation, including audits, assessments, etc.

It addresses also the virtual aspects of "virtual development" and "virtual testing or verification". They are also addressing the given standards for safety, which should provide the methodology framework.

One major aspect is the timing aspects for an entire function and the related signal chain. A paper from the Japanese delegation gives a good insight into the targets (Fig. 4.86).

In low-speed ALKS scenario, the avoidance capability required for the driver model is braking control only. This driver model is separated into the following three segments (Fig. 4.87):

"Risk perceive situation", "delay in time" and "deceleration degree and Max. G-force"

They consider a significant time difference between "normal braking and emergency braking" (Fig. 4.88).

"Delay in time" occurs in each driving process of human: "perception-decision-reaction".

The paper is not to argue that automated driving is faster and safer than human drivers. It should demonstrate, how to develop a driving scenario and how to make validation criterion for the assessment available for virtual development and tests.

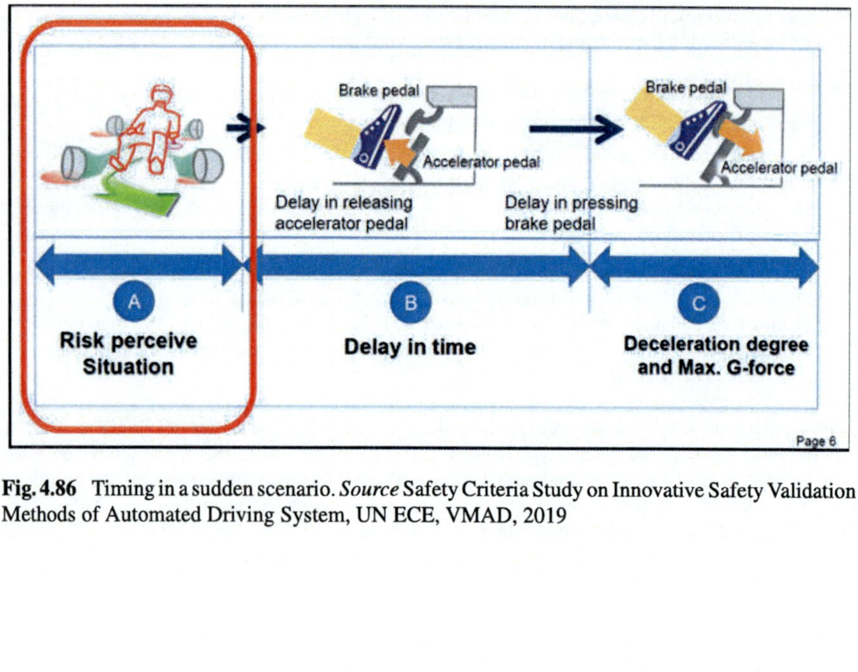

Fig. 4.86 Timing in a sudden scenario. *Source* Safety Criteria Study on Innovative Safety Validation Methods of Automated Driving System, UN ECE, VMAD, 2019

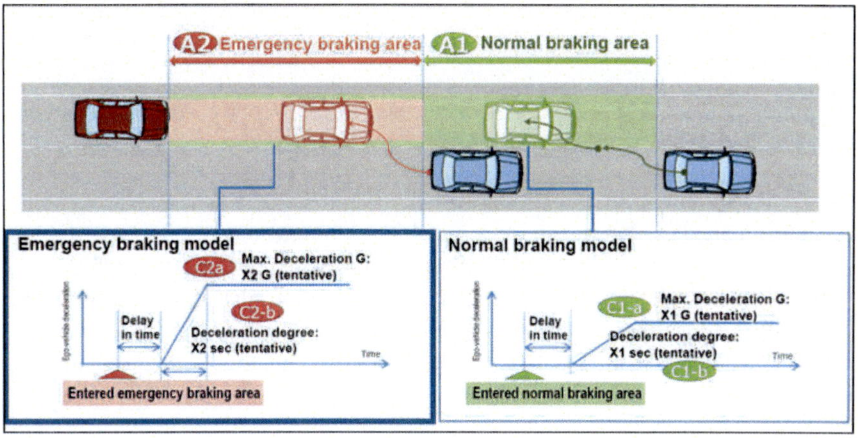

Fig. 4.87 Timing breakdown of scenario. *Source* Safety Criteria Study on Innovative Safety Validation Methods of Automated Driving System, UN ECE, VMAD, 2019

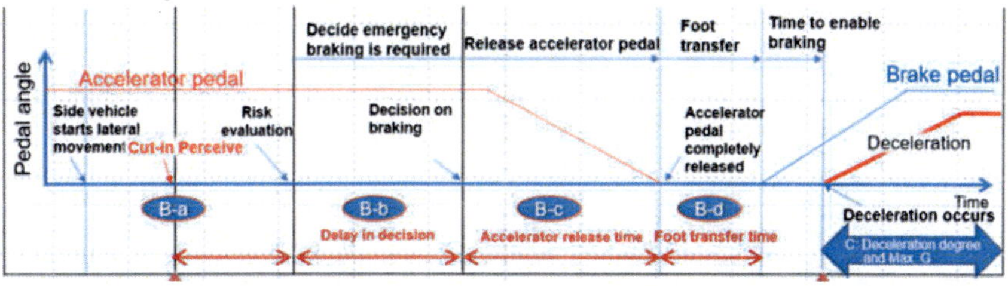

Fig. 4.88 Timing chain and related action of a human driver. *Source* Safety Criteria Study on Innovative Safety Validation Methods of Automated Driving System, UN ECE, VMAD, 2019

4.9.4 Safety Validation in ISO 26262

ISO 26262 considers examination for correctness, consistency of requirements as verification activities. For example, examining whether a goal has been achieved, this includes the validation of the safety goals in ISO 26262. A test, for example, if the SW fits into the microcontroller would be a verification after integration. Methodically, you checked but against the requirements of the higher level through the intermediate step, whether they were consistent, correct and completely achieved. A second step would be whether the goal of a safe integration of hardware and software has been achieved. Thus, it is also in the second interpretation, the request itself is really what is required, according to ISO 26262 in deriving the functional safety concept for technical safety concept through the component requirements. Here ISO 26262 calls the activity also for the verification of requirements.

The major goals for validation based on underlying verifications such as:

- The goals are achieved.
- The demands were right.
- The requirements were implemented correctly and adequately.

If this leads to a consistency and completeness of reasoning, creating a sufficient transparency should be no problem. If the activities and the methods used are extensively documented, so that an auditor could confirm the adequateness of the activities and an assessor should be able to confirm the achievement of functional safety.

This means that all verifications are receipt confirmations that all relevant requirements and specifications are correctly implemented for a certain system on vehicle level for a specific vehicle or vehicle class. The safety validation should provide the final argument for functional safety on vehicle level.

4.9.5 Validation Phase

Validation based on many validation criteria and based on that it is always the question how many intermediate "goals" are necessary (Fig. 4.89).

The validation phase is the last column after the integration phase. The arrows show that validation provides feedback for all the previous phases. Therefore, the

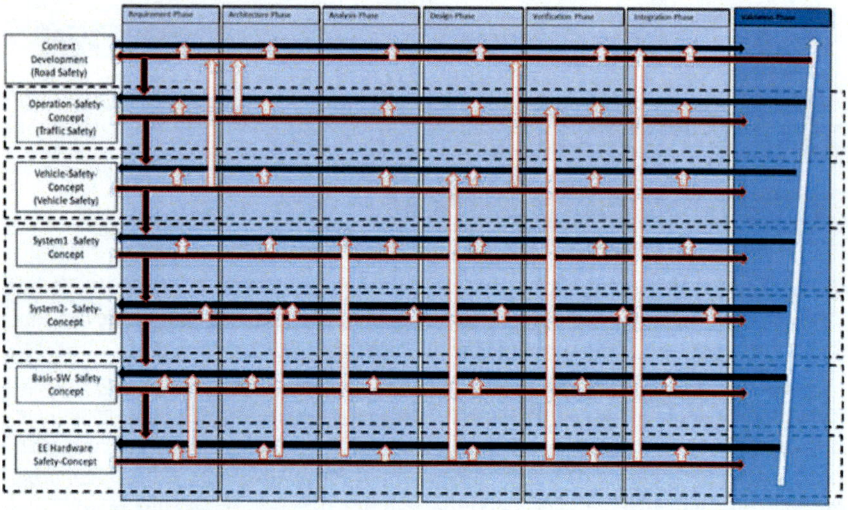

Fig. 4.89 Validation phase

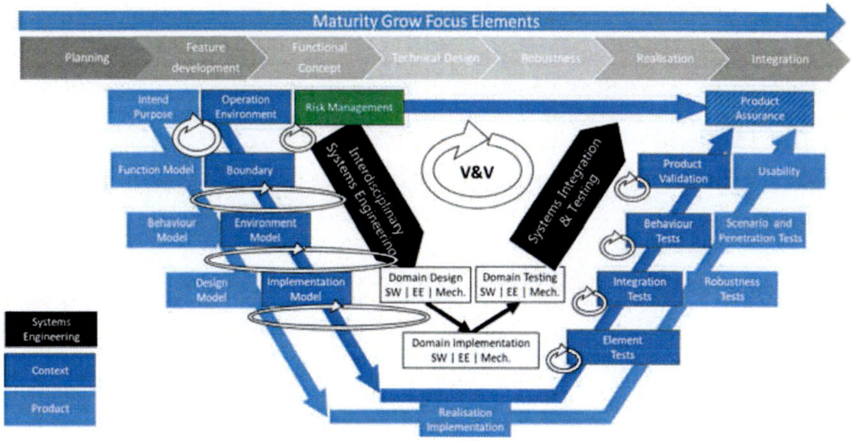

Fig. 4.90 Development process including V&V and risk management

verification and validation activities need to be iteratively integrated into the product development process (Fig. 4.90).

In a "real-world" project, during development, the system and the context need to be derived from top-down to the realisation and the implementation. Verification and validation measures are required in any new level of abstraction in order to assure traceable results. According the V-model approach, the integration and parallel verification measures follow as a bottom-up approach. During the entire development process, the risk management evaluates the current degree of risk reduction. The final "overall validation" and the risk or safety assessment confirm the achievement of the "goals".

Chapter 5
Organisational Viewpoint

Accidents are a class of adverse events. They have many causes. What may appear to be bad luck for the individual (being in the wrong place at the wrong time)? Analysis often shows a chain of events, failures and errors that lead almost inevitably to the specific adverse event.

These causes can be classified as follows:

- immediate causes: the agent of injury or ill health (the blade, the substance, the dust, etc.),
- underlying causes: unsafe acts and unsafe conditions (the guard removed, the ventilation switched off, etc.),
- root causes: the failure from which all other failings grow, often remote in time and space from the adverse event (e.g. failure to identify training needs and assess competence, low priority given to risk assessment, etc.).

To prevent adverse events, you need to provide effective risk control measures, which address the immediate, underlying and root causes.

The role of human impact in major disasters has been widely acknowledged, with studies concluding, that human error is responsible for a high average of accidents. Accident analyses of industrial disasters like Seveso, Three Mile Island, Bhopal and Chernobyl confirm the view. Two types of human error are said to cause systems disasters:

- active errors
- latent errors.

While the effects of active errors are almost immediate, the consequences of latent errors may lie dormant in the system for many years before they combine with active failures and local triggers to create an accident opportunity.

Active errors are more likely to be caused by people directly involved in the process. Latent errors tend to be made by designers, managers and other personnel who are sometimes not even aware of their errors and their consequences.

H.-L. Ross, *Safety for Future Transport and Mobility*,
https://doi.org/10.1007/978-3-030-54883-4_5

Causes of latent errors during operation are often:

- time pressure,
- understaffing,
- inadequate equipment or
- fatigue of people (fatigue of material is another class).

Other latent errors had been their causes,

- during development,
- production
- or instruction.

Fatigue of material is development failure, and causes could be as follows:

- Wrong material had been selected.
- The user profile and the stress profile are not aligned.
- Unexpected environmental impacts increase ageing of material.
- Inadequate production or processing of material,
- damaging of goods during maintenance,
- etc.

This is one of the reasons why today's development processes for safety relevant equipment require a lifecycle approach.

5.1 Accident Research

We mainly learnt to build safe vehicles and define the necessary road traffic regulations by experiences.

Unfortunately, for experiences, similar classifiers are relevant than for the risk:

- Occurring damages, accidents and noticed incidents are the basis for the severity scaling. A number of people died due to accidents are strong arguments for improvement.
- Exposure is only an argument unless nothing happened. If there are no injuries and no incidents, there is no claim for compensating damages. Noticeable incidents and how frequent such incidents happen are a matter of field monitoring and how preventive the field monitoring is implemented. The factor likelihood strongly depends on the quality of field monitoring.
- In countries with a focus on common law, like the USA, damages in road traffic are often controlled on local legislation (city, county or state law) and even private (injured party and party causing the damage) level. Those countries provide recommendations and not so detailed regulation like the German road traffic act and the derived legislation. Preventive legislation and recommendations of consumer protection organisations are fluent and sometimes redundant and could

not be traced to their origin. Many safety measures are just best practices and their "why" and "how good for what" are not anymore present.

These have strong impact to the organisation of an engineering team. The vehicle manufacturer is responsible for the entire vehicle in the road traffic context and not only obliged to assure the roadworthiness of the vehicle. Primarily, the vehicle manufacturer is also responsible for the integration of various systems or items into the vehicle. Usually, the entire breakdown of responsibility of risk is due to vehicle design, instruction and production based on contracts between the vehicle manufacturer and their supplier and subcontractors.

Very early accident research became an important science for road traffic (Fig. 5.1).

In the early thirties of last millennium, a significant increase of accidents had been already noticed. Rescuing in road traffic became an important measure to reduce the consequences of accidents and saving life of injured people. Zero accidents were at that time already the target, but the society learnt to accept the accidents as an unpleasant fact (Fig. 5.2).

Frank E. Bird and others had expanded the pyramid from Heinrich. Especially, the extension from Bird addressed the unsafe acts and near misses. Incidents and accident prevention become with that measure more in the focus, because unsafe acts and nearly accidents became also scope of research.

The figures had been often discussed in the pyramid. Is the number of accidents or number of dead people a fixed indication of the layer below? A lot of researches have shown that we could play with the definition of the horizontal interfaces, like what is sever and what is minor to make the model representative, but the main stream always confirmed.

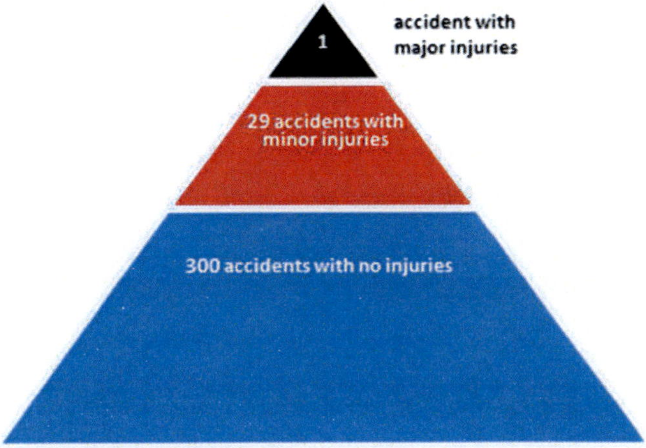

Fig. 5.1 Relationship as first proposed in 1931 by Herbert William Heinrich in his "Industrial accident prevention: a scientific approach"

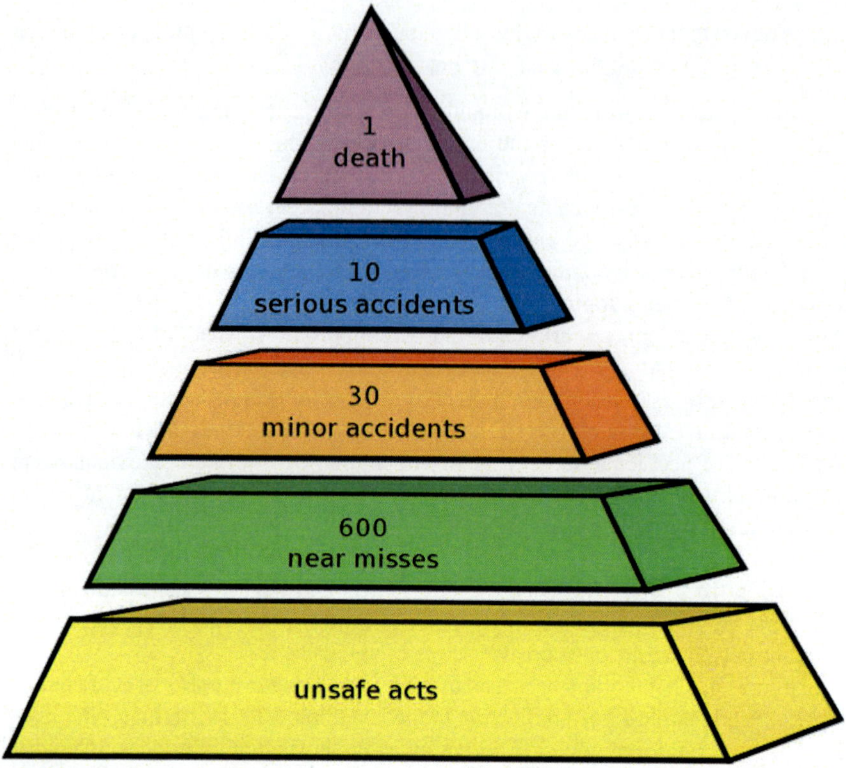

Fig. 5.2 Further developed theory by Frank E. Bird in 1966, based on the analysis of 1.7 million accident reports from almost 300 companies

Also the theory of "Iceberg" beginning from the accidents research of Titanic to today's accidents shows similar factors as visible or invisible below the water level (Fig. 5.3).

Many issues are addressed in various known Iceberg theories, but they classify always in the same groups, as we know them from the technical risk management.

- product failure and their integration,
- handling failure,
- misuse,
- systematic failure during development,
- wrong application of regulations,
- Over ambitioned traders or producer and gaps in the delivery chain. (Fig. 5.4)

There are various interpretations of the "Swiss Cheese Model" from Raeson. In the accident research, there are various interpretations, which address the causes of accidents. An interesting example considers the following slices:

- technology,
- organisation,

- communication and information,
- qualification, experiences and knowledge,
- attitude and behaviour.

The idea of the "Swiss Chess Model" is that any slice means a barrier for causes of dangers, unless the holes in the slices do not allow a complete pass through of a cause of danger.

The technology or the technical product is the basic root cause and as closer; the risk of danger could be controlled or even avoided as better for the product liability. The further causes of risk should provide barriers but those are also own sources of

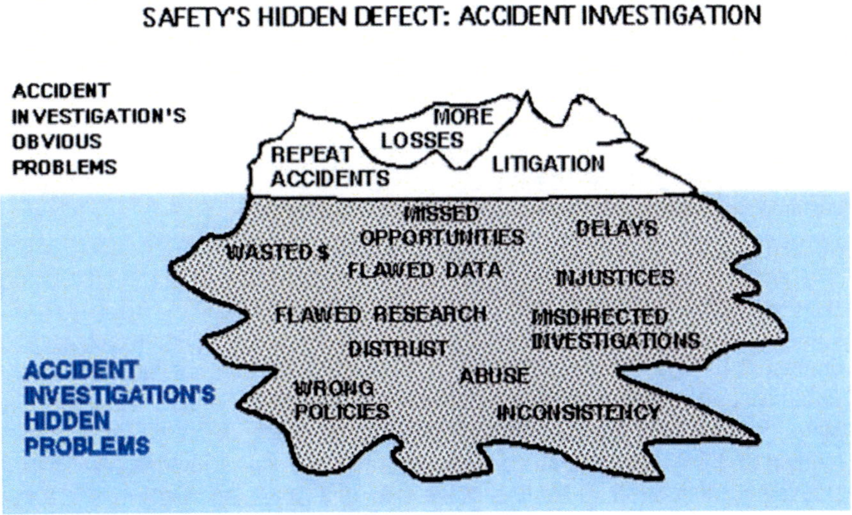

Fig. 5.3 Typical iceberg in accident research

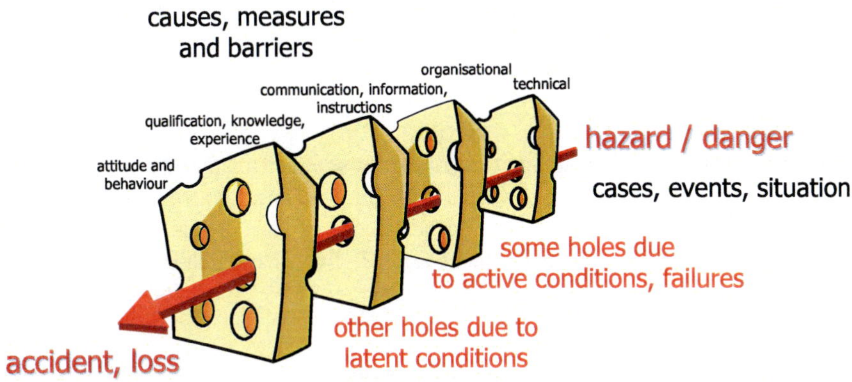

Fig. 5.4 Swiss cheese model similar to Raeson

risk in case they are not adequately developed for the considered products, technology or use cases.

The main characterises following approaches:

- analysing of causes, procedures and consequences of accidents,
- retrospectively, based on historical data,
- providing evidences for future regulations, approaches, methodology and
- measures and their effectiveness scaling.

Potential stakeholders are authorities (e.g. regulation, legislation), automotive industry (OEMs and suppliers), infrastructure planners, any road users or the society (e.g. education, behaviour), rescue forces (medicine, trauma surgery) and police (enforcement). The common goal of accident research could be defined as follows:

- safe transportation for all kind of road users.

Safe could be defined with attributes like

- tolerable risk (robust limits),
- clean (e.g. environmental friendly),
- sustainable (stabile and continuously crowing of safety level).

Typical accident research databases are as follows:

- International Road Traffic and Accident Database (IRTAD). IRTAD is a data collection maintained by the Organisation for Economic Cooperation and Development (OECD) and the International Transport Forum (ITF) in Paris, covering safety data in countries within and outside of Europe.
- State Traffic Safety Information (STSI). STSI reports contain additional from Federal Highway Administration: Highway Statistics Series, United States Census Bureau: population data. National Center for Statistics and Analysis (NCSA) manages STSI, they are an office of the National Highway Traffic Safety Administration and are responsible for providing a wide range of analytical and statistical support to NHTSA and the highway safety community at large.
- Institute for Traffic Accident Research and Data Analysis, Japan (ITARDA) was founded in 1992 for this purpose, being promoted by the National Police Agency, the Ministry of Transport, the Ministry of Construction, Japan Automobile Manufacturers Association Inc. and the General Insurance Association of Japan and many other bodies.
- German In-Depth Accident Study (GIDAS) is accident data collection project in Germany. GIDAS project has been collecting around 2000 traffic accidents involving personal injury per year in the Hannover and Dresden metropolitan areas. The project is supported by the Federal Highway Research Institute (BASt) and the Forschungsvereinigung Automobiltechnik e.V. (Automotive Research Association, FAT).

Typically, the databases provide statistics about

- road user deaths by age, road network area or road use,
- traffic accidents with personal injuries by road network area.

Accident research and access to the relevant databases are essential for the development of safety-related products. They provide.

- orientation for state of the art,
- support for risk evaluation,
- guiding and benchmarks for risk assessment,
- reference for effectiveness and effectivity of measures, etc.

Managing risk and experiences how to manage risk is a core competence for company, which wants proving services or safety-related products.

5.2 Quality Management

Quality management and an adequate quality management system ensure that an organisation, product or service is consistent. Without that consistency, a company is no reliable partner for service user or product customers. Quality management focuses not only on product and service quality, but also on the means to achieve it. Quality management, therefore, uses quality assurance and control of processes as well as products to achieve continuous improvement in consistent quality. What a customer wants and is willing to pay for it determines quality. It is written or unwritten commitment to a known or unknown consumer in the market. Thus, quality is often described as fitness for intended use or, in other words, how well the product performs its intended function. The quality management and the implemented quality management system provide process, methods, qualified tools and adequate educated staff. Without those implicit knowledge into the organisation in development and production products could not maintained, no continuous improvement is achieved and sustainable knowledge is determined on low level. A learning organisation entering new markets and new technology approaches starts always on a basic level. All progress of products or improvements for better or new products depend on personal knowledge and motivation of employees. If key people leave the company, it always means lost of knowledge for the company. A best practice for a high quality in production process requires maintaining the quality of the product independent from any human influence. In production, human impact is always a difficult calculating risk.

Quality management systems provide not only the typical activities like

- quality planning,
- quality assurance,
- quality control and
- quality improvement.

It also provides the management, development process and any kind of supporting process and related basis for implicit methods, which are the basis of the know-how of a company.

Risk management, business reporting, escalation principles, release management and field monitoring are important basis for a trustful company.

In safety standards, a safety culture and an operational quality management system provide the basis for an organisation that is capable of developing safety-related products.

The basic processes in a process landscape show the identity of a company. Process landscapes are coherent sets of interconnected processes within an enterprise, business or organisation. They show the structure, grouping, modularity, functionality and technology of chain processes, business processes and working processes.

In the automotive industry, an addition to ISO 9001 has been established. ISO TS 16949 had been continuously improved and is now published as an IATF standard. The IATF 16949 standard combines existing general requirements for quality management systems in the (mostly North American and European) automotive industry. They were jointly developed by the IATF members and published as enhancement of EN ISO 9001. More than 100 existing car manufacturers agree with these harmonised requirements of the nine IATF members (BMW, Chrysler, Daimler, Fiat, Ford, General Motors, PSA, Renault, VW), however the big Asian car manufacturers have differentiated own requirements for the quality management systems of their group and their suppliers. It describes additions especially to the product development and production, which developed into standards in this industry. Manufacturers from Asia still refer to different standards, based on historical reasons. Especially in Japan, quality requirements focus more on the ideals of the six sigma philosophy (e.g. DFSS, design for six sigma).

The manufacturing process design output shall be expressed in terms that can be verified against manufacturing process design input requirements and validated. The manufacturing process design output shall include according to IATF 16949 the following:

- specifications and drawings,
- manufacturing process flow chart/layout,
- manufacturing process FMEAs,
- control plan,
- work instructions,
- process approval acceptance criteria,
- data for quality, reliability, maintainability and measurability,
- results of error-proofing activities, as appropriate and
- methods of rapid detection and feedback of product/manufacturing process nonconformities.

The standard requires a well-organised, well-established and well-applied quality management system. Such requirements are in any safety standard the basis to develop safety-related products. The American manufacturers particularly Ford, GM or Chrysler met at AIAG to define joint requirements for quality management. In Germany, similar standards and requirements were developed under the umbrella of VDA. The aim was to define processes for the development as well as planned advanced product quality improvement measures, advanced product quality planning (APQP according to AIAG). VDA and AIAG published a series of documents, which are considered to be the foundation for VDA or AIAG members. Those various volumes of these documents are often mandatorily referenced in the contract documents for supplier. Unfortunately, these documents are not highly consistent. For example, both organisations describe different FMEA methods (or several FMEA methods); in 2019, a common FMEA approach has been published. In addition, these organisations also developed milestones or maturity level concepts, which were primarily used for the synchronisation of automotive manufacturers and supplier (Fig. 5.5).

AIAG defined APQP with five "milestones" as follows:

- The first phase "concept, initiation, approval" is a mere planning phase.
- In the second phase, before the program approval, the planning as well as the product and process development should have a certain maturity. The feasibility of the product is then verified as part of the program approval.
- The third phase focuses on the development of the first prototypes, the verification (often prototype tests) and the product and production process validation. At this point, the product design should be almost finalised.
- In the fourth phase, the first series development (close-to-series production, pilot) products are produced. Those products should already be produced with the series production tools.
- The product launch initiates the series production. This requires the development of supply chains, and the production needs to be able to guarantee a sufficient quantity and quality.

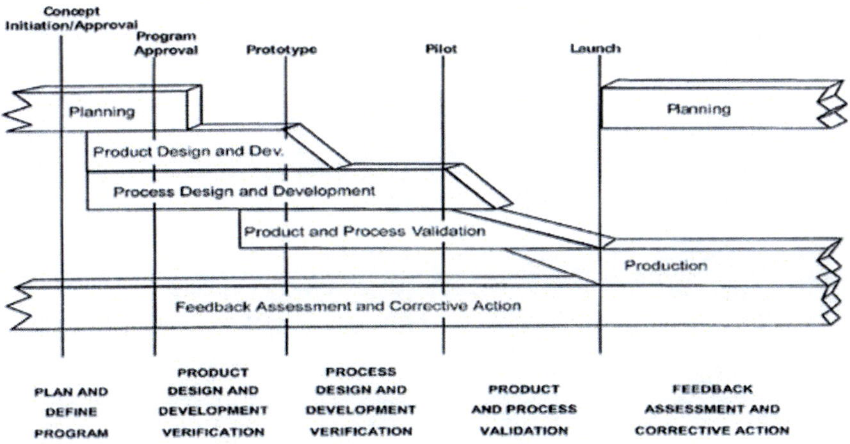

Fig. 5.5 Advanced product quality planning (Reference: APQP AIAG 4th edition). *Source*: APQP AIAG 4th edition

After the product launch, an assessment of the product development and appropriate corrective actions are expected. All activities are continuously monitored, and necessary corrective actions need to be implemented when field findings arise.

AIAG proposes the following process examination steps for production parts:

- plan and define program,
- product design and development verification,
- process design and development verification,
- product and process validation,
- feedback assessment and corrective actions.

During planning and concept development, the development process is based on cognitive work. The verification of the product is an earlier phase because the process design does not only consider the development process it also considers the production process (Fig. 5.6).

Similar to the AIAG framework, also VDA focuses on production parts. The entire supply chain bases are on the idea that the production parts at different stages of maturity also adapt continuously at their interfaces. At an early stage, it is only functional and technical interfaces, but also mechanical components show ageing effects over their lifetime and have tolerances due to their design but also due to how they are produced (Fig. 5.7).

The VDA maturity model shows the challenges during product development. A cross-functional team with experts from.

- development
- purchasing

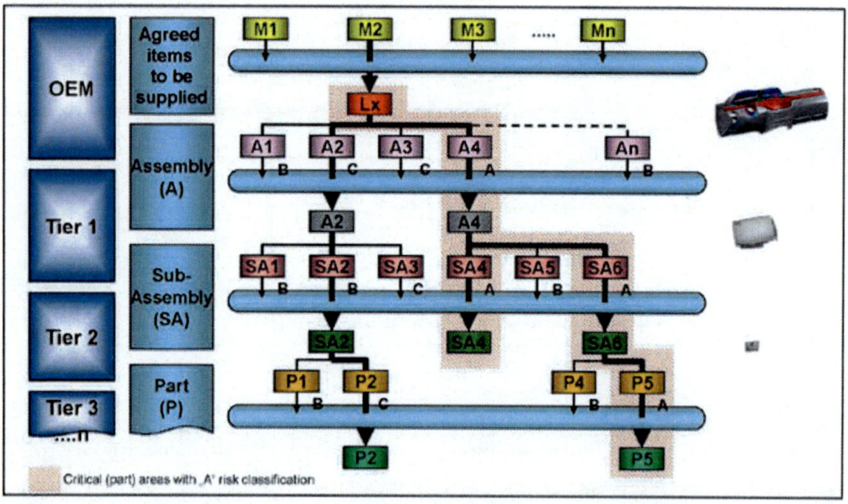

Fig. 5.6 Supplier management—specifying the critical path. *Source* VDA publication „maturity level assurance for new parts ", 2nd edition, 2009

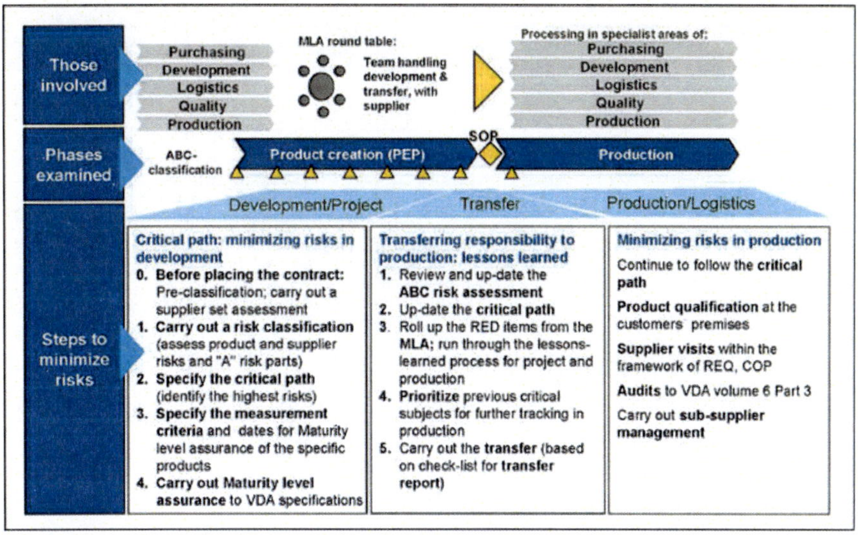

Fig. 5.7 VDA maturity model, maturity hedge for new parts. (*Source* VDA maturity for new parts)

- logistics
- quality and
- production.

should be involved in several activities. A three-step risk reduction process is proposed for the phases:

- development,
- production planning and transferring the product from development to production,
- during production.

They propose a typical risk management process with measures and adequate monitoring of the key risk factors.

5.3 Software Quality

The process for software-intensive products is usually very different, since software provides the key quality factors and not the average from the production part (Fig. 5.8).

ISO 12207 "ISO/IEC 12207 Systems and software engineering—Software lifecycle processes" is still considered as the basis of all lifecycle approaches for software-intensive products. It covers the necessary activities for software lifecycle processes covering for developing and maintaining software. As a lifecycle standard, it includes processes and activities applied in acquisition and configuration of the

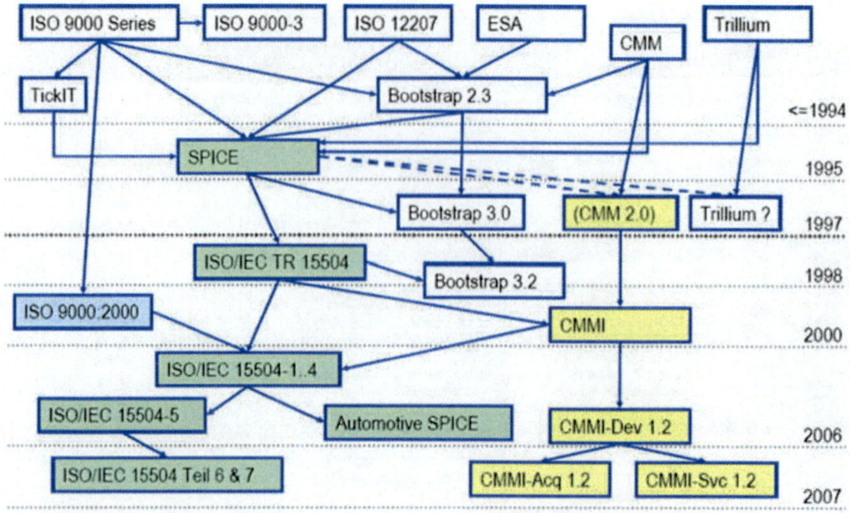

Fig. 5.8 Quality management and ISO 12207 as root for software process standards

services of the system. It defines 43 system and software processes, which are the basis for tailoring of software projects.

Processes are classified in three types:

- basic
- support
- organisational

The support and organisational processes must exist independently of the organisation and the executed project. The basic processes are instantiated according to the situation.

The major sections of lifecycle are the system context processes:

1. Agreement:
    1. acquisition process
    2. supply process
2. organisational project enabling:
    1. lifecycle management
    2. infrastructure management
    3. project portfolio management
    4. HR management
    5. quality management
3. project:
    1. project planning process
    2. project assessment and control process

    3. decision management process
    4. risk management process
    5. CM process
    6. information management process
    7. measurement process

4. technical:

    1. stakeholder requirements definition process
    2. system requirements analysis process
    3. system architectural design process
    4. implementation process
    5. system integration process
    6. system qualification testing process
    7. software installation process
    8. software acceptance support process
    9. software operation process
    10. software maintenance process
    11. software disposal process

5. software implementation.

    1. software implementation process
    2. software requirements analysis process
    3. software architectural design process
    4. software detailed design process
    5. software construction process
    6. software integration process
    7. software qualification testing process

6. software support.

    1. software documentation management process
    2. software configuration management process
    3. software quality assurance process
    4. software verification process
    5. software validation process
    6. software review process
    7. software audit process
    8. software problem resolution process

7. software reuse.

    1. domain engineering process
    2. reuse asset management process
    3. reuse program management process.

The list shows that software development requires very different disciplines and education of the project members.

SPICE, Automotive SPICE, CMMi, etc., all those standards have their roots in ISO 9001. However, none of the software quality standards considers the path, which had been developed for production parts, for example, IATF 16949. Many approaches had been always defined which could see software as a stand-alone product. However, value adds functionality could only provide customer benefit if it fits into the target system. In most software standards, a well-specified system is the entry point of any software development, but is this a realistic assumption?

5.4 Process Models

There are many ideas and approaches to find also a lean software development process. The waterfall process model is a very basic understanding of process models for software development. (Fig. 5.9).

A typical waterfall process model uses the following steps:

- Initialisation: It covers all the system requirements and constraints as well all stakeholder aspects. In addition, the organisational set-up and the tool infrastructure are strong driver for the product development.
- Analysis: After elicitation of all requirements and constraints, an analysis phase initiates feasibility studies and provides feedback about completeness, correctness, applicability and consistency of given requirements and constraints.
- Drafting: After the first analysis, a draft realisation shows the feasibility. The basic design principles and the architectural assumptions are evaluated.
- Realisation: The realisation phase leads to the product.

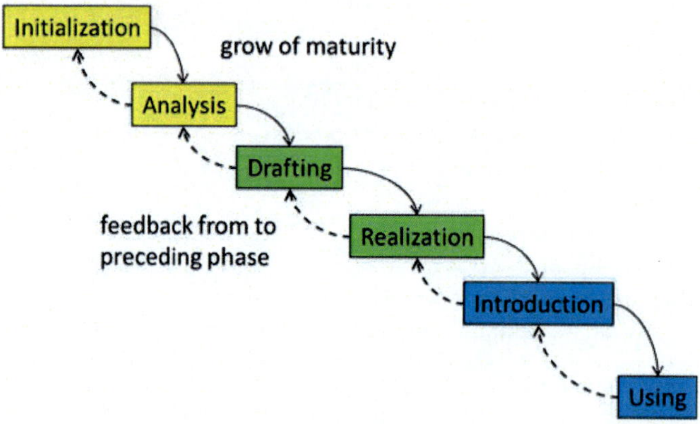

Fig. 5.9 Waterfall with feedback from preceding activities

- Introduction: The introduction phase assures that the product meets the stakeholder requirements. The product and the principles behind are communicated for example also to the users.
- Using: User experiences and handling analysis provide experiences with the product in the intended application context.

This very short interpretation shows the weaknesses of a waterfall process. There are many steps before you get the user feedback, the stakeholder feedback and the suitability of the entire development environment. Most of the software tools are developed according to a waterfall process, and anybody ever got tools from his IT department knows the weaknesses of the process (Fig. 5.10).

The V-Modell is a process model for IT development projects in the Federal Republic of Germany.

V-Modell is a protected trademark of the Federal Republic of Germany. The "V" in the term "V-Modell" stands for visualises the idea of contrasting specification and decomposition with realisation and integration. This general procedure was proposed as the V-Modell in the late 1970s. In the meantime, V-Modell XT supports not only this procedure, but also many other approaches (e.g. agile system development). A governmental owned company developed the first V-Modell in 1986. Initially, it was intended for IT projects in the public sector in Germany, but in the meantime, it is also used in the private sector. In contrast to classical phase-oriented process models, the V-Modell only defines activities and results and does not require a strict chronological sequence. In particular, it lacks the typical acceptances that define the end of a phase. Nevertheless, it is possible to map the activities of the V-Model-XT to a waterfall model or a spiral model, for example.

The V-Model-XT considers the following main activities:

- project management,
- quality assurance,
- configuration management,
- problem and change management.

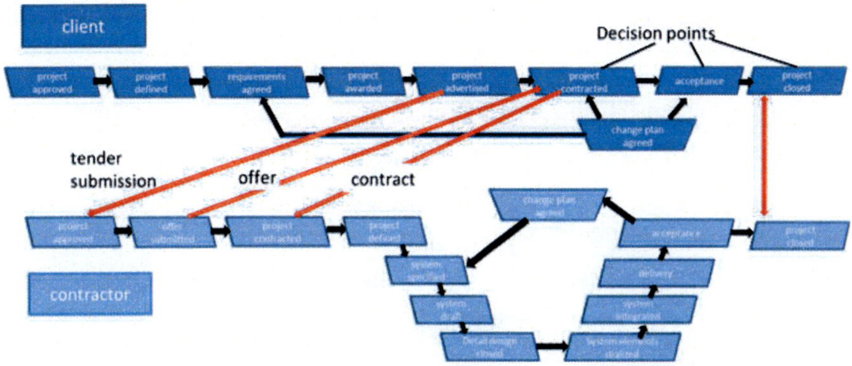

Fig. 5.10 Interface V-Modell customer–supplier according to V-model XT. *Source*: V-model XT 1.2

These are also typical support processes as considered in Automotive SPICE. There are typical system engineering activities like

- System creation,
- specification of requirements.

A major set of activity is the contract agreement with the following main activities:

- delivery and acceptance by supplier,
- delivery and acceptance by customer,
- conclusion of contract by supplier,
- conclusion of contract by customer.

The V-part itself is a typical system process with

- system specification,
- draft system,
- detail design.

the descending branch starts with

- system elements realisation,
- system integration,
- delivery,
- acceptance.

The V-model XT seems to be quite close to the waterfall model because the final customer acceptance is based on the delivered product.

The V-model XT proposes an acceptance with a specific feedback (Fig. 5.11).

This acceptance cycle from V-model XT provides various steps before the management or the customer would accept the product or design. There is a short cycle, which verifies the considered element after realisation and a wider management cycle to provide the product or design acceptance by responsible organisation (Fig. 5.12).

The system design information flow based usually on a heterogeneous set of requirements and consequences, constraints, etc., is derived from the architecture, design and the context of operation during runtime. Whenever there is a change within the system, design, it could lead to major or minor impact to components or subelements and it could lead to changes in the derived requirements, architecture, design and the implementation or realisation.

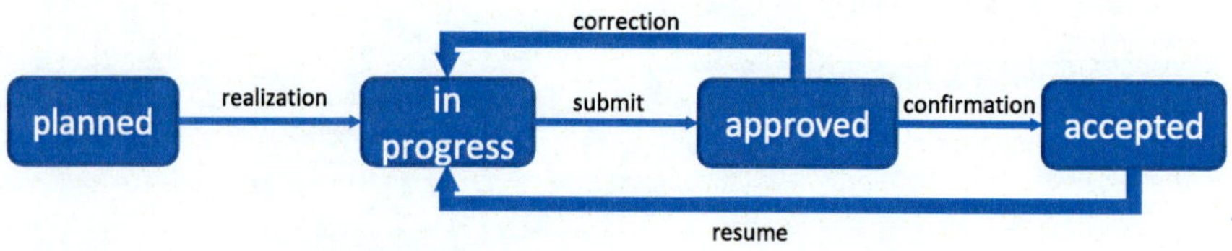

Fig. 5.11 Acceptance feedback loop similar as proposed by V-model-XT

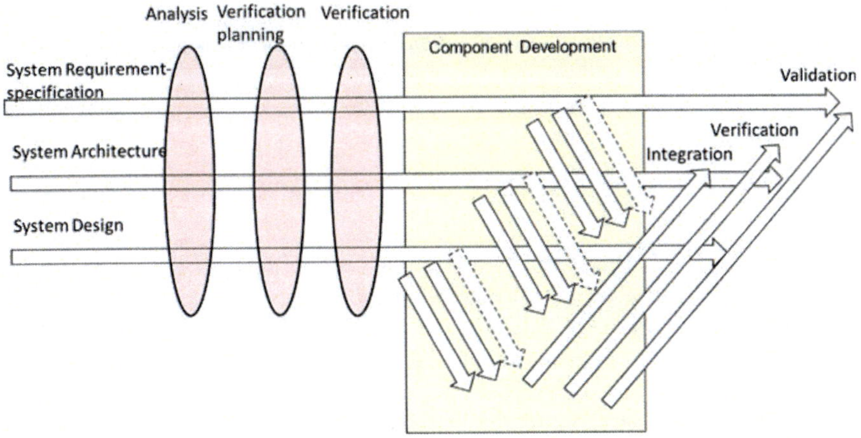

Fig. 5.12 Process flow and verification

ISO 26262 does not address many of the requirements for the alignment of safety activities (processes) to the product development in general. In addition, the confirmation measures in the standard are not described precisely so the verification of the safety activities, which can result from the interleaving of safety activities, has never been described. Since the norm never described this process, the chains could be destroyed during project planning. Therefore, verifications of safety activities should not be neglected. Verification is needed on all horizontal development layers and between all activities related to breakdown requirements, viewpoints from architecture and design, etc. However, verifications had been done during the functional and technical safety concept as well as the horizontal architecture interfaces in the components. Each input and output of a safety-related activity should be verified. Caution, the safety process is due to systematic failure of course also in the intended function. Consequently, a QM attribute or no ASIL attribute does not mean any verification of the activities that are needed. For multiple horizontal system levels, the verification is required after each interface. The benefit is that the work results once again become input for the next phases after the verification, which ensures the confidence and resulting safety for the input and the predeceasing activities.

The purpose of the V-Modell is to communicate the process model or to explain how to link the individual development activities. Therefore, you pass the V-Modell at least twice, once during the planning of the activities and once during the realisation of the product. During planning, the specific activities are thought out in advance, which are then applied during implementation.

A further cycle through such a V-Modell is the verification and validation process. Verification is the means by which the correctness, consistency and completeness of the respective requirements are examined. Validation questions the objectives and helps to assess the achievement of the objectives with regard to function, handling, applicability, producibility, maintainability and technical feasibility, etc.

The realisation of the product is also a validation of the development plan, regarding the quality and feasibility of the previously planned process and methodology.

5.4.1 Cyclic Process Models

In automotive industry, since the fifties of the last century, there are sample phases as a cyclical process. The goal here is to achieve a higher degree of maturity in customer acceptance in each phase (Fig. 5.13).

Today, traditional sample names such as A-, B-, C- and D-sample are only referenced in certain company standards (e.g. in Daimler's). In the APQP standards from Automotive Industry Action Group (AIAG) or German: Verband der Automobilindustrie e. V. (VDA), all samples refer back to the initial sample. The sample groups for different customers are mostly aligned with the requirements of the vehicle development.

Today, traditional sample names such as A-, B-, C- and D-sample are only referenced in certain company standards (e.g. in Daimler's). In the APQP standards from AIAG or VDA, all samples refer back to the initial sample. The sample groups for different customers are mostly aligned with the requirements of the vehicle development.

Phases of the Spiral: Prototypes

The dream of every process developer and design engineer is that a complete specification is available right at the beginning of product development. In reality, requirement development became a lifecycle approach. For classical mechanics, size and

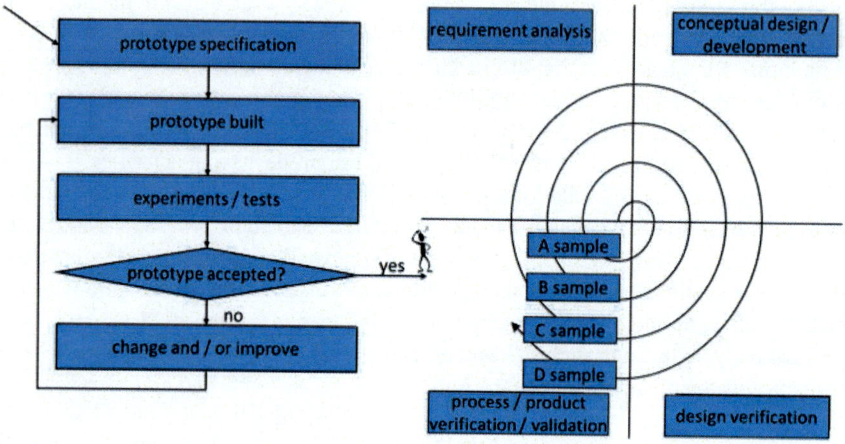

Fig. 5.13 Typical spiral process for prototype samples

geometrical data are the essential parameter for the product. Functionality crows during the development. There is often only a specification, which is defined on a higher level of abstraction for the first iterations.

Previously, in the early stage of automobile engineering, the A-sample could be made out of wood since the focus at the beginning was the question how to arrange all the products and components in the vehicle. In the CAN bus age today, the connectivity is an early target for the first samples.

Construction of Prototypes

In order to verify samples have to be delivered to the customers, they have to be produced, and communication interfaces should be aligned with the vehicle parameter. In the following iterations, the degree of maturity in terms of functionality, precision, adaptability, etc., increases continuously. The D-Sample, first sample or initial sample is the basis for series production.

Experimenting—Trial/Acceptance

The sample has to be tested according to the given requirements. The sample will be tested, in the first iterations under laboratory conditions, and then later in the customer environment.

All parties involved in the process hope to be able to figure out all necessary requirements in the first shot and that the sample returns with a positive test result. In the real world, the prototype is an essential input factor for the requirement analysis. The acceptance phase is often a multi-step approach. Single elements, subcomponents, etc., are often tested in advance in a simulation environment before it should be delivered to the customer. The customer considers that the delivered sample meets the agreed specification.

Changes, Modifications or Enhancements

After the feedback from tests, verifications, validations or change requests and agreement on the next sample changes initiate the next maturity level and the start of the next cycle.

5.4.2 *PDCA and CIP*

A comparable approach from quality assurance is the Deming cycle or Deming wheel, Shewhart cycle or PDCA cycle. It describes an iterative three- or four-phase process for learning and improvement. PDCA stands for "Plan - Do - Check - Act".

Walter Andrew Shewhart described a three-step approach:

Specification → Production → Inspection.

Shewhart wrote: "*These three steps must go in a circle instead of in a straight line, as shown ... It may be helpful to think of the three steps in the mass production process as steps in the scientific method. In this sense, specification, production and inspection correspond respectively to making a hypothesis, carrying out an experiment, and testing the hypothesis. The three steps constitute a dynamic scientific process of acquiring knowledge."*

Deming added a further step to Shewhart's three-stage process to illustrate evolutionary quality development. This gave the process the four-stage form of presentation that is customary today. A Shewhart cycle became a stage in the continuous improvement process of a system. In accordance with the human relations approach preferred by Deming, he focused on the work system (Gemba): "Go to the place of action" and above all, he put the local employees with their exact knowledge of the situation at the workplace at the centre of the planning. Due to the American occupation in Japan, the idea of the development cycle came to Japan. Ishikawa Ichiro took up the theme.

The PDCA cycle consists of four elements:

Plan.

The respective process must be planned before its actual implementation: "Plan" includes the identification of improvement potentials (usually by the employee or team leader on site), the analysis of the status and the development of a new concept (with intensive involvement of the employee).

Do.

Contrary to widespread belief, Do does not mean the introduction and implementation on a broad front, but rather the trial and error or testing and practical optimisation of the concept with quickly realisable, simple means (e.g. temporary fixtures) at a single workplace with strong involvement of the employee according to Gemba. (Gemba walks denote the action of going to see the actual process, understand the work, ask questions and learn. It is also known as one fundamental part of lean management philosophy).

Check.

The process flow realised on a small scale and its results are carefully reviewed if successful, generally released for implementation on a broad front.

Act.

In the phase "act", the new general requirement is introduced on a broad front, laid down and following by regularly checks for compliance (audits). This is indeed a "major action" which, in individual cases, can involve extensive organisational activities (e.g. changing task lists, NC programs, master data, carrying out training courses, adapting the organisational and operational structure) as well as considerable investments (at all comparable workplaces, in all plants). The improvement of this standard starts again with the plan phase (Fig. 5.14).

The PDCA cycle is often combined with the idea of continuous improvement (CIP), which also targets a growing quality maturity.

In the industry and in the service sector, PDCA cycles and CIP are basic components of the standard families DIN EN ISO 9000, ISO 14000, ISO/IEC 20,000 and

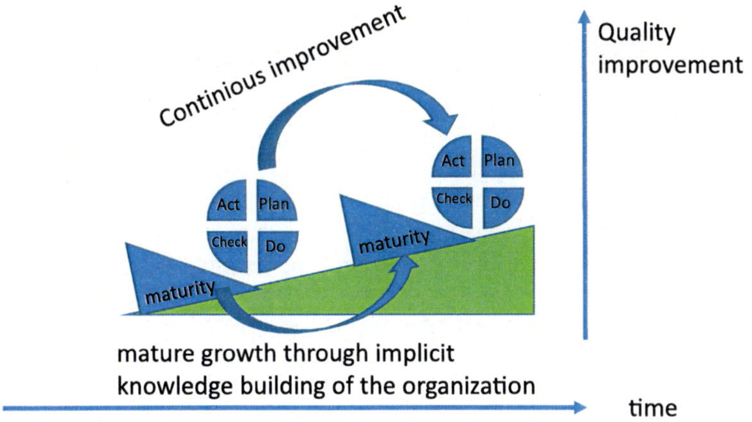

Fig. 5.14 PDCA combined with continuous improvement (CIP)

ISO/IEC 27,001, etc. Another cyclic approach from the domain of process management is the **DMAIC cycle** (***D****efine –* ***M****easure–****A****nalyse–****I****mprove–****C****ontrol).* Here the approach of key performance parameter is considered so that the improvement potential could be measured.

Another approach derived from Walter Andrew Shewhart the idea of cycles is the statistical process control "SPC".

His book "Economic Control of Quality of Manufactured Product" described the scientific principles of SPC. Shewhart started from the assumption that the quality of the final product depends essentially on the combination of the scatter of the parameters of the individual parts. He found two fundamentally different mechanisms as the cause for this scatter:

- scatter due to general causes random deviations from the mean value, which result from a stochastic process noise (physics) and
- scattering due to special causes (material defects, machine defects, design defects, etc.).

Shewhart's second important finding was that two mistakes could be made when trying to minimise this scatter:

- Error 1: Assigning a deviation to a special cause, even though it was caused by a general cause.
- Error 2: Assign a deviation to a general cause, even though it was caused by a particular cause.

Either one or the other error can be avoided completely, but never both at the same time. Therefore, a way had to be found to minimise the costs of error avoidance. Extensive statistical research and theory building finally led Shewhart to develop control charts as an optimal tool to put the knowledge gained into daily practice. William Edwards Deming recognised that these insights and tools could be applied to all types of processes (business processes, administrative processes, etc.) with the same positive results. This teaching fell on fertile ground especially in Japan, where it was further developed within the Toyota Production System, among others.

Today, statistical process control is seen as a component of a quality management system and, as a service process, accompanies the core process of production or service. All statistical methods that serve to monitor and optimise the core process are summarised under the term statistical process control. These methods go beyond the various control chart techniques and include, for example, the methods of statistical design of experiments (DoE), FMEA or also the method collection six sigma. SPC variables are incorporated into customer–supplier relationships as process capability indices.

All that ideas consider a cyclic process and measurement of the key parameter in order to prioritise, optimise and select the necessary measures for effective improvement.

5.5 Complexity Due to Organisation

A typical vehicle manufacturer organisation defines of course a management layer responsible for a specific type of vehicle. In order to find synergies with several vehicle domains, they are organised in a matrix. Such a matrix organisation leads already to very complex interfaces (Fig. 5.15).

The three circles represent basic types of interfaces.

- organisation interfaces (other department or even more separated organisation unit),
- technical interfaces, based on the technical elements that need to be integrated into a vehicle,
- project interfaces are not independent, but depend also on development progress, availability of elements, historical and external influence factors.

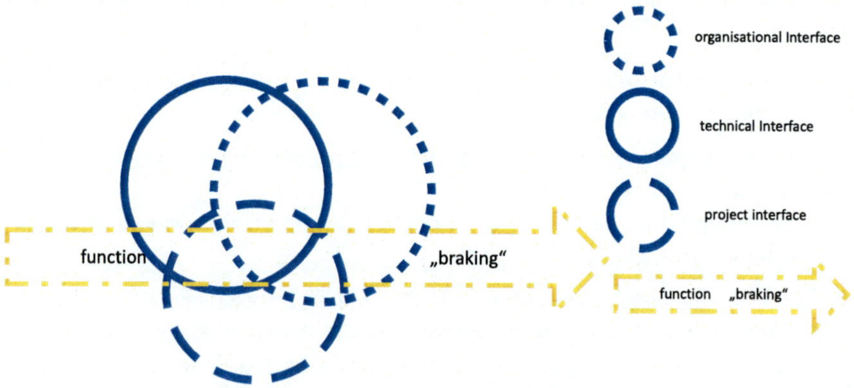

Fig. 5.15 Organisational interfaces in a typical technical development project

It is obvious that a basic function like braking provides several organisational interfaces, and if the interfaces could not be harmonised, the complexity increases exponential.

From a quality management point of view or derived from ISO 12207 all the activities needed for software intensive products, we get the following organisational units.

Just for the technical processes, ISO 12207:2008 defines the following:

- business or mission analysis,
- stakeholder needs and requirements definition,
- systems/software requirements definition,
- architecture definition,
- design definition,
- system analysis,
- implementation,
- integration,
- verification,
- transition,
- validation,
- operation,
- maintenance,
- disposal.

5.6 Vehicle Structure and Organisation

A vehicle, form homologation point of view, is a heterogeneous device. In order to reduce the complexity, a generic structure is required. Those structures should provide at least as low as possible interfaces or interfaces, which are industry wide accepted.

Many engineers state, "If you see a vehicle EE-Architecture, you recognise the organisation responsible for the development of that vehicle".

Typical organisational structure considers the following three basic segments:

- chassis,
- powertrain and
- interior.

Chassis covers the main basic vehicle control functions like braking and steering. Powertrain provides the energy management and propulsion. In interior, the focus is on any kind of human–machine interface (HMI) and all functions needed for the basic operation of the vehicle. All kind of driver information, maintenance and comfort functions and also occupancy safety are typically seen as part of interior. In many organisations, the crash management is separated, since there are major legal aspects and various safety-related interfaces to chassis and powertrain functions. All legal

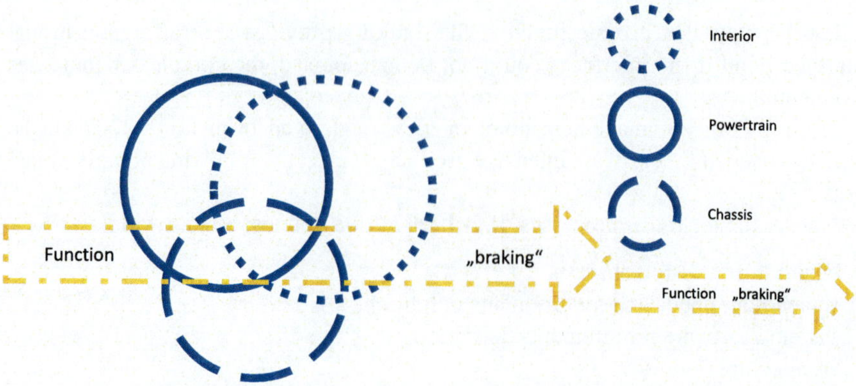

Fig. 5.16 Organisational interface type for a basic vehicle function

requirements assigned to view, illumination and driver´s activities belong often to the body domain (Fig. 5.16).

The EE architecture reflects the organizational interfaces of the business units for technical development. They show the structure, grouping, modularity, functionality and technology of chain processes, business processes and working processes (Fig. 5.17).

Even in the two thousand twenties, nearly all conventional vehicle platforms use various CAN segments for

- chassis,
- powertrain and
- interior.

The chassis CAN provides communication segment for

- EPS: electric power steering,
- ESC: electronic stability control,
- EPB: electronic park brake,
- CCS: chassis control systems.

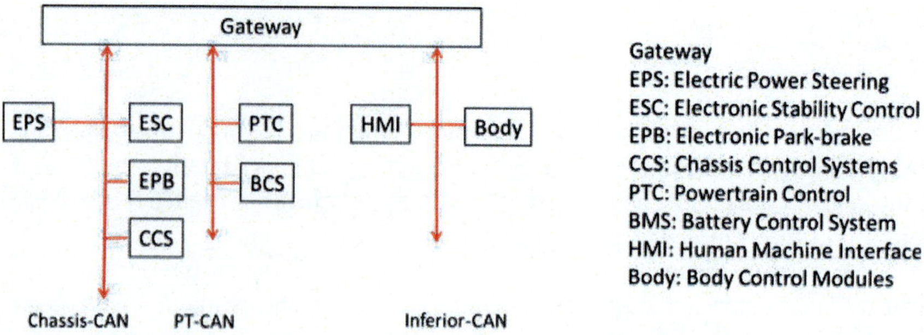

Fig. 5.17 Typical ComNet architecture

Due to the yaw rate sensor, the airbag control unit is often, as well as, a separate steering angle sensor in the chassis CAN.

The powertrain CAN provides communication segment for

- PTC: powertrain control and
- BMS: battery control system.

For an electrical powertrain, the entire energy management is often divided into various control units. Charging, heating and cooling systems are typically also part of the powertrain segment.

The interior CAN provides communication segment for

- HMI: human–machine interface and
- Body: body control modules.

The interface with the driver except pedals is usually part of various HMI control units. All the functions related to enter the vehicle, climaxing, locking doors, booth, etc., belong to the body segment. Mobiliser for theft control, etc., is usually also part of the body segment.

This distribution is not brand-or manufacturer-specific but reflects the historical development of the respective vehicle platform. If a German and a French platform are based on joined developments, it could be traced many model cycles later. Even in electric vehicles, the history of the com-network from the combustion predecessors could be recognised.

However, not only the ComNet is the reason for the complexity, all functions and features are assigned on historical reasons. Whenever there are time-critical communications between sensors, control units or actuators, an additional CAN bus needs to be added, because data networking with more than two control units is not possible.

Due to the power-up, wake-up or energy management, the sequences for the control devices depend on the ComNet. The fuse and terminal concept need to be adapted accordingly. Any degradation requires respecting the given architecture, because nearly all control units need signals from other control units. All the state machines for vehicle entry, mobilising, etc., depend of course also from the given ComNet architecture. In case of communication failure due to EMC, etc., the entire vehicle functions could be confused and corrupted. This is typically the case if all the warning lights come on and the motor powered down.

SAE 3049 provides a basic framework, how to model a road vehicle. They decompose the so-called ground vehicle system into different subsystems.

The body-frame subsystem comprised the following five subsystems, which are highlighted in figure (Fig. 5.18):

1. body-frame supervisory control subsystem,
2. under body subsystem,
3. upper body subsystem,
4. cabin interior subsystem and
5. cabin climate subsystem.

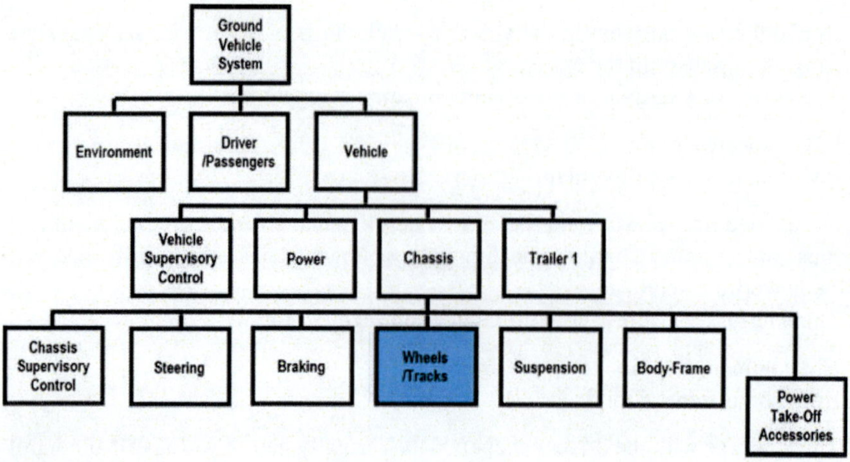

Fig. 5.18 Vehicle decomposition according to SAE J3049

> These subsystems are interrelated and work together to provide an enclosed space that is protected from the elements of the weather, with seating for transporting people, storage areas for transporting goods and subsystems for controlling the climate of the enclosed space. The body-frame subsystem supervisory control subsystem supervises and coordinates the control of all of the subsystems in the body-frame subsystem. In addition, it provides a flexible controller information bus structure for coordinating all of subsystem controllers in the body-frame subsystem with each other and with controllers throughout the vehicle subsystem, as needed. A high-level controller coordinates the operation of all of the subsystems of the body-frame subsystem by monitoring the capabilities of these subsystems and coordinating the requests of the human controlling the vehicle with the capabilities of all of the other subsystems of the body-frame subsystem at this level of the hierarchy.
>
> *Source* Model Architecture and Interfaces Recommended Practice for Ground Vehicle System and Subsystem Dynamical Simulation, SAE J3049, 2015

This standard provides a basic structure of the vehicle elements, which are typical for the entire automotive industry. The standard provides many aspects in order to develop a model of a standard road vehicle.

5.7 Dimensions of a Capable Organisation

An organisation needs many control mechanism in order to control the enterprise risk. Even for financial reasons, a company must set up a business risk process. In most countries, for financial reasons, a company is obliged to make a profit. Profit is also required for a company to be able to bear potential liability risks (Fig. 5.19).

The cube from Nancy C. Leveson shows three dimensions.

- 6 levels for the intent specification,
- Part whole shows system in the context, human impact and process artefacts.
- Refinement provides a dimension to product and activity details.

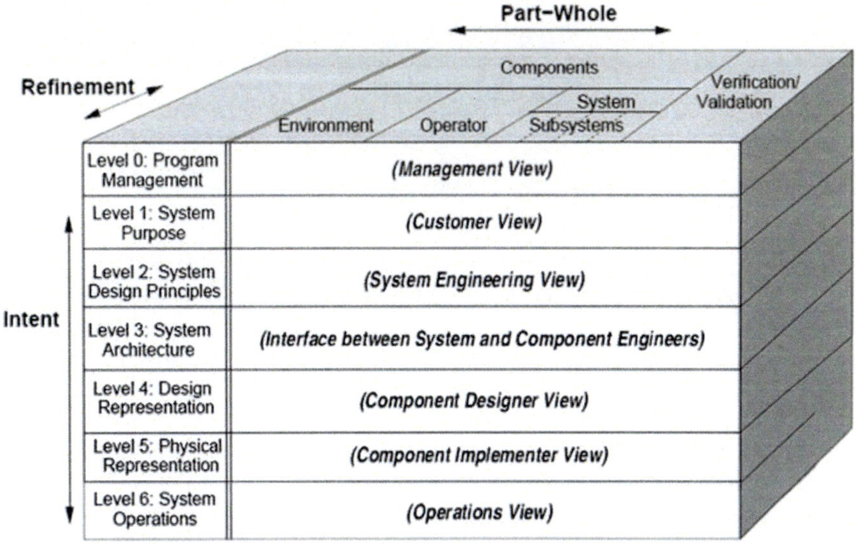

Fig. 5.19 Structure of an intent specification. *Source* Nancy C. Leveson, Engineering a Safer World, Multi-dimensional structure of specification (Source: Fig. 10.1, Nancy G. Leveson Engineering a Safer World)

The intent specification dimension covers similar to V-model XT also the program management aspects, and this particular covers all platform-related topics.

- Level 0, Program management is addressed as an management view.
- Level 1, System purpose addresses the customer view.
- Level 2, System design principles address system engineering views.
- Level 3, System architecture addresses the interface between system and component engineers.
- Level 4, Design representation provides the component designer view.
- Level 5, Physical representation provides the component implementers view.
- Level 6, System operation provides the operations view.

In system safety engineering, the cube would be considers other dimensions but all the aspects need to be addressed by a responsible organisation.

5.8 Structure of Organisations

An organisation is usually based on a historical evolution, but in modern technology companies. There are typical research, pre-development or advance development and the typical product development organisation (Fig. 5.20).

In most companies, you see a typical hierarchical org chart, which shows project or product management in a typical matrix to the management hierarchy. Program

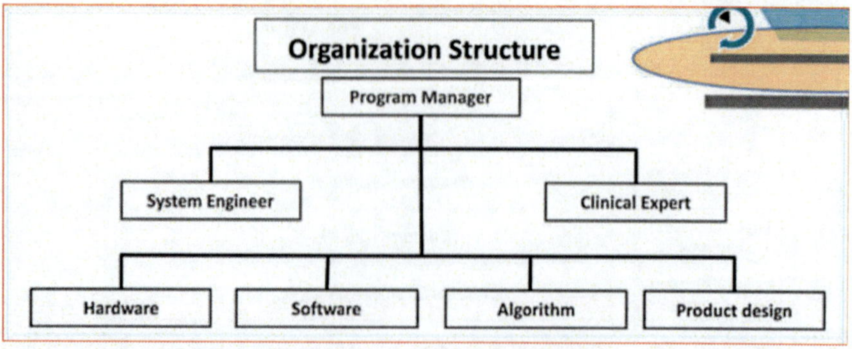

Fig. 5.20 Matrix organisation for program and project management

management is usually the link between sales and the development, and the project manager controls various projects for example a dedicated product development.

There are the typical value-adding processes that generally represent the product development process. Around there are the typical supporting processes as known from ISO 12207. During research, there are many uncertainties and insights need to be developed so that feasible pattern could be evaluated. During concept phase, the "noise" from research should be "filtered" for the product or platform concept (Fig. 5.21).

Management needs to provide targets for the product idea, which provides orientation, and framework for the entire product development process. Research does not need to be completely unstructured due to uncertainties, it should lead systematically in the concept phase (Fig. 5.22).

The typical product development cycle starts stakeholder analysis with all necessary interactions. The V-model typically starts with requirement, architectural development of the branch down to realisation or implementation. All the integration activities via all necessary validations and verifications lead to the product release.

Fig. 5.21 Typical transition from research to concept development

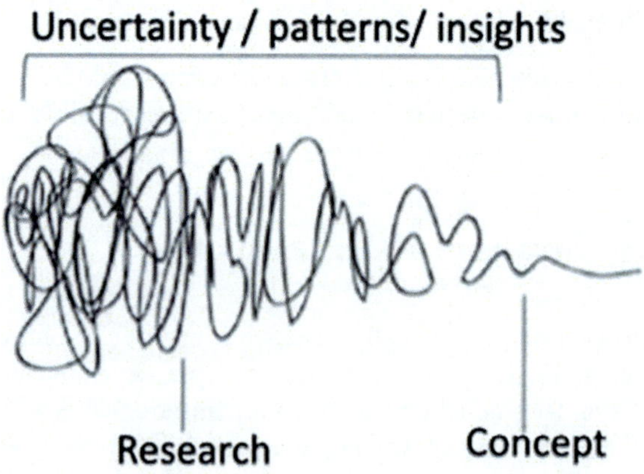

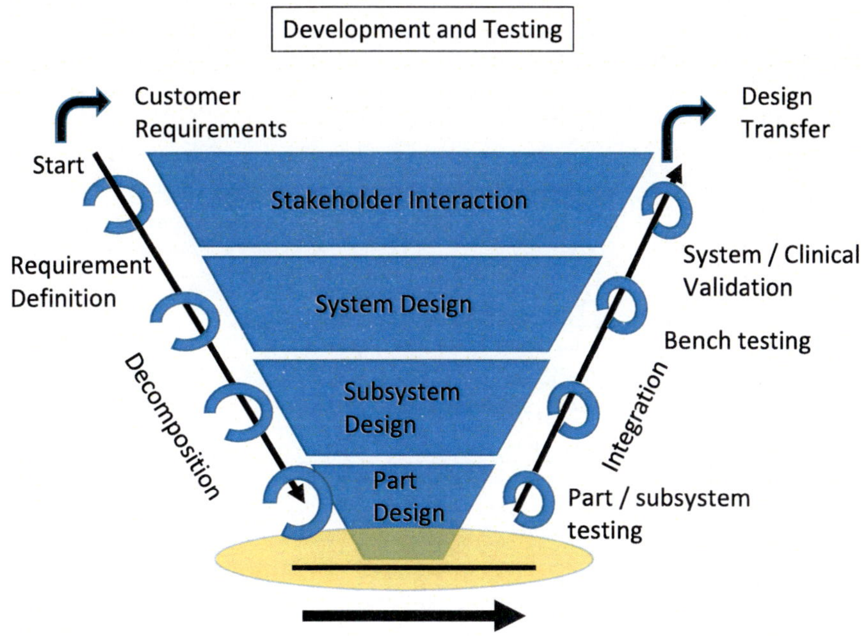

Fig. 5.22 Typical development process including stakeholder interaction

The process is supported by many subprocesses, which run in parallel, sequences and very often in iterations (Fig. 5.23).

A typical organisation works still in a hierarchical structure. As bigger a company is as bigger is also the organisational overhead, and it became more difficult to identify synergies. Without a structural organisation also common resources for plants, laboratories and supporting infrastructure, systematic control of an organisation on enterprise level is very difficult. The first interface in case of product liability is the CEO of a company. Usually, the CEO assigned a specific organisation to manage such interfaces, which is most probable, a quality organisation. The quality management organisation is usually also assigned to develop the company processes and controls the maturity based on the product lifecycle.

Organisational failure, which led to accidents, could not be traced to special business segment or functions in the organisation, like development or production. The quality organisation usually just manages the tasks like complain management, but the individual quality manager could not be made responsible since his personal cross-negligence is not an identified cause for the accident.

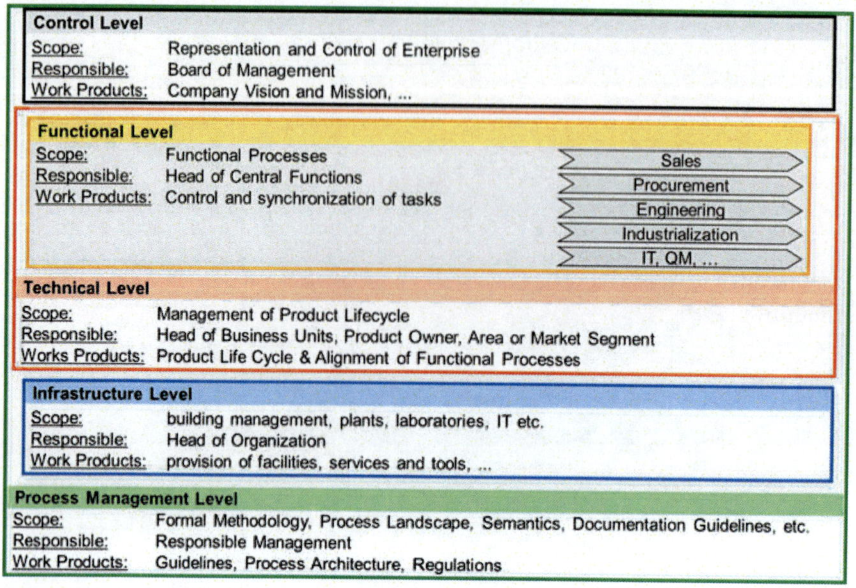

Fig. 5.23 Organisational structure of a capable organisation

5.9 Organisational Aspects for Future Mobility

For future mobility, the owning of a motor vehicle particularly with a noisy combustion engine will be not anymore in the favour of the society. There are two mayor business models in the focus of future mobility:

- services for availability of mobility and transportation and
- operating of transportation systems for people and goods.

How the two models developed, if they devolve somehow overlapping or just hand in, are questions for futurists.

Other fields that will play a significant role in international mobility are dealers and importers. Due to the trade laws and product liability, there are also further aspects to be considered for these sectors in terms of safety.

The term "mobility as a service (MaaS)" becomes a slogan for the future mobility ideas. It describes a shift away from personally owned vehicles for transportation, travelling, etc., and towards mobility provided as a service. This combines transportation services from public and private transportation providers through a unified gateway that creates and manages the trip, which users can pay for with a single account. Users can pay per trip or a monthly fee for a limited distance. The key concept behind MaaS is to offer travellers mobility solutions based on their travel needs. Specialist urban mobility applications are also expanding their offerings to enable MaaS, such as the car-sharing providers or transfer providers like Transit, Uber and Lyft.

Travel planning typically begins in a journey planner, which could be an app on a smartphone. For example, a trip planner can show that the user can get from one destination to another by combinations of trains, busses, aeroplanes and other transportation services like shuttles, etc. The user can then choose their preferred trip based on cost, time and convenience. At that point, any necessary bookings (a seat on a long-distance train) would be performed as part of the service. Such service requires roaming functions, so that end-user app should work in different cities, without the user needing to become familiar with a new app or to sign up to new services.

Many new mobility concepts will be developed which operate just with a limited permission or homologation, the number of typical mass production car, which is owned by the driver or his family, will decrease. At least, the private owned car will not be fitted with the high-end automation features. Most probable new protection systems for pedestrians, cyclists and occupancies will dominate the design and type of assistance systems and system to control unauthorised use. The same is true for car-sharing models; they will use concepts to return to charging or parking facilities with support from automated function. Nevertheless, the customer does not have a direct benefit from those necessary services; the driver and the occupancy require just a safe transportation from one point to another point. Parking and charging become a task and burden for the operator of the car-sharing or rental car service (Fig. 5.24).

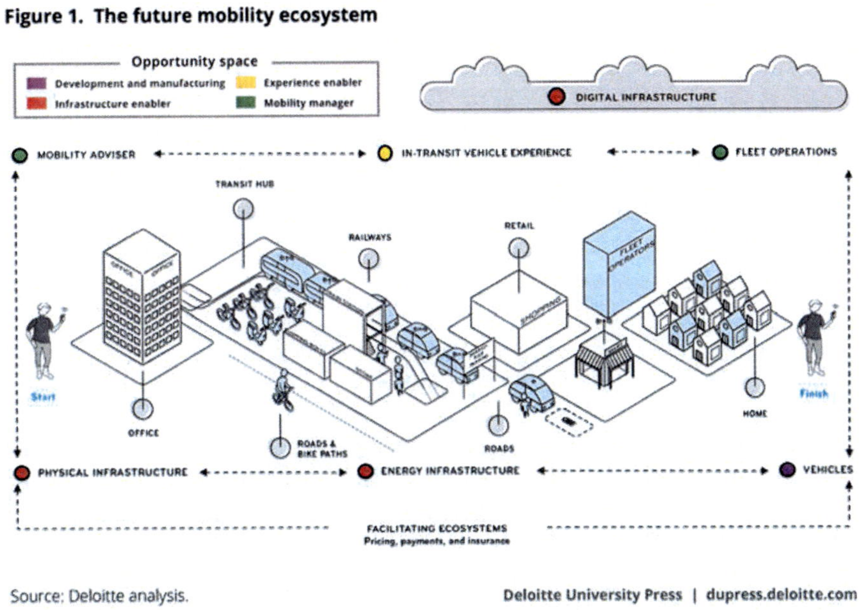

Fig. 5.24 View on a future mobility ecosystem *Source* Deloitte, https://www2.deloitte.com/us/en/insights/deloitte-review/issue-21/transportation-ecosystem-future-of-mobility-reshaping-work.html, Making the future of mobility work: How the new transportation ecosystem could reshape jobs and employment

By Burt Rea, Stephanie Stachura, Laurin Wallace, Derek M. Pankratz, downloaded 4–2020.

Deloitte sees the connected mobility between railway, motor vehicles, bike cyclists and pedestrians, but also airports, heliports or even air taxies need to be considered in the future.

Air systems operate in a very special controlled area; the air corridors are well-controlled by the authorities. Unauthorised use by cyclists or pedestrians is more or less impossible.

Typically, there is an infrastructure organisation which provides

- the road for road vehicles,
- the railway for trains,
- air corridors for aeroplanes.

The infrastructures for various transport systems are usually defined by legislation.

The European Community for Mobility and Transportation provides the European Civil Aviation Handbook.

It consists of the following topics:

European Civil Aviation Handbook:

Part I. Regulations and Directives.

Title 1. Air transport and market issue.

Subtitle 1. Internal Market.

- licensing of air carriers [Regulation (EC) No 1008/2008],
- access for community air carriers to intra-community routes [Regulation (EC) No 1008/2008],
- fares and rates for air services [Regulation (EC) No 1008/2008],
- insurance requirements for air carriers and aircraft operators [Regulation (EC) No 785/2004],
- code of conduct for computerised reservation systems [Regulation (EC) No 80/2009],
- statistical returns [Regulation (EC) No 437/2003],
- implementing rules of statistical returns regulation [Commission Regulation (EC) No 1358/2003],
- subtitle 2. Relationship with third countries,
- air services agreements [Regulation (EC) No 847/2004],
- protection against subsidisation [Regulation (EC) No 868/2004].

Title 2. Passenger rights.

- air carrier liability [Regulation (EC) No 889/2002],
- denied boarding cancellation or long delay of flight [Regulation (EC) No 261/2004],
- right of disabled persons [Regulation (EC) No 1107/2006].

Title 3. Safety.

- harmonisation [Regulation (EC) No 216/2008],
- investigation of civil aviation accidents and incidents [Council Directive 94/56/EC],

- common rules—EASA establishment [Regulation (EC) No 216/2008],
- occurrence reporting in civil aviation [Directive 2003/42/EC],
- rules for the airworthiness [Commission Regulation (EC) No 1702/2003],
- continuing airworthiness [Commission Regulation (EC) No 2042/2003],
- safety of third country aircraft using community airport [Regulation (EC) No 216/2008],
- board of appeal of the EASA [Regulation (EC) No 216/2008],
- community list of air carrier subject to an operating ban [Regulation (EC) No 2111/2005],
- implementing rules for the banned air carrier list [Commission Regulation (EC) No 473/2006],
- fees and charges levied by the EASA [Commission Regulation (EC) No 593/2007],
- EASA working methods for standardisation [Commission Regulation (EC) No 736/2006],
- collection and exchange of information on the safety of aircraft [Commission Regulation (EC) No 768/2006],
- common rules—EASA establishment [Regulation (EC) No 216/2008],
- list of banned air carrier [Commission Regulation (EC) No 1043/2007].

Title 4. Security.

- common basic standards on aviation security [Commission Regulation (EC) No 820/2008],
- national civil aviation security quality control programmes [Commission Regulation (EC) No 1217/2003],
- security restricted areas at airports [Commission Regulation (EC) No 1138/2004],
- procedures for conducting inspections in the civil aviation security [Commission Regulation (EC) No 1486/2003],
- civil aviation security [Regulation (EC) No 300/2008].

Title 5. Environmental protection.

Subtitle 1. Noise emission.

- limitation of noise [Council Directive 89/629/EEC],
- operation of aeroplanes covered by Part II, Chap. 3 Vol.1 of Annex 16 [Directive 2006/93/EC],
- introduction of noise-related restrictions [Directive 2002/30/EC],
- management of environmental noise [Directive 2002/49/EC].

Subtitle 2. Gas emission.

Title 6. Airport.

Subtitle 1. Slots.

- allocation of slots [Council Regulation (EEC) No 95/93],
- subtitle 2. Ground handling,
- access to ground handling market [Council Directive 96/67/EC].

Title 7. Air traffic management.

- framework for the creation of the Single European Sky (SES) [Regulation (EC) No 549/2004],

- provision of air navigation services in the SES [Regulation (EC) No 550/2004],
- organisation and use of the airspace in the SES [Regulation (EC) No 551/2004],
- interoperability of the European ATM network [Regulation (EC) No 552/2004],
- requirements for the provision air navigation services [Commission Regulation (EC) No 2096/2005],
- rules for the flexible use of airspace [Commission Regulation (EC) No 2150/2005],
- air traffic controller licence [Directive 2006/23/EC],
- airspace classification and access of flights [Commission Regulation (EC) No 730/2006],
- automatic systems for the exchange of flight data [Commission Regulation (EC) No 1032/2006],
- procedures for flight plans in the pre-flight phase for the SES [Commission Regulation (EC) No 1033/2006],
- common charging scheme for air navigation services [Commission Regulation (EC) No 1794/2006],
- establishment of a joint undertaking to develop SESAR [Council Regulation (EC) No 219/2007],
- requirement for the application of a flight message transfer protocol [Commission Regulation (EC) No 633/2007],
- safety oversight in air traffic management [Commission Regulation (EC) No 1315/2007].

Title 8. Personnel and social issues.

- mutual acceptance of personnel licences [Regulation (EC) No 216/2008]

Title 9. Competition rules.

Subtitle 1. Antitrust practices law.

- implementation of the rules on competition [Council Regulation (EC) No 1/2003],
- application of Art. 85–3 in the air transport sector [Council Regulation (EC) No 487/2009].

Subtitle 2. Economic concentration law.

- control of concentration [Commission Regulation (EC) No 802/2004],
- subtitle 3. State aids,
- transparency of financial relations [Commission Directive 2006/111/EC],
- other related texts.

Source https://ec.europa.eu/transport/modes/air/internal_market/handbook/part1_en, download Mai 2020

The European Community for Mobility and Transportation provides also "Safety Regulations and Standards for European Railways".

They provide a summary of the "Directive 2004/49/EC—safety on the EU's railways ". They define the following aims and key points:

What is the AIM of the Directive?

It Aims to Establish a More Competitive and Safer Railway System Which Covers the Entire European Union (EU) Market Area Instead of Confining Itself Mainly to National Markets.

Key Points

Scope

The directive applies to the railway system of the EU countries and covers safety requirements for the system as a whole, including infrastructure and traffic management, and the interaction between railway undertakings* and infrastructure managers*.

In This Connection, the Directive Focuses on ***4 Major Aspects***:

- the setting up, in each EU country, of an ***authority responsible*** for supervising safety,
- the ***mutual recognition of safety certificates*** delivered in the EU countries,
- the establishment of ***common safety indicators*** (CSIs) in order to assess that the system complies with the common safety targets (CSTs) and facilitate the monitoring of railway safety performance,
- the definition of ***common rules for safety investigations***.

Development and Management of Safety

Safety Rules and Standards, Such as Operating Rules, Signalling Rules, Requirements on Staff and Technical Requirements Applicable to Rolling Stock Have Been Devised Mainly Nationally.

These national safety rules, should gradually be replaced by rules based on common standards, established by technical specifications for interoperability (TSIs)—the ability of equipment or groups to operate in conjunction with each other. The European Commission has the power to suspend the implementation of a national safety rule for a maximum of 6 months.

In This Connection, the EU Countries Must Ensure that:

- Railway safety is generally maintained and continuously improved, taking into consideration the development of EU legislation.
- Safety rules are laid down, applied and enforced in an open and non-discriminatory manner.
- Infrastructure managers and railway undertakings must take responsibility for the safe operation of the railway system and the control of risks associated with it.
- Information is collected on common safety indicators through annual reports in order to assess the achievement of the CSTs and monitor the general development of railway safety.

Source https://eur-lex.europa.eu/legal-content/EN/TXT/HTML/?uri=LEGISSUM:l24201a&from=EN, download Mai 2020

Similar to typical contents of national Road Safety Acts, the focus is on:

- basic requirements on the transport ways (airways, railways, roads),
- regulations for the coexistence of the traffic participants,
- requirements on the transport conveyance (vehicle, train aeroplane), which are the typical homologation standards which describe the roadworthiness for vehicle or airworthiness for aeroplanes, etc.,
- licencing or certification for parts, equipment and transport conveyance,
- licencing for operator, driver, etc. (could be a person or an organisation).

Today, the basic principles are as follows:

- A driver drives the road vehicle.
- A pilot controls and operates a plane.
- A train driver drives a train.

Since in the past, the operator was mainly responsible to provide the infrastructure and an organisation or an individual driver owned the rolling or flying transport conveyance. Since in 2016, the change even in the "Vienna Convention" changed, a system could be considered to operate, control or "drive" a road vehicle (Fig. 5.25).

Usually, the following four parties are usually considered:

- The driver gets a driving licence based on road traffic training during driving school. Age and fitness could influence the capability, therefore renewal of licence is in discussion.
- The infrastructure provides quality and capacity of transport ways. They are responsible for repair and adaption of road capacity based on the growth of traffic.
- The operator is usually the government; they provide limitations and the necessary traffic environment. For the dynamic conditions and rules, the government provides the road traffic regulations.
- The vehicle provides feature and attributes, which are provided by the vehicle manufacturer including procedures for inspection, maintenance and modifications.

These were the aspects, which had been during "U.S. Automotive Industry Coalition" meeting with the target of harmonisation for worldwide harmonisation of homologation standards.

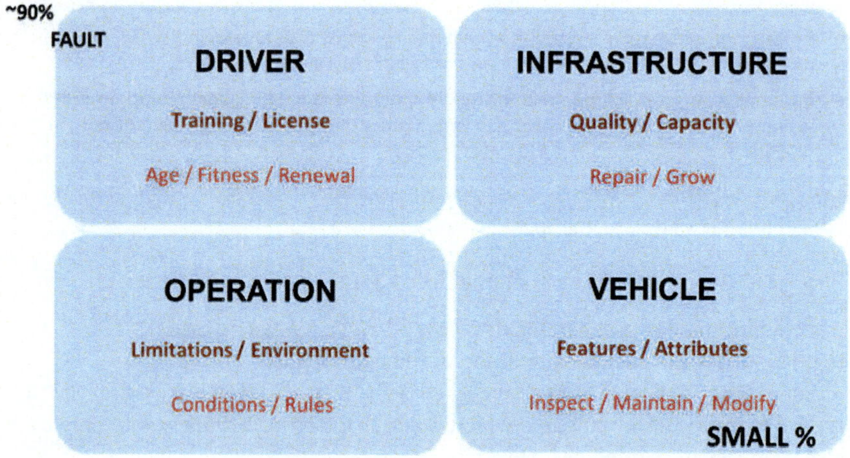

Fig. 5.25 Responsibility of stakeholder in road traffic. *Source* U.S. Automotive Industry Coalition Meeting, Andean/Mexico Delegation, December 7, 2016

One of the major sources of accidents had been analysed and the question arise, who is responsible for the compensation of occurring damages? The research of the history showed that nearly 90% of the damages occurred during accidents had been claimed to the driver.

Usually, the driver is responsible by law for the following:

- proving of vehicles drivability,
- adequate driving in road traffic,
- maintaining and assessing of his physical capability to drive the vehicle,
- adaption of driving environmental conditions.

The role for individual mobility already changed, since we are dealing with more car sharing, or other rental car business models. That the driver of a car is usually the owner of the car and consequently covers all that obligations, just to have the authority about the use of a specific vehicle becomes more and more negligible way of thinking. The burden of owning a car becomes too high in comparison with the advantages of owning a vehicle. As a consequence, the typical insurance business model will change.

5.10 Organisational Structure in Product Development

In order to manage a whole lifecycle, the classical safety lifecycle must change. The lifecycle according ISO 26262 starts with the item definition, since we like to consider the total vehicle as an item, various considerations about non-EE systems and their protection or safety function need to be re-evaluated. In order to address automated functions, an operation layer of system engineering based on an environmental-centric development approach is necessary. Considering not only traffic scenarios, but also the protection demand of other road traffic participants, a breakdown of the environmental-centric approach to a human-centric approach is necessary. The drivers, HMI, are one important interface, nearly as old is the need and the design of the compartment and the necessary occupancy protection mechanism. The typical misuse of other traffic participants was not typical part of the car manufacturer, it was more the job of driving schools and the governmental executives, like the police. Since today, most of the accidents are traced back to misuse by the driver and other road traffic participants. Safety-in-use focuses more on design issues, like sharp rims or ergonomically issues, like the view (of the driver), seat position, pedal arrangement, etc. System failures could be in the past well-managed and the lobbyists were strong enough, so that the vehicle design very seldom comes into public discussions. Failure of airbag-system led to the first public discussions and recalls with a high attention of the press. Field monitoring became very early a strong domain for the quality departments at manufacturer and TIERs, but most of the recalls were silent or garage repair measures.

5.10.1 Classical Organisation Until the Last Millennium

Logistic and marketing were the domain of the vehicle manufacturer.

Tier 1 provides the main parts such as steering , braking, propulsion and other interior systems.

A typical brake supplier was divided into the following main area:

- foundation brake (the basic hydraulic or pneumatic brake system),
- actuation, where the brake booster and the tandem master cylinder were the main product (the brake pedal and park brake lever, etc., were often supplier parts from secondary TIERs),
- electronic brake system,
- utilities such as pipe, hosts, discs and brake pads.

The main understanding in the legislation was that it was necessary to have two independent actuation systems. One for the normal operation or service brake, and the alternative or auxiliary brake system was the park brake which had the following basic design requirement:

- The vehicle could be decelerated with its overall allowed weight to a still stand and
- hold the vehicle with its allowed weight at a specific hill (e.g. 25–35%).

The basic brake system had in the early version of UN ECE R13h, to be designed in a way, that the foundation brake system could hold the vehicle against the engine torque. Later fixed data had been defined which were particularly calculated to achieve a specific stopping distance. The requirements for the brake booster grow, and the requirements for the foundation brake decreased.

The development of the electronic brake function starts as protection functions. The active brake force distribution was an adaption of the chassis systems with the upcoming brake disc, and it was realised that the front brake could provide shorter brake distances. The rear axle brakes were recognised to be important for the vehicle stability, with too high brake force the rear axles wheels tend to blocking, since the full weight of the vehicle moved to the front axle. Another effect was detected, wheels that tended to lock can no longer follow the steering request in steering systems according to the Ackermann principle. This led to the need and the development of anti-lock braking system (ABS). The ABS is primarily a protective mechanism to maintain steerability. First attempts to reduce the blocking effect by mechanical solutions such as orifices and throttle valves did not lead to sufficient results. Thus, the microcontroller and with it the software systematically moved into the essential safety systems in the vehicle.

The electronic brake departments separated into a hydraulic part, an electronic part, a sensor part and a continuously growing software department. Since software safety became a significant part of the software-intensive systems, the software departments distributed into base software development, application software development and the fail-safe or software safety department.

5.10.2 Future Driver of a Development Organisation

Since the vehicle systems became more software-intensive and nearly all major systems provide their key features based on software functions, the software came more and more in the focus. We will all wish for generalists such as Leonardo Da Vinci, but we will not find them in the future for such complex relationships. Thus, one is forced to develop whole groups of products according to certain process models. Today, it is still a common practice that a vehicle is considered finished after the start of production and then the manufacturers are paid according to the number of units delivered. It is already apparent today, through systems such as "Firmware over the Air" or the apps we know from our mobile phones that the systems and the vehicles only provide the carrier of functionality. The money is earned with the services that can be booked. The more modular the system platform, the better the business model is. For a successful product, other parties will also have to offer corresponding applications. Amazon makes no money by producing the products itself. Such a trading platform earns its money by relating customers and manufacturers. Such a platform also provides the basis for explaining the benefits of the product and promoting the capabilities of the manufacturer. Thus, the car itself will probably only become an essential carrier platform and the other services and services to be brought closer to the user. To ensure this, the vehicle must meet some basic requirements:

- a modular and adaptable platform,
- it must be safe and secure by design,
- open standardised interfaces for further value-added products and functions,
- rules and standards need to be defined, how to enhance the functionality,
- a business platform needs to be established.

This means that today's vehicle manufacturers are becoming producers of a platform whose purpose is to provide the means of transport.

In most countries, it is reserved for the state to provide the basic infrastructure, although here too Alan Musk has already positioned himself in the market with the Boring Company, a prominent advocate for private transport routes and infrastructure. In addition to the means of transport and infrastructure, there will have to be operators. These will offer services in such a way that people and goods are transported from A to B. They will have the original interest that the efficiency of the transport is in the first place.

In order to ensure operation over a longer period of time, maintenance is required for the means of transport and the infrastructure, as well as for the technical equipment. For the systems and components in question, there will again be manufacturers who guarantee maintenance and optimisation and also the adaptation of their systems.

Comparing this with aviation, there are manufacturers like Boeing and Airbus. Airlines like Lufthansa, Air China or Delta Airlines etc., ensures that the customers and goods get from A to B. The airport is operated by another company, and the air traffic area is usually a public authority which, together with service companies such as air traffic control, ensures the traffic area. The airlines decide which service the customer is actually provided with which means of transport. If one applies the ratios to road mobility, in future the operator will decide which vehicle will take the customer from A to B.

To this end, there will be regulatory authorities which, together with other government bodies, will be responsible for road safety. These will, of course, then also formulate the socially accepted risk level.

Chapter 6
Automated Driving and Control

Automated driving (AD) is another engineering discipline as development of driver assistant systems. The idea evolutionary development like five-level classification from SAE, VDA, etc., from manual driving via continuous enhancements of driver assistant system failed (Fig. 6.1).

Automation system requires a cybernetic approach, with a systematic breakdown of methods for communication and control strategy (Fig. 6.2).

> The introduction of conventional cars and trucks capable of operating without this active monitoring will represent a significant technical and conceptual leap. This threshold between partial automation (Level 2) and conditional automation (Level 3) corresponds to the line that several US states have drawn between non-automated and automated vehicles. Because of its assumption that the human driver will resume actively driving shortly after being prompted to do so, conditional automation raises particularly difficult issues of human-machine interaction that have not been satisfactorily solved. A more normative taxonomy developed by the U.S. National Highway Traffic Safety Administration arguably passes over conditional automation altogether on its way to high automation. (NHTSA 2013)
>
> *Source* OECD, Automated and Autonomous Driving Regulation under uncertainty, 2014.

6.1 Vehicle Behaviour

Any observer could describe any real-world scenario based on his perspective. Whenever the observer changes the viewpoint, or the context changed for the observation, the observed behaviour could be perceived differently. In road traffic scenarios, we need to find a perspective, which describes the behaviour of the vehicle in the road traffic context. Therefore, we need a standard perspective for the description of the behaviour of road vehicle. ISO 8855 provides basic terms and conditions, which are harmonised with SAE J670.

H.-L. Ross, *Safety for Future Transport and Mobility*,
https://doi.org/10.1007/978-3-030-54883-4_6

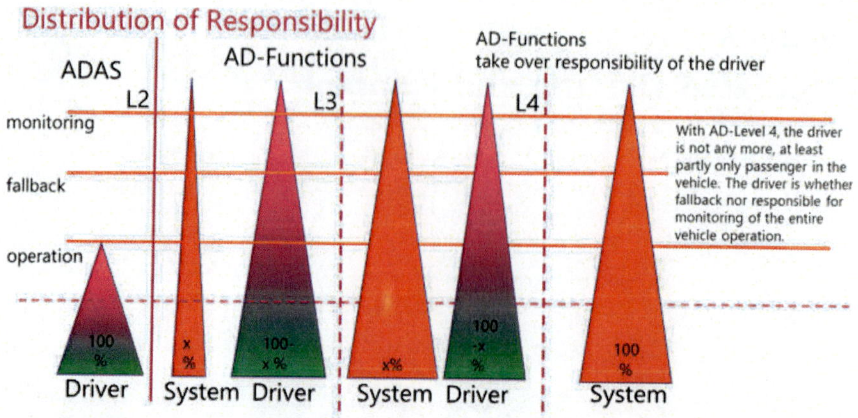

Fig. 6.1 Level of automation enhanced by automation functions

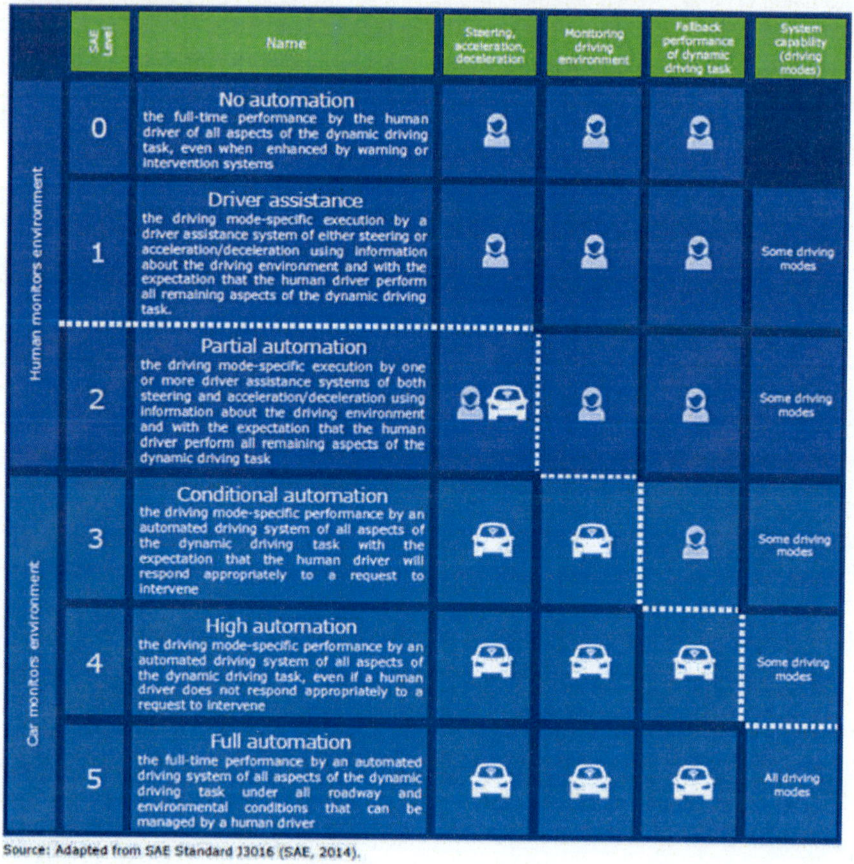

Source: Adapted from SAE Standard J3016 (SAE, 2014).

Fig. 6.2 Classification according to SAE J3016

ISO 8855:2011, "Road vehicles—Vehicle dynamics and road-holding ability—Vocabulary" provides the following Introduction and Scope:

> **Introduction**
>
> This International Standard defines terms appertaining to road vehicle dynamics, principally for use by design, simulation and development engineers in the automotive industries. This second edition has been prepared in response to a requirement to update the first and to harmonize its contents with that of the comparable standard published by SAE International (SAE J670:JAN2008). This revision extends the scope to include provision for separate tyre and wheel axis systems, inclined and non-uniform road surfaces, tyre forces and moments, multiple unit commercial vehicles, and two-axle vehicles possessed of four-wheel steer geometry.
>
> The vocabulary contained in this International Standard has been developed from the previous edition, and SAE J670, in order to facilitate accurate and unambiguous communication of the terms and definitions employed in the test, analysis and general description of the lateral, longitudinal, vertical and rotational dynamics of road vehicles.
>
> ***1 Scope***
>
> This International Standard defines the principal terms used for road vehicle dynamics. The terms apply to passenger cars, buses and commercial vehicles with one or more steered axles, and to multi-unit vehicle combinations.
>
> *Source* ISO 8855:2011

For the road vehicle, the standard focuses on the vocabulary to describe the behaviour of the vehicle on public roads. The holding forces are considered based on road surface conditions and road geometry.

6.1.1 Degree of Freedom of Road Vehicles

Driving on public roads is a closed-loop control function. Usually, the road provides the basic context, road traffic regulations, road markings and signs, etc., and provides the basic functionality. In order for a vehicle to drive on such a road, certain characteristics, properties, etc., are required for a vehicle to be sufficiently adequate for such road traffic environment. Usually, the homologation standards define the roadworthiness, for the dedicated area of their validity.

The essential nominal functions of a vehicle are:

- driving
- steering
- braking.

Without these basic control functions, the vehicle cannot perform its intended function at all. Without assuring such basic control functions a risk assessment is not possible because the reference is missing. The moving vehicle has a kinetic energy that already represents a basic risk. Steering particularly influenced the lateral movement, and the propulsion assures the forward driving. The basic brake functionality is the control of longitudinal movement, by reducing speed. Driving and braking are just a common mechanism to control lateral movement. Usually by gear selector, the driving direction during vehicle still-stand is preselected (Fig. 6.3).

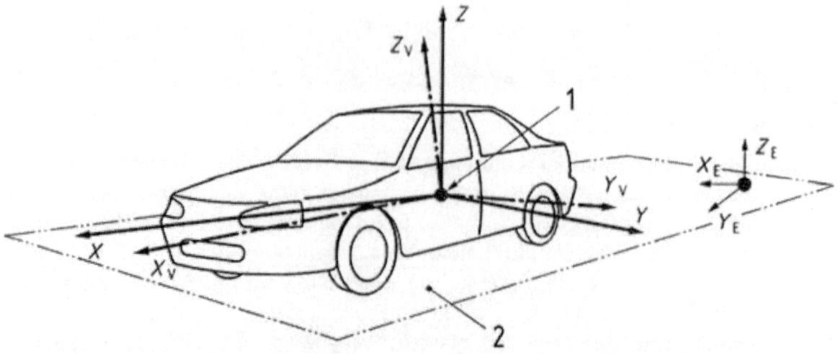

Fig. 6.3 Degree of freedom of road vehicles. *Source* DIN ISO in 8855 road vehicle dynamics and performance

In the DIN ISO in 8855, the vehicle parties become a coordinate system by the axes X_V, Y_V and Z_V stipulated, the horizontal coordinate system by *X*, *Y* and *Z*. The stationary coordinates are a *XE*, *YE* and *ZE*.

The following axes are considered:

- the vehicle moves on the *X*-axis in longitudinal direction,
- on the *Y*-axis in lateral direction and
- on the *Z*-axis in vertical direction.

The *Z*-axis is not influencing directly from the typical operation point of the vehicle. Suspension system would provide functions in *Z*-axis, but those are usually to compensate road design impacts. In order to improve the side stability, there are functions to improve the lane stability by the suspension system.

Mission Phase

The functional failure (malfunctions) is derived from the possible degree of freedom of a generic 4-wheel car. It is the major source of hazards during the mission phase of a car.

Longitudinal: Acceleration/deceleration

Malfunction
- Driving torque is too high/low/unintended
- Brake torque is too high/low/unintended

Hazard
- Side collision, pedestrian/bicycle accident due to sudden acceleration
- Rear collision, instable or not steerable vehicle due to locking wheels

Lateral: Left/right; steering/destabilisation; and yawing

Malfunction
- Yaw torque is too high/low/unintended
- Too less or too much steering angle/steering torque
- Oscillating steering

Hazard
- Insufficient vehicle stability control, under-/oversteering is possible
- Lane dereliction
- Accident with oncoming traffic.

In addition, the rotation effects of the vehicle could provide impact on the lateral and longitudinal axis of the vehicle. Usually, it should be part of the vehicle design to assure limited influence of those effects to vehicle stability.

DIN ISO in 8855 considers besides street design also the impact on the vehicle like weather. For example, wet road usually reduces the friction from the road, and the standard does not address direct influences to a specific sensor.

Also in a hazard analysis and risk assessment according to ISO 26262, malfunctions could derive from unintended deviations the degree of freedom of the vehicle. Considering a vehicle as an ITEM according to ISO 26262 could allow finding a bridge to the EUC approach (EUC—equipment under control) from IEC 61508.

According to ISO 26262, the degrees of freedom around the other axes, such as nodding, rolling and pitching, are only considered if EE functions have the ability to influence the behaviour of the vehicle.

Due to nodding and pitching, a vehicle seldom gets into extreme driving situations for chassis stability. What can mean, however, very intensive performance in extreme situations is known from the racing sport, and here there are accidents repeatedly where the air constructs a so immense pressure by the journey wind that vehicles take off or overturn even.

Hazards or dangerous situations typically occur when the vehicle behaves differently than expected. There are various classes of people who may be endangered by the unintended behaviour of the vehicle, such as the driver, occupancies and other road traffic participants like people in other vehicles, cyclists, pedestrians, etc. Also wrong handling of the driver or wrong behaviour by other road traffic participants could lead to dangerous situations. Malfunctions and failures are a potential source of harm, but many other classes of risk could lead to dangerous situations in road traffic.

There is a major experience; we learnt from avionics, airplanes, etc., that are safe by design. A plane flies safely due to its design. Due to the design of the fuselage and wings, an aircraft needs only one propulsion to fly towards it. When it is at a certain height, it can use the kinetic energy to fly or sail in a descent without any propulsion. A fighter plane needs a permanent electronic control system to be able to fly. Therefore, the design requirements for passenger transport aircraft are different from those for military aircraft or transport aircraft.

Today's road vehicles are also largely safe due to their design. The passenger compartment, all crash measures, but also the steering and the entire chassis design follow certain safety criteria. We use four wheels and steering system according to the Ackermann principles because we do not know any safer design. The car and the homologation regulations are arranged so that a responsible driver drives the car. If the vehicle is controlled by a system, many of the learned functions and abilities of the driver have to be recorded and evaluated and must be mapped by the system technology. A very big challenge is to transfer the vehicle control from a system to a human driver. This transfer requires a safe collaboration between the system and the human driver. Today, many researchers stated that a safe handover to human driving is only possible during still-stand of the vehicle.

6.1.2 Framework for Inertial System of Space and Time

A "frame of reference" is a standard relative to which motion and rest may be measured; any set of points or objects that are at rest relative to one another enables us, in principle, to describe the relative motions of bodies. A frame of reference is therefore a purely kinematical device, for the geometrical description of motion without regard to the masses or forces involved. A dynamical account of motion leads to the idea of an "inertial frame" or a reference frame relative to which motions have distinguished dynamical properties. For that reason, an inertial frame has to be understood as a spatial reference frame together with some means of measuring time, so that uniform motions can be distinguished from accelerated motions. The laws of Newtonian dynamics provide a simple definition: an inertial frame is a reference frame with a timescale, relative to which the motion of a body not subject to forces is always rectilinear and uniform, accelerations are always proportional to and in the direction of applied forces, and applied forces are always met with equal and opposite reactions. It follows that, in an inertial frame, the centre of mass of a closed system of interacting bodies is always at rest or in uniform motion. It also follows that any other frame of reference moving uniformly relative to an inertial frame is also an inertial frame. For example, in Newtonian celestial mechanics, taking the "fixed stars" as a frame of reference, we can, in principle, determine an (approximately) inertial frame whose centre is the centre of mass of the solar system. Relative to this frame, every acceleration of every planet can be accounted for (approximately) as a gravitational interaction with some other planet in accord with Newton's laws of motion.

It originated in a profound philosophical consideration of the principles of relativity and invariance in the context of Newtonian mechanics. Further reflections on it, in different theoretical contexts, had extraordinary consequences for twentieth-century theories of space and time.

The entire light electric effects which we use for lidar or camera systems do not follow in all effects and conditions Newton's laws. Here, we need the "theory of special relativity". The theory of special relativity explains how space and time are linked for objects that are moving at a consistent speed in a straight line. One of its most famous aspects concerns objects moving at the speed of light.

For collaborative automation systems, with many independent moving objects, vehicles and people in the same space, such a unique reference frame is necessary. By luck, in general we are on the same planet, with nearly the same gravity, which could be considered for all actors, items, obstacles, etc.

The deviation and impact of gravity due to height and position are only relevant to some measuring principles like pressure or rotation measurement. Nevertheless, since such measurements based on different references we get context failure, which could not be detected during examination of the device under development on its own. Context-specific validation becomes necessary, and correctness could be only claimed for the same context condition.

ISO 8855 considers different vehicle reference points such as:

- C.G. (centre of gravity)—a point in the vehicle reference frame that coincides with the centre of mass of the entire vehicle when the suspensions are in equilibrium and the vehicle is resting on a flat-level surface,
- Aerodynamic reference point—a point in the vehicle reference frame that lies in the intersection of the vehicle plane of symmetry and the ground plane, midway between the front and rear axles, when the suspensions are in equilibrium and the vehicle is resting on a flat-level surface.

It is based on the following axis:

- vertical position *Z*—*ZE* coordinate of the C.G.,
- *X* position *X*—*XE* coordinate of the C.G.,
- *Y* position *Y*—*YE* coordinate of the C.G.

The standard describes the following axes of motion for a vehicle:

- lateral acceleration A_y—*Y* component of acceleration vector of the C.G.,
- lateral velocity V_y—*Y* component of velocity vector of the C.G.,
- longitudinal acceleration A_x—*X* component of acceleration vector of the C.G.,
- longitudinal velocity V_x—*X* component of velocity vector of the C.G.,
- vertical acceleration A_z—*Z* component of acceleration vector of the C.G.,
- vertical velocity V_z—*Z* component of velocity vector of the C.G.

For the front of the vehicle, there are definitions, which are also important for the positioning of sensors:

- pitch centre—imaginary point in the XVZV plane through the lateral centre of the vehicle reference frame at which a longitudinal force applied to the vehicle body is reacted without causing suspension jounce (front or rear),
- roll centre—imaginary point in the YVZV plane containing the two wheel centres of an axle, at which a lateral force applied to the vehicle body is reacted without causing suspension roll angle (Fig. 6.4).

The yaw, pitch and rolling angles are important for the chassis design; the yaw rate is the basis for electronic stability control (ESC) systems. Rolling and pitching influence not only in the mathematical dependency the yaw-rate, especially, if the sensors are not exactly installed in the vehicle centre of gravity, all movements of the vehicle lead to deviations and need to be compensated (Fig. 6.5).

A camera, radar or lidar sensor is usually not implemented in the centre of gravity or any other central position in the car, and they are implemented, for example, at the 2 corners symmetrically in the vehicle front. Their direction is typically zero degrees from the horizontal in order to have a traceable reference to the driving direction. Depending on the distance targets for the sensor, the angle could be higher or lower than zero degree from the horizontal line. Due to pitching effects, the sensor centre line moves up and down depending on the road geometry (Fig. 6.6).

The rolling of the vehicle leads to up and down and to deviations from the horizontal line for the environmental sensing. For the vehicle front, the rolling centre

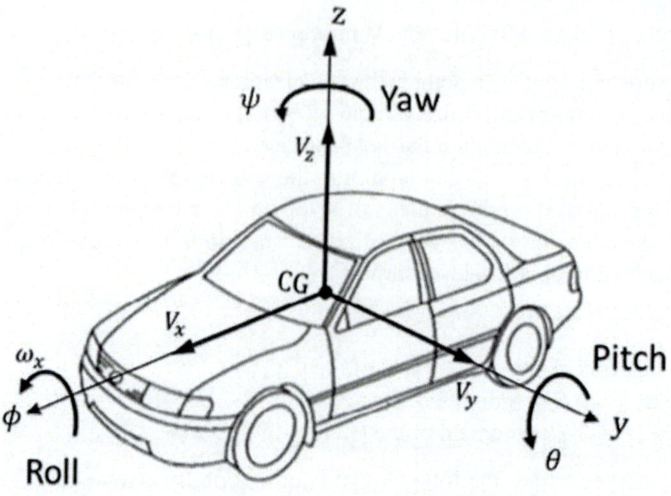

Fig. 6.4 Axes of vehicles according to ISO8855. *Source* ISO 8855

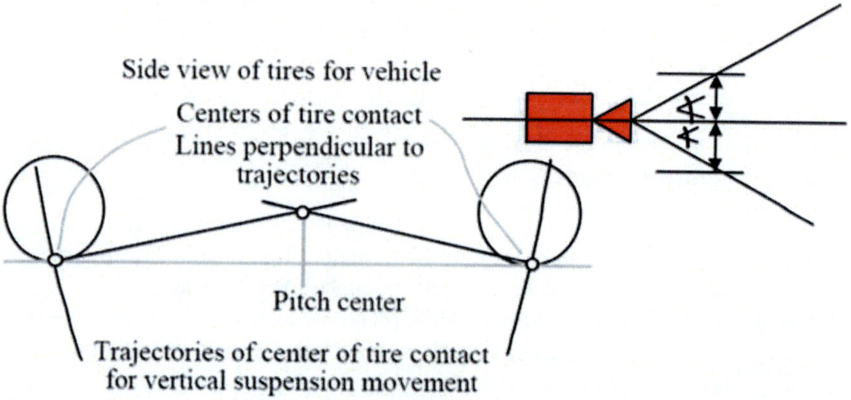

Fig. 6.5 Influence of pitching for the sensor measurement

Fig. 6.6 Influence of rolling for the sensor measurement

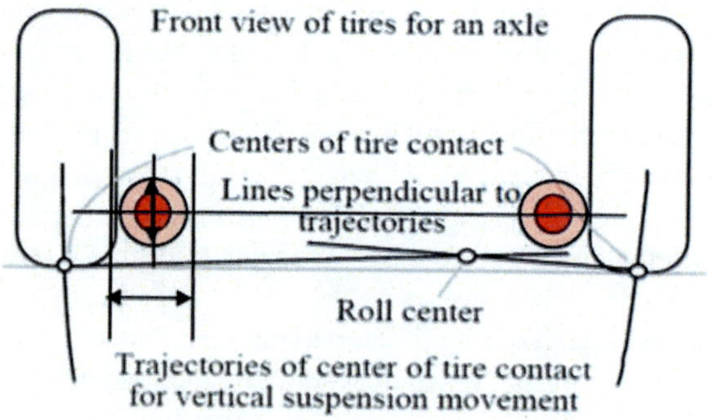

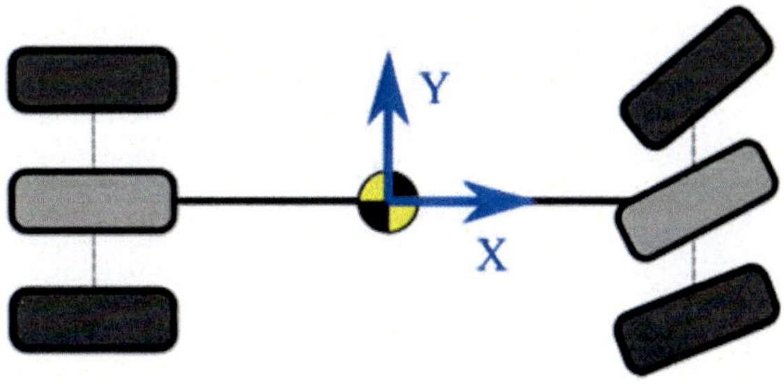

Fig. 6.7 Base of single-track model

is depending on the front axle geometry and so strongly depending on the current steering angle of the vehicle.

For the brake system design, usually a single-track model is sufficient and the reference for such a model is the rear axle, since this is independent from the steering angle (Fig. 6.7).

In a typical single-track model, there are two tires in single axle combined to a single virtual tire per axle. The centre of gravity built the reference for longitudinal and lateral movement.

Integrating a single-track model into a real-world coordinate system uses in the first glace also the rear axle as a reference point (Fig. 6.8).

The wheelbase of the single-track model considers a fixed rear-axle and synchronous wheel angle. The wheel slip angle is a resulting unit based on many vehicle data in dynamic mode. The vehicle moves based on a fixed steering angle in a specific distance around a turn point.

6.1.3 Road Vehicle in "Real World"

In case of integrating the single-track model into a world coordinate system, using the rear axle centre is an often used reference (Fig. 6.9).

The world coordinate system is often called also a Cartesian coordinate system. A Cartesian coordinate system is an orthogonal coordinate system. It is named after the Latinised name Cartesius of the French mathematician René Descartes, who introduced the concept of "Cartesian coordinates". In two- and three-dimensional space, it is the most frequently used coordinate system, since many geometric facts can be described clearly and concisely in this system. It that specifies each point uniquely in a plane by a set of numerical coordinates, which are the signed distances to the point from two fixed perpendicular oriented lines, measured in the same unit of length. Each reference line is called a coordinate axis or just axis (plural axes) of the system, and the point where they meet is its origin, at ordered pair (0, 0). The coordinates can also be defined as the positions of the perpendicular projections of the point onto the two axes, expressed as signed distances from the origin. Typically, an abscissa is the X-value of a point in the Cartesian coordinate system. The ordinate is the Y-value (Fig. 6.10).

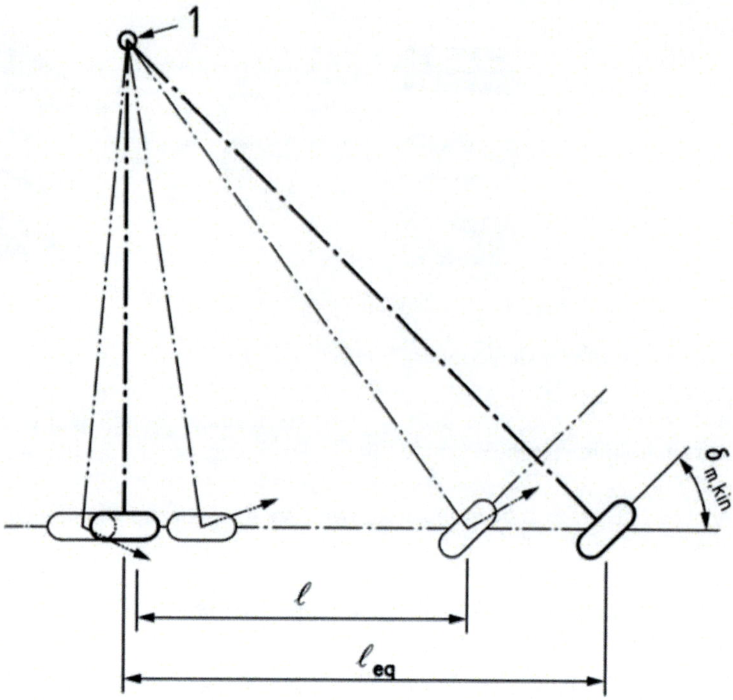

Fig. 6.8 Wheel distances and angles in a single-track model (equivalent wheelbase). *Source* ISO 8855, 2013

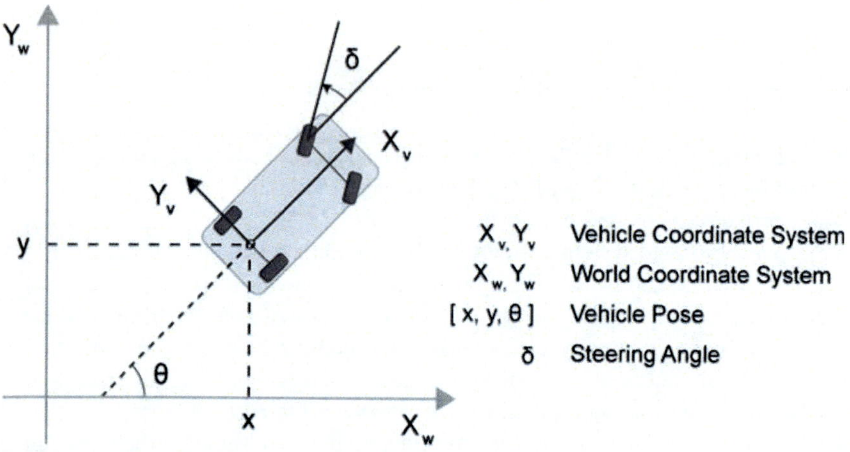

Fig. 6.9 Vehicle coordinates in the world coordinate system

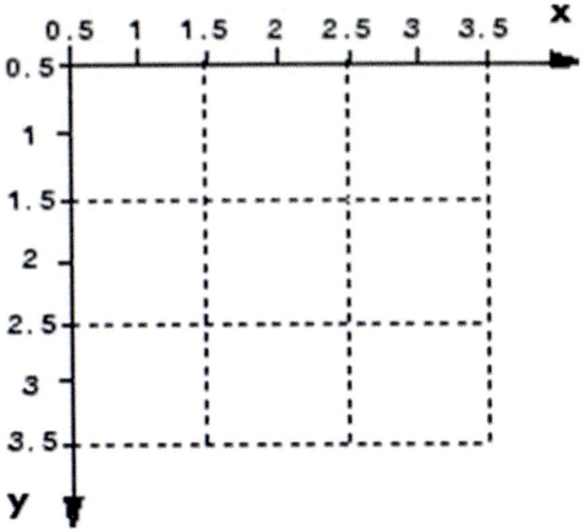

Fig. 6.10 Typical Cartesian coordinate system

Based on that understanding, we consider the world coordinate system as a spatial coordinate system. Another spatial coordinate system is a polar coordinate system or spherical coordinate systems (Fig. 6.11).

Spherical coordinate systems are coordinate systems for three-dimensional space where three numbers specify the position of a point:

- the radial distance of that point from a fixed origin, its polar angle measured from a fixed zenith direction,
- the azimuthal angle of its orthogonal projection on a reference plane that passes through the origin and is orthogonal to the zenith, measured from a fixed reference

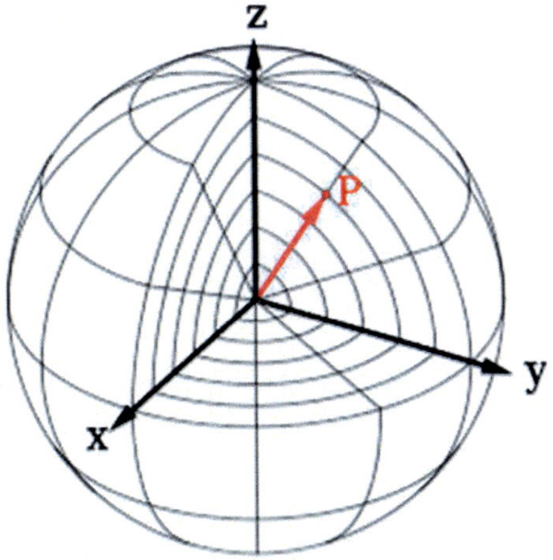

Fig. 6.11 Spherical or polar coordinate system

direction on that plane. It can be seen as the three-dimensional version of the polar coordinate system.

There are further coordinate systems which bring the earth orientation into the framework, e.g. for navigation systems.

Local tangent plane (LTP) coordinates, sometimes named local vertical, local horizontal (LVLH) coordinates, are a geographical coordinate system based on the local vertical direction and the earth's axis of rotation. It consists of three coordinates: one represents the position along the northern axis, one represents the position along the local eastern axis, and one represents the vertical position. Two right-handed variants exist: east, north, up (ENU) coordinates and north, east, down (NED) coordinates. They serve for representing state vectors that is commonly used in aviation and marine cybernetics (Fig. 6.12).

There are 2 typical variants:

- Local east, north, up (ENU) coordinates; in many targeting and tracking applications, the local east, north, up (ENU) Cartesian coordinate system is far more intuitive and practical than ECEF or geodetic coordinates. The local ENU coordinates are formed from a plane tangent to the earth's surface fixed to a specific location, and hence it is sometimes known as a "local tangent" or "local geodetic" plane. Typically, the east axis is labelled with "x", the north with "y" and the up with "z".
- Local north, east, down (NED) coordinates; in an airplane, most objects of interest are below the aircraft, so it is sensible to define down as a positive number.

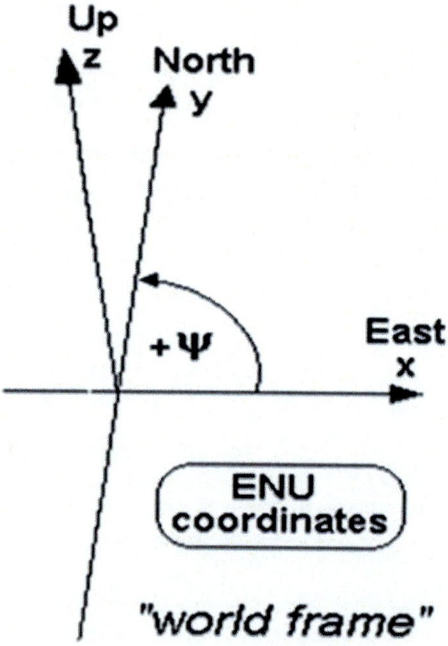

Fig. 6.12 Local east, north, up (ENU) coordinates

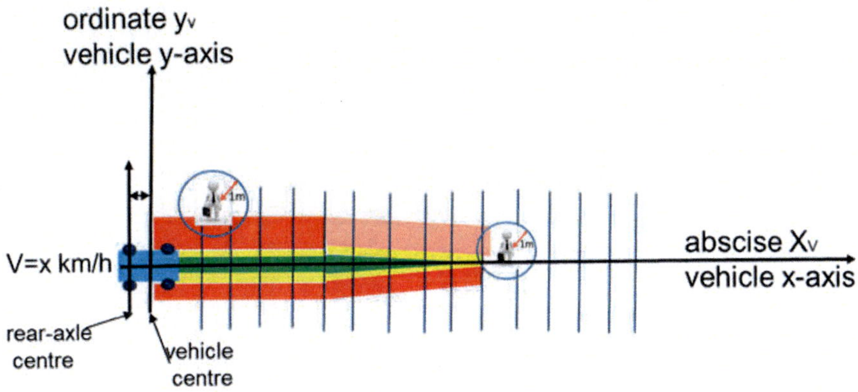

Fig. 6.13 Vehicle ordinate and abscise

The origin of this coordinate system is usually chosen to be a point on the surface of the geoid below the aircraft's centre of gravity. However, care must be taken since, if the aircraft is accelerating (turning or accelerating linearly), then the NED coordinates are no longer inertial coordinates (Fig. 6.13).

For brake system simulation, the alignment with rear axle centre is typical reference for longitudinal movements. Steering system goes for the vehicle centre, due to the dependency of longitudinal movement and lateral forces, e.g. centrifugal forces. In the longitudinal reference, the difference is fixed for any vehicle. For the simulation of mobility concepts, all elements need to consider the same reference frame.

6.1.4 Criticality Depending on Distance

A vehicle has a local risk area, and due to its speed, the risk is speed depending. Considering only the longitudinal axis, we need to consider that stopping distance increases based on the square of the speed. In driving school, we learn that the brake distance from 100 kph is about 100 m. Based on physical equations, the braking distance or stopping distance is the distance a vehicle covers from the time of the full application of its brakes until it has stopped moving. This is often given as a "100–0 kph" distance, e.g. 56.2 m, and is measured on dry road. Some $M1$ class vehicle makes it below 35 m; in order to get good voting from NCAP, a vehicle should make the test below 40 m. In case of human driving, we need to add the driver reaction time. The distance the vehicle travels is also called thinking distance. A vehicle travels the thinking distance, when the driver applies the brakes after realising they need to stop. The thinking time needs to be added to the braking distance, which means the distance a vehicle travels in the time after the driver has applied the brake. Perception distance is the distance your vehicle travels from the time your eyes see a hazard until your brain recognises it. The perception time for an alert driver is about

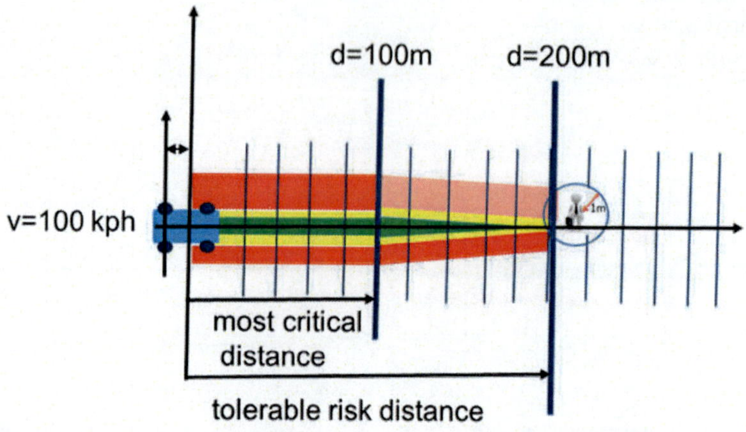

Fig. 6.14 Critical distance

usually considered as one second. At speed of 100 kph, we travel about 40 m in those second.

Human distance to still-stand = perception distance + thinking distance + stopping distance

For a speed of 100 kph during good weather conditions, a very conservative calculation leads to the following data:

stopping distance: 40 m,
thinking distance: 40 m,
perception distance: 40 m.

This leads to a distance to standstill of 120 meters from the vehicle at a speed of 120 kph. The normal perception time and reaction time of the driver is 1 second. However, if the driver is not fully engaged in the driving process, for example because he is distracted, the time can increase significantly. The stopping distance is mainly depending on road and weather conditions and on vehicle capabilities.

From ISO 26262 and from accident research, we learn that a collision speed of 40 kph could lead to dead, which leads in the average that we need to consider a high severity, in the area of 100 m in front of a vehicle driving with a speed of 100 kph (Fig. 6.14).

Unless we are on a motorway, we consider no people in the road area. On country roads, where law allows speed up to 100kph, we need to consider people on road. Even during an accident, we need to consider people on a motorway. Based on the assumptions, we need to consider a distance in front of 200 m, to enter a corridor of tolerable risk.

Lateral Movement

The other dimension of the moving vehicle is the lateral movement. Consider a perfect road, a perfect steering system and a perfect driver would lead to no lateral movement

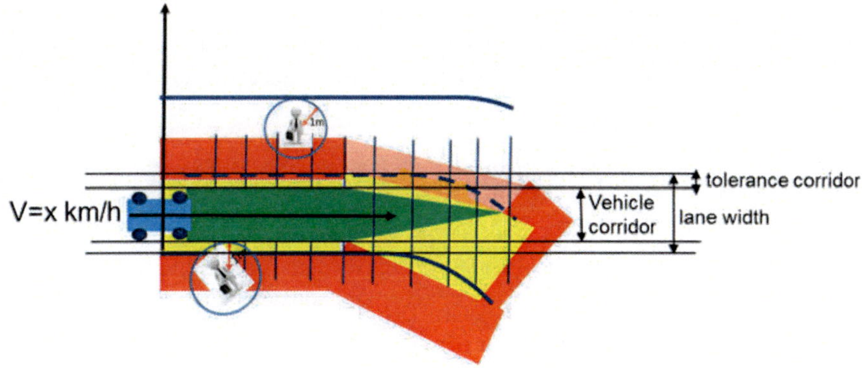

Fig. 6.15 Lane in relation to lateral movement

on a straight road. The average vehicle has a width of more than two metres (2 m + mirrors). According to the design guidelines for motorways, the corridor for the road width increases with the permissible vehicle speed. The maximum vehicle width on European roads is 2.5 m. The right lane of German motorways is usually 3.5 m. This width allows only half a metre spare on each side of a truck. Any deviation of the steering angle could lead within one second to a significant dissipation from the lane. Many judicial practices show the following risk tolerance corridors:

- A person feels endangered if even at low speed a car passes the person at a distance of less than one metre.
- A cyclist is considered to be at risk if a vehicle passes at a distance of less than 1.5 m (Fig. 6.15).

On a completely straight road, the corridor for a vehicle can be easily determined. In this case, the tolerance corridor can be well determined by the driver and the steering. Modern electronic steering assistance systems also correct slight steering effects by the driver. If the road is uneven or curvy, it becomes very difficult to predict the lane. A slow and cautious driver will try to take the curves exactly in the middle of the lane, and a fast sporty driver tends to cut the curves.

For urban roads, such distances require a lane change. For pedestrians on the pavement, such distances represent a continuous risk if they walk along the edge of the road. Any risk of unexpected behaviour of any actor in public road traffic means a risk. Therefore, all road traffic laws require special care and consideration for weaker road users. These subjective limits are weighed by people and are weighed differently in different situations. Man drivers simply drive with a higher willingness to take risks. Such behaviour cannot be implemented in a technical system, which is why more and more lobby groups are demanding clear traffic rules for automated driving. It is massively doubted, whether in the current situation, society and politics are prepared to do this at the expense of pedestrians and cyclists. Today we always discuss a "human-centred design" in urban road traffic environment. There are several research activities, which deal with the coexistence of road vehicle and other road traffic participants. However, even in Asia, urban city planner considers the need of the entire planning of new living quarters. They put special care on handicapped people and other groups that require higher protection such as children.

6.2 Driver–Vehicle Interface

For any mobility scenario, we should not forget the people. They could act in various classes. Typically, the following classes of actors are considered:

- drivers (of the AD vehicle, or driver of other vehicles)
- occupancies
- pedestrians
- cyclists, scooter driver, skateboarders, etc.

While drivers and occupancies are usually inside of the vehicle, pedestrians, cyclists, skateboarders, etc., are outside of the vehicle.

For the driver inside the controlled vehicle, we consider a typical human–machine interface, but for all the other road traffic participants, the automated vehicle is a counterparty as any human-controlled vehicle as well. They need to share the same space, but only the road traffic regulation provides the rules for periodisation and condition.

The usual operation scenario sees the vehicle under control just as a black box, with its possible degree of freedoms. In a HARA, you would never consider that a vehicle would lift up to the sky by its own forces. The driver and the occupancy are not visible on that layer of abstraction. Consequently, the interface of the driver to the vehicle becomes only visible on a lower level of abstraction, where you see the pedals, steering wheel, gear lever, etc. (Fig. 6.16).

In order to describe the interface to the driver, the driver interfaces need to be identified. Braking, propulsion and steering are an evolution since more than hundred years of vehicle development. Since Bertha Benz drove with a three-wheel motor vehicle, today all vehicles have four wheels using a steering system based on the Ackermann principle. There are rear-wheel, front-wheel and all-wheel propulsion today available; only similar to the beginning of automobilisation, we try to use again an electric powertrain. Due to the changed torque characteristics of the e-drives, the requirements for the chassis design change. This leads to advantages and disadvantages also for the vehicle interior. It provides new room in the compartment, and usually less vibrations leads to higher comfort during driving. However, the different E-drive concepts and their characteristics lead also to chassis design changes due to the heavy weight of the traction battery.

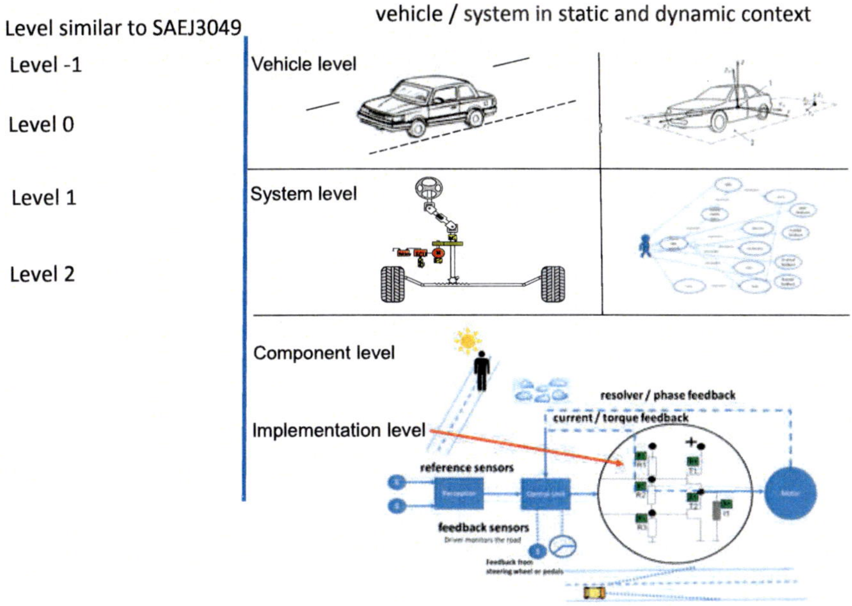

Fig. 6.16 Hierarchy and various viewpoints

6.2.1 Controllability by the Driver

In higher level of automation, the driver interface is only a demand interface; up to SAE Level 3, the driver interface plays a significant role. In driver assistance systems, the steering system only provides support for the driver's intended behaviour, while in automation systems the driver interface must be adapted to the different system states in the various operating situations.

In normal assistant system, it is necessary to analyse the acceptable degree of influence by the system, which allows the driver to control any driving situation. Depending on the torque of the power steering system and the speed of the vehicle, the possible impact on the reaction and the vehicle behaviour is different. At a higher speed, an unintentional steering intervention has more effect on lane keeping than at a low speed. Consequently, tolerable disturbances of the system to the steering system are speed-dependant, which lead to lower ability of the driver to control the failure of the power steering system (Fig. 6.17).

Since ISO 26262 considers only an ITEM, which is for example a steering system, the vehicle context and possible safety-relevant impacts are not considered by the standard. The

- driver
- environment and
- other (road) participants

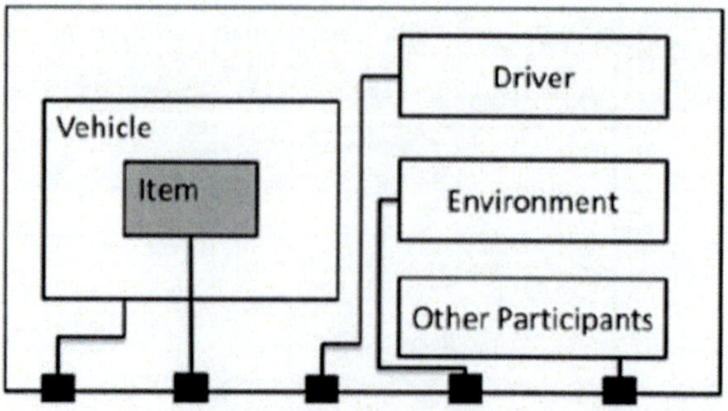

Fig. 6.17 Example similar to ISO 26262, where the ITEM is part of the vehicle

affect the ITEM on very heterogeneous ways. There are many influence factors between the driver and the steering system (Fig. 6.18).

The ASIL distribution for power steering systems follows often a hyperbola. The typical exposure rating for a power steering system is $E = 4$. The occurring damages are leading to death when a significant departure of the lane would be considered; the controllability is depending on the possible forces at the steering wheel. The ordinate in the diagram shows the possible unintended forces or the performance. Another aspect is that the kinetic energy of the vehicle grows with its speed. On the ABS ice, the possible intensity of the malfunction provides the relation to the possible performance of the power steering system. The following failure modes could be considered:

- too strong intervention of the power steering system, degree of intensity of possible malfunctions,

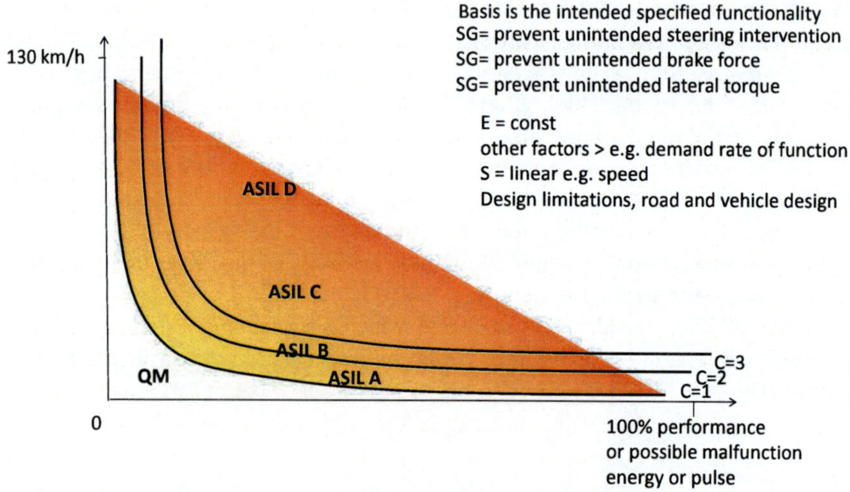

Fig. 6.18 ASIL hyperbola covered to the controllability by the driver

- too strong or asymmetric braking at one wheel,
- too strong deceleration due to propulsion malfunctions; usually only in curves, the propulsion leads to unwanted lateral interventions.

As higher the intensity of the unintended lateral effect, as lower a driver could control possible failure effects of the systems. It depends of course on driving situations and road conditions, if and when such malfunctions lead to lane dissipations.

The following effects could be considered:

- accident with oncoming traffic, by lane dissipation,
- accident with vehicles on other lanes with the infrastructure, by vehicle destabilisations, e.g. yawing effects, etc.
- wrong intervention by the driver due to unexpected system or vehicle impact.

The driver could also react wrong on system effects or on unexpected vehicle movements.

Typical causes for wrong driver interventions are:

- Malfunction of a vehicle system like braking or steering leads to an effect at the steering wheel, and the driver tries to compensate this effect in an unexpected manner.
- The vehicle drives through a pothole, and the driver reacts too intensive to compensate the bump.
- The driver gets information from other systems like navigation or other and reacts wrong.
- The driver reacts wrong due to the behaviour of other road traffic participants, swerving of a sudden animal into the oncoming traffic.

Especially the evaluation, how the driver reacts on the various sources of system and vehicle impacts or on the various sources of information, is very difficult to assess.

In the ITEA2 funding project "Safe", work package 3 looked at steering systems. The influence on the driver was simulated, and the limits of controllability were analysed.

An electrical steering assistance is used to demonstrate the variability approach. The steering wheel in driving direction commands a change of the steering angle. In addition, the driver receives feedback from the road at any moment in time unless he holds the steering wheel. The following requirements and interactions with the environment apply to steering systems:

- The required hand forces of the driver should be as low as possible, but the force gives the feedback so that the driver is sufficiently involved in the steering process.
- The rotation and number of turns of the steering wheel are linear equivalent to the steering angle and the intensity of the steering torque initiated by the driver.
- The steering system returned to the straight-ahead position without any force from the driver while driving.
- The feedback of the road condition is noticeably communicated to the driver.

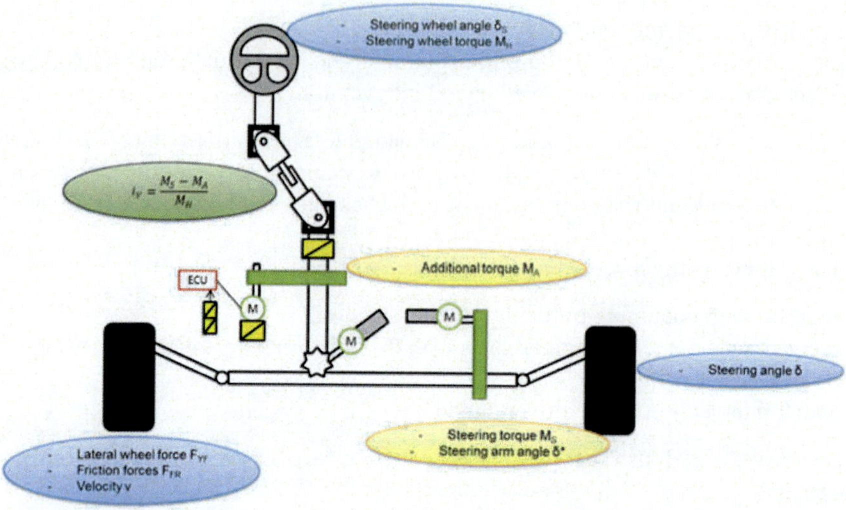

Fig. 6.19 Steering system and system interfaces. *Source* Safe, ITEA2, WP4, Deliverable D3.4.a, "Guideline for description of variant management techniques"

- The legal requirements for maximum operating force and the duration of the steering assistance provide a constant steering feel for the driver (Fig. 6.19).

The electric power steering system is based on the following basic principles:

- The rotation of the steering wheel is detected by a torsion torque sensor, which transmits the value to an electronic control unit (ECU). The ECU calculates the necessary supporting torque, the desired power amplification in the respective time interval.
- The power amplification is transmitted to the steering system via a transmission, the electric motor and the power electronics at the control unit. The design definition of the elements also provides us with the technical view of the architecture including the technical aspects of the internal and external interfaces.

The identified risk in this scenario is unintended torque when no torque or such an intensive torque is requested:

- context: speed higher than 50 kph and curved road with a steady radius,
- risks: within 100 ms, torque MaxTorque(x) exceeds higher than specified for a duration higher than x,
- MaxTorque(x) gives the at most tolerable torque during a period of 100 ms (Fig. 6.20).

The engine torque and the temporal course of the moments that the driver perceives as disturbing or even uncontrollable steering wheel torque depends on the duration and the intensity of the deviation from the desired correct (at that moment in time) value.

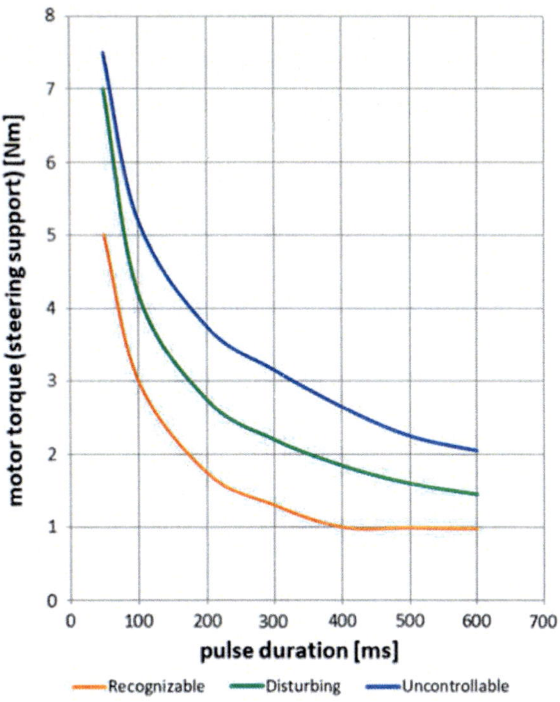

Fig. 6.20 Motor torque and duration of a steering system. *Source* Safe, ITEA2, WP4, Deliverable D3.4.a, "Guideline for description of variant management techniques"

Based on those assumptions, the possible failure modes on vehicle level are identified and resulting safety goals according to ISO 26262 could be formulated in the way

- "Avoid that any malfunction leads to a hazardous situation".

During the next steps, the possible design limits for the power steering system should be specified. The following questions (e.g. derived from a HAZOP) are providing criterion for such limits or constraints:

- What are the correct steering torque characteristics from the power steering pack, so that the driver still could safely operate the vehicle?
- What limits of steering torque are acceptable in terms of duration and frequency?
- What maximum transient and torque limits could be considered for the implemented mechanical part of the steering system? Does the power steering pack provide the necessary characteristic and performance?
- Could the power steering system provide the necessary reaction and performance to adapt the torque characteristics to sudden events, due to bad road conditions or any other external event?
- Could the power steering system provide the right reaction and sufficient performance in case of electrical failure?

In order to design a steering system, not only the electrical failure could lead to hazardous situations. The entire power steering system needs to be safe, and safe means much more than controlling of EE failure.

Typical safety goals according ISO 26262, would not sufficiently cover the design constraints of the entire steering-system, the following kind of safety mechanism need to be considered as well:

- What kind of protection mechanism is necessary for the mechanical part of the steering system? What limits and constraints are relevant?
- What safety measures are necessary to compensate credible sudden effects from the road or any other external impacts?
- What design limits, constraints and limits are to be considered, so that the driver could control the driving situations?
- What torque characteristics allow a takeover by the driver after termination of lane-keeping functions?
- etc.

Such design limits or constraints could be considered as systematic faults during development of the item "electric power steering system", but the activities as provided by ISO 26262 do not lead to adequate measures. An ASIL tailoring for such design-related failure is not possible, since a cause is not an EE failure.

A very concise example is the necessary performance of braking systems. The nominal performance of a brake system refers to the possible maximum pressure gradients and the volume flow of brake fluid into a brake calliper. This maximum pressure gradient can be sufficient to brake a car up to 3.5 ton, but not for a 40-ton truck. Where we will end up with a battery electric vehicle and whether the brake systems will provide sufficient brake performance for vehicles with a weight of more than 3.5 tons could be difficult. A pneumatic brake system would not be a solution.

The wheel speed sensors may be able to detect the grip on the asphalt during braking, but three is not adequate measuring technology to measure the grip in advance before braking. In automotive technology, nearly no sensor does measuring the targeted physical effect. There is always a physical effect, which is linear or in any other deterministic relation to the measuring target, that will be calibrated or configured in a way, so that we have a sufficient confidence for the measurement. The torsion bar in a power steering system is a strain gauge, which is calibrated to the torque difference between drive hand torque and the torque affected by the road friction and the steering system stiffness.

A wrong calibration, configuration or design could not be traced to EE faults, and the fault is a systematic error usually of the design engineers or at least of the product development process and the applied methods.

"Indirect measurement" is normal today to measure the nominal or functional performance or the necessary system capability.

The next challenge, is the design of the entire actuator system to control the possible critical scenarios by EE-functions, does it need protection mechanism of other technology, or there well-proven design principles, which are State-of-the-Art and they could not change.

The question is the design world-proven?

These are all aspects to determine a nominal function for situations and certain states and transitions of states as correct. Aspects of occupant protection or conceivable faults in the steering or braking system that require special protective functions have not yet been considered. An important example is the active safety or protective function of anti-lock braking system (ABS), which has the task of avoiding excessive pressure gradients in the braking system that lead to locked wheels.

Nevertheless, what is the danger behind it?

- With blocked wheels, the vehicle is no longer able to steer, and the vehicle pushes straight ahead over the front wheels despite the steering angle.
- The rear axle no longer guides the vehicle, and it loses lateral stability or other effects.

However, the ABS protection function is only a brake protection function. The ABS function could not avoid blocked wheels or drive axles by the propulsion.

A blocked drivetrain could be also caused by:

- The rear axle commands forward, and the front axle commands backward or vice versa.
- excessively recuperation or is mechanical blocking.

Usually, all that failure could not be avoided and very seldom controlled by ABS or a traction control function in the brake system.

There is also a big discussion on how much detection from the real world is enough by the environmental sensing. The radar by its measuring principle is perfect to detect moving objects, and any speed camera has caused a lot of pain, due with its correctness in speed measurement. However, is the radar the right device for object detection? Is a camera the perfect device for 3D measurement? The radar is a perfect distance control and warning system even more for dynamic objects. A stereo camera has no huge distance between the two imagers, very similar like the human eyes. Similar to our eyes, the stereo camera can recognise many patterns, also 3D pattern, but it remains at least a recognition system. The lidar is a bit better for shapes, but even if it is that expensive, it is not a perfect sensor for everything.

The inadequacy of functions could be considered within the framework of IEC 61508, as this standard has defined its reference to "EUC, equipment under control" (the system to be protected). In most cases, the EUC would be the vehicle, so according to IEC 61508 the decisive question in validation would be whether the vehicle can steer and brake sufficiently safely by means of the defined braking or steering system, if it functions correctly. However, can the item according to ISO 26262 (the system under consideration) not be regarded as control function of the vehicle?

IEC 61508 always refers to the EUC when considering a system. This means that in the context of IEC 61508 a correctly functioning machine (example of an EUC) would never lead to harm or accidents, unless the system performs, works, operates and behaves as specified. In a second step, possible failure in the entire equipment, mechanical and electrical equipment, will be analysed and the risk based on this analysis will be assessed. On the operation level, adequate protection measures and usability concepts are elaborated. After a technical risk analysis, adequate measures are defined based on electrical or electronic technology but also other technologies like hydraulic, pneumatic or other mechanical solutions. For any deriving EE system, there are the following classes of systems

- EE nominal control function, like the operational brake system or the basic steering system,
- protective control systems, which control operational risk or system risks,
- EE safety mechanism that had been developed according to a safety integrity standard like ISO 26262 or IEC 61508.

IEC 61508 distinguished between "on-demand mode" and "continuous mode" of operation. On the operation layer, a vehicle distinguished the following principles:

- A braking system is an on-demand system (it must intervene effectively when requested).
- A steering system is a continuous mode system (it acts continuously).

However, we should understand that any protection system or safety mechanism usually is just on-demand systems, but sometimes the time interval or the demand rate is so low that a continuous mode of operation should be considered. For the entire event or failure, monitoring a cyclic diagnostic is necessary anyway.

Analysing accident scenarios and their cause is usually also very heterogeneous and is an experienced-based science. Forecasting of accidents fails very often.

6.2.2 Accidents and Measures Against Their Root Causes

Accidents are a class of adverse events. They have many causes. What may appear to be bad luck (being in the wrong place at the wrong time)? By analysing, be seen as a chain of failures and errors that lead almost inevitably to the adverse event?

These causes can be classified as:

- immediate causes: the agent of injury or ill health (the blade, the substance, the dust, etc.),
- underlying causes: unsafe acts and unsafe conditions (the guard removed, the ventilation switched off, etc.),
- root causes: the failure from which all other failings grow, often remote in time and space from the adverse event (e.g. failure to identify training needs and assess competence, low priority given to risk assessment, etc.).

To prevent adverse events, you need to provide effective risk control measures, which address the immediate, underlying and root causes.

The role of human impact in major disasters has been widely acknowledged, with studies concluding, that human error is responsible for a high average of accidents.

Accident analyses of industrial disasters like Seveso, Three Mile Island, Bhopal and Chernobyl confirm the view. Two types of human error are said to cause system disasters:

- active errors
- latent errors.

While the effects of active errors are almost immediate, and the consequences of latent errors may lie dormant in the system for many years before they combine with active failures and local triggers to create an accident opportunity.

Active errors are more likely to be caused by people directly involved in the process. Latent errors tend to be made by designers, managers and other personnel who are sometimes not even aware of their errors and their consequences.

Causes of latent errors during operation are often:

- time pressure
- understaffing
- inadequate equipment or
- fatigue of people (fatigue of material is another class).

Other latent errors had been their causes,

- during development
- production
- or instruction.

Fatigue of material is development failure, and causes could be:

- Wrong material had been selected.
- The user profile and the stress profile are not aligned.
- Unexpected environmental impacts increase ageing of material.
- Inadequate production or processing of material.
- Damaging of goods during maintenance.
- etc.

This is one of the reasons why today's development processes for safety-relevant equipment require a lifecycle approach.

For an entire vehicle, such relations become very heterogeneous and could not be traced to specific features of a system in the vehicle. Without a systematic engineering process including adequate simulations, we are not able to make the challenges and the possible systematic failure transparent.

6.3 Control and Communication

Future mobility means also communication to various stakeholders. Communication could not remain inside a vehicle and just build up the driver–vehicle interface. Many publications and advertisements see the motor vehicle as a "smartphone on wheels". For such solutions, it is necessary to integrate the vehicle into the consumer networks.

6.3.1 Event-Driven Control

Human driving means the driver is in the centre of all control task and implementing a system that operates as a driver leads to a high complex control system, which have to be a deterministic digital twin of a human driver, including sensors, nervous system and brain, as well as the reflexes of experienced drivers.

In automation engineering one of the first questions are,

- what is the potential of current technology,
- what needs to be improved?
- do we need disruptive approaches or could we stay with current design principles?

The human system could be compared more with an event-driven control system, rather than a deterministic maybe time-triggered control system.

The VDA safety concept provides a boundary of the motor management control system in the centre of engineering (Fig. 6.21).

The EGAS system includes the engine control unit, the acceleration pedal and the throttle valve. Sensors, fuel injection, other μC, the brake system or other control lever is considered as external device.

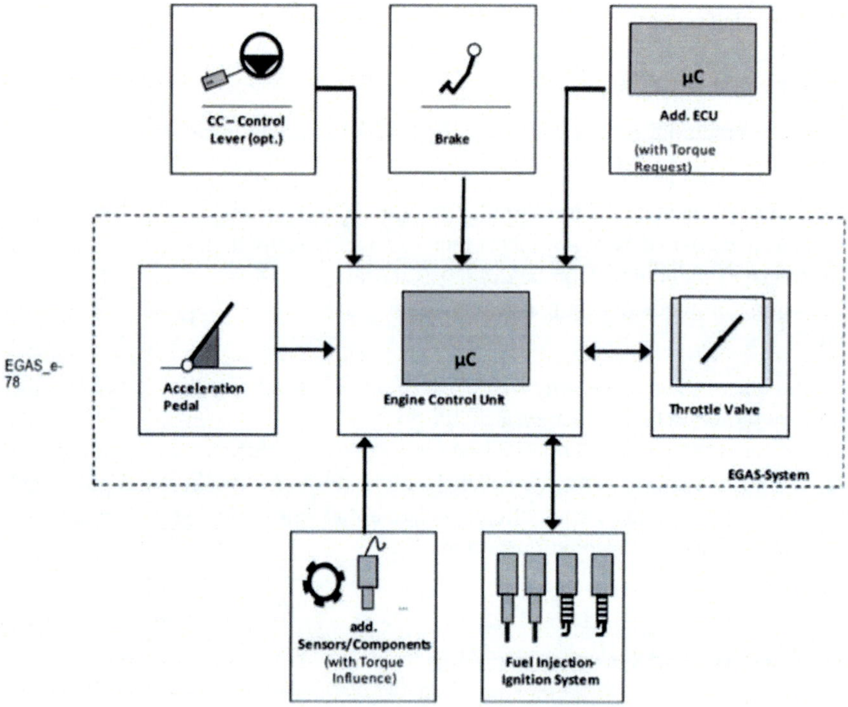

Fig. 6.21 VDA EGAS boundary. *Source* VDA EGAS, 6.0, 2013

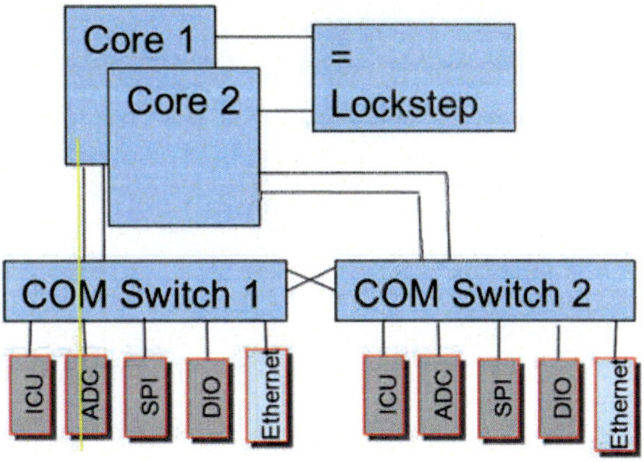

Fig. 6.22 Typical lockstep core with two com-switches

The mechanical engine, considered as the equipment under control (EUC) is not even part of the EE-system at all. However, the engine does have no electrical interfaces, but is significant for the behaviour of vehicle. Nevertheless, the engine can provide torque or not, for that it needs energy and communication to the devices that command.

Any device needs access to the communication channel and energy for its own control and for the actuator.

A typical solution for the EE-architecture could consist of a dual core lock-step controller with 2 independent Ethernet switches (com-switches) (Fig. 6.22).

The disadvantage of the lockstep controller is missing availability in case of detected faults. Most lockstep principles switch the core off, in case of detected failure by the lockstep comparison. Switching-off just one core needs sufficient diagnostic information, what was the defective core or what core caused the identified deviation of the comparison result. The advantage is it runs with the same software and the synchronisation and comparison of the two cores requires microseconds (Fig. 6.23).

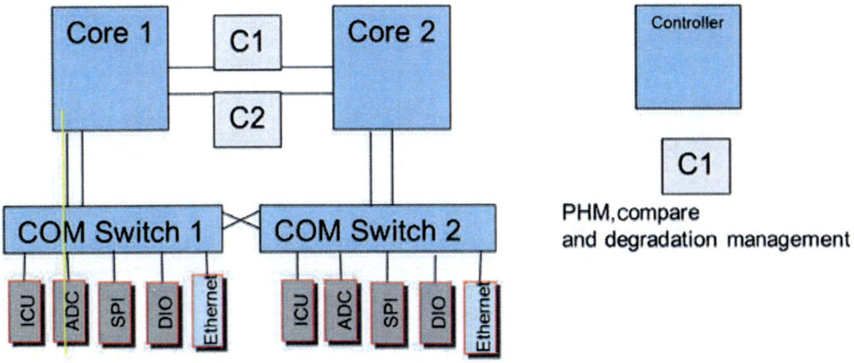

Fig. 6.23 Dual core with PHM and compare and degradation management

Using an independent dual-core principle with asymmetric multi-processing requires more engineering effort, but measures against systematic faults need to be analysed and implemented anyway. Together with a predictive health monitoring (PHM), the cause of failure is transparent and the two cores could run based on dissimilar software or even asymmetric mathematical models or logical principles. Since a continuous comparison is provided, both cores could provide equivalent result to other control units, and first data could be used for processing. Power supply could be added for each core and peripheral element independently, so that also external disturbances would not lead to a complete device shutdown.

If we assume a control network, which could get data from whatever source it needs, we ultimately in communication technology find the idea of ring communication (Fig. 6.24).

Sensors and actuator need of course also a redundant access to the redundant control network. Sensors could be used as shared resources between ring communication lines. Sufficient independence for the two ring networks is the highest design goal for sensor and actuator integration. The easiest way is that the vehicle words are with one ring completely independent; consequently, each ring has its own power supply, sensors and actuators. As long as both rings are designed and realised with the same electronic hardware, any random hardware fault could be detected by cross-comparison between the two rings. The cross-comparison units should be galvanic isolated, so that overvoltage, EMC, etc., do not provide any common cause. Systematic failure of any control unit needs its own predictive health monitoring (PHM). If sensors are assigned to control units, they should be integrated into the PHM of the control unit. Any actuator electronic should be completely independent redundant and integrated on the mechanical design. PHM should be designed in a way that all monitoring data are available in both ring networks. Any degradation should be initiated by other ring network by graceful degradation. Any degradable unit needs to be contained in a way that such a device could not cause any fault, when degraded.

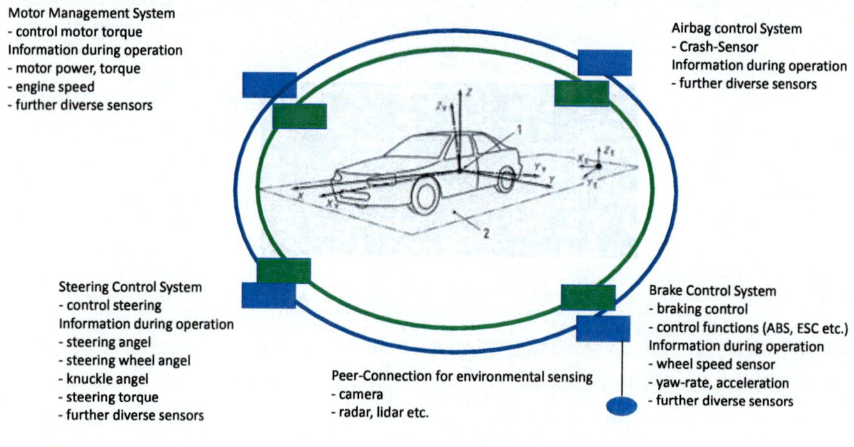

Fig. 6.24 Ring architecture for the entire vehicle com-net

Performance degradation could be done by mechanical design, for example two independent sets of coils on an electric motor. Only if both sets of coils could provide the full performance, the actuator also could provide the performance according to the design limits. For a steering actuator, 50% performance is enough for most driving manoeuvre, but losing vehicle lateral control during driving even just in a limp home mode means a significant risk. Even with a speed of 50 kph, losing lateral control means a significant risk. Consequently, for the steering actuator we need further measures, unless we are not considering a human driver as fallback. For the propulsion engine, such a 50% performance degradation could be tolerable (Fig. 6.25).

The control network should work as a client–server network. Any sensor and any actuator should be a client. All clients should be designed in such a way that all necessary data can be retrieved from the network or be continuously made available to the network with time stamps, sequence counter, etc. The associated servers should be available redundantly in the network, whereby in the event of a server failure, a backup server should also be available in each ring network. As long as both clients are fully functional in both networks, the clients can receive the control and monitoring data from both ring networks at any time. At the actuator, the first valid data set can be used to control the actuator. The second data set, which receives at the actuator by the second ring network, can be used for monitoring or for performance enhancement. The redundant actuators should be mechanically coupled in such a way that minor time differences are compensated. In this case, the synchronisation effort could be significantly reduced. The necessary real-time window for the redundancy comparison is considerably larger and making the entire system much more fault-tolerant. As long as the lockstep comparison does not degrade the own control device, the lockstep would not limit the availability. However, the synchronisation effort, time

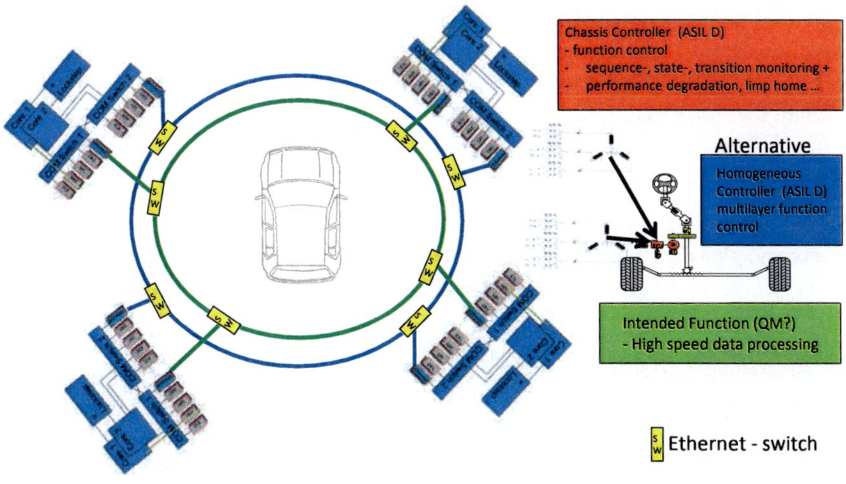

Fig. 6.25 Redundant ring network

delays and lack of controller resources are disadvantages for the lockstep design. Nevertheless, if the controller is provided by different suppliers, any supplier could be made responsible to control their own random hardware faults. Environmental perception, trajectory planning, etc., should be considered by dissimilar control and different logical principles that should be implemented redundantly in both ring networks. Any monitoring software should also be implemented as part of the PHM in a separate partition of the controllers.

Systematic failure needs to be avoided by an adequate design process, which requires systematic design assurance for the entire EE architecture of the vehicle. The understanding of DAL for the entire development process (development assurance level) as well as the component design process according to the specific design assurance level like in the avionics industry is of course a hard burden for the development process. The aviation industry would not have established such highly complex development processes if there were leaner and more efficient solutions. In detail, there may be simpler solutions, but every development process should allow this freedom to tailor according to the level of risk. However, such a development process must be verifiable and the results in the safety case must remain transparent. As long as the safety argumentation is comprehensible, old, well-worn process regulations should not be a barrier to new safe solutions.

6.3.2 Publish–Subscribe Networks

Publish–subscribe describes communication systems for asynchronous management of information. It is a messaging facility. Users, consumers of information, present the system with continuous queries and coined subscriptions. Sources of data generation (producers) present the system with data-carrying publication events. The publish–subscribe system infrastructure is responsible for (asynchronously) matching the publication events to all relevant subscriptions. Hence, in essence, this infrastructure filters all available information for every user and presents to each user only the information units, which has defined as relevant.

Publish-Subscribe uses the following principles:

- Software components do not necessarily know to whom they are communicating.
- Producers of data publish that data to the system as a whole.
- Consumers of data subscribe to and receive data from the system as a whole.
- Information is labelled so that software modules can identify the available information. Such labels are also called a topic.

Queue managers (QMs) control the filtering and interactions between publishers and subscribers. Each queue manager matches messages published to a topic with the locally created subscriptions that have subscribed to that topic. You can configure a network of queue managers so that messages published by an application connected to one queue manager are delivered to matching subscriptions created on other queue managers in the network. This requires additional configuration over simple channels between queue managers. Most messaging systems support both the publish–subscribe and message queue models in their Application Process Interface (API), e.g. Java Message Service (JMS).

A distributed publish–subscribe configuration is a set of queue managers connected together. The queue managers can all be on the same physical system, or they can be distributed over several physical systems. When you connect queue managers together, subscribers can subscribe to one queue manager and receive messages that were initially published to another queue manager.

IBM provides in their knowledge centre an example of a single queue manager publish–subscribe river configuration (Fig. 6.26):

> In order to illustrate this, the following figure adds a second queue manager to the configuration described in Example of a single queue manager publish/subscribe configuration.
>
> – Queue manager 2 is used by Publisher 4 to publish weather forecast information, using a topic of Weather, and information about traffic conditions on major roads, using a topic of Traffic.
> – Subscriber 4 also uses this queue manager, and subscribes to information about traffic conditions using topic Traffic.
> – Subscriber 3 also subscribes to information about weather conditions, even though it uses a different queue manager from the publisher. This is possible because the queue managers are linked to each other.
>
> You can manually connect queue managers in a parent and child hierarchy, or you can create a publish/subscribe cluster and let IBM® MQ define much of the connection detail for you. You can also use both topologies in combination, for example by joining several clusters together in a hierarchy.
>
> *Source* https://www.ibm.com/support/knowledgecenter/SSFKSJ_8.0.0/com.ibm.mq.pro.doc/q005120_.htm.

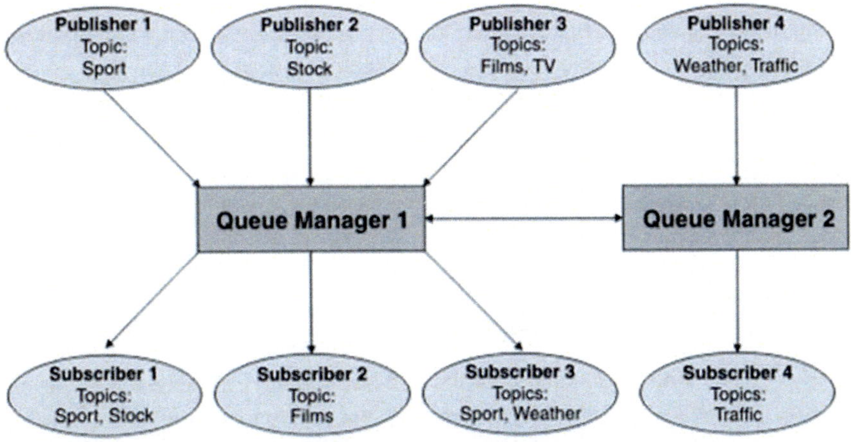

Fig. 6.26 Publish–subscribe example with two queue managers

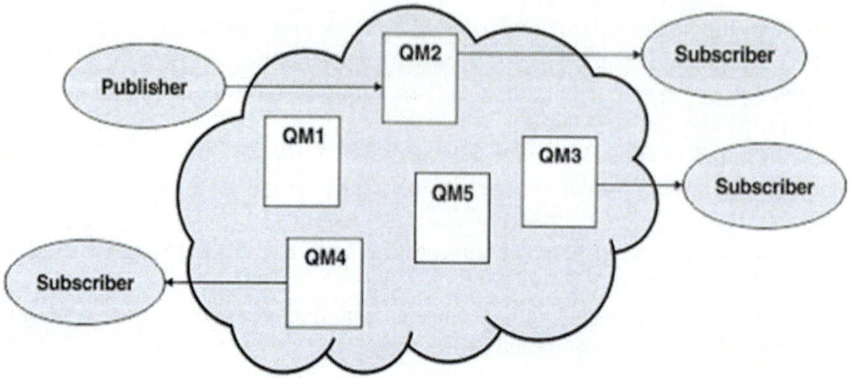

Fig. 6.27 Publish–subscribe cluster. *Source* https://www.ibm.com/support/knowledgecenter/SSFKSJ_8.0.0/com.ibm.mq.pro.doc/q005120_.htm

Such a basic framework provides major basic communication feature. Such a publish–subscribe network provides the basic needs for complex networks, with different stakeholder needs (Fig. 6.27).

The basic configurations could be applied how messages are routed in a cluster:

- direct routing,
- topic host routing.

These two configurations form the basic building blocks for any communication architecture, however heterogenic. Several query managers (QMs) could be implemented in the same context, e.g. traffic management, and second topic within such a context could be weather or road conditions. This provides opportunities such as combining weather forecast, traffic jam assist and navigation system with active vehicle control for automated transport systems (Fig. 6.28).

In order to provide various viewpoints from different stakeholders, it is necessary to bring such a network in a hierarchical structure. In the example from IBM, QM1 is in the highest hierarchical level and QM3 is a level higher than QM4 and QM5. This could be a viewpoint for road traffic conditions, and for weather or traffic control the hierarchical viewpoints lead to another hierarchy of the query managers (Fig. 6.29).

In such a network, typical publisher (PM-x) could be used to provide information on

- weather conditions,
- road conditions,
- road traffic condition and/or
- the current map of the road environment.

The weather, traffic jam and routing forecasts could be implemented as a subscriber, which get always the latest data from the different publishers. The query managers are configured so that the subscribers get the right data in time, depending on conditions and situations.

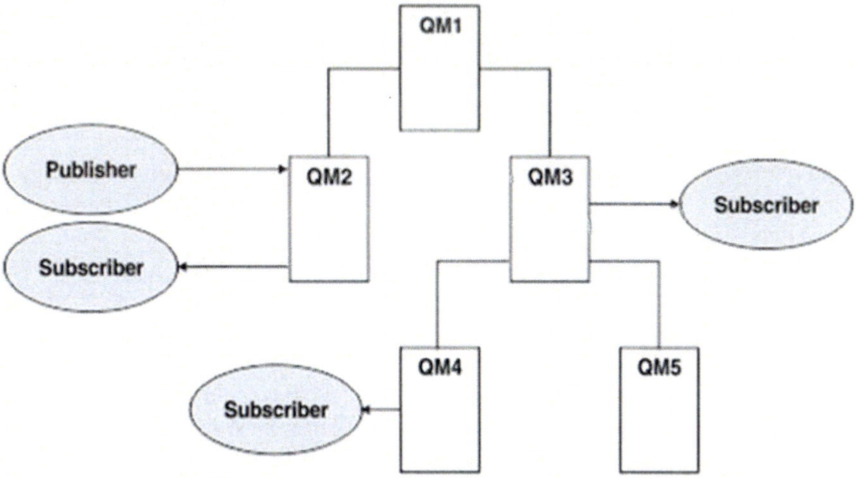

Fig. 6.28 Publish–subscribe hierarchical view. *Source* https://www.ibm.com/support/knowledgecenter/SSFKSJ_8.0.0/com.ibm.mq.pro.doc/q005120_.htm

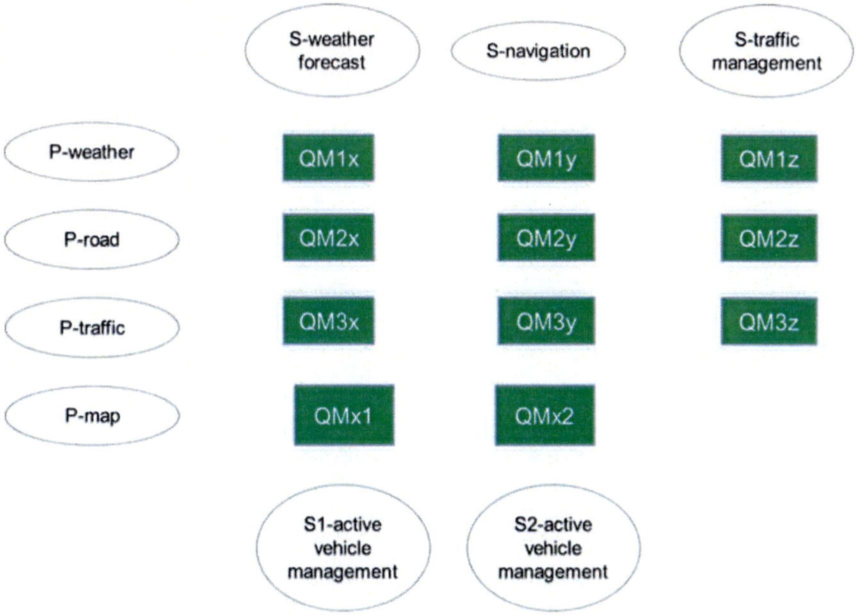

Fig. 6.29 Traffic management based on publish–subscribe network

An active vehicle management could be implemented based on a redundant subscriber, so that in case of failure, the necessary functionality or any kind of degraded mode could be applied (Fig. 6.30).

Similar to the human reflexes, road traffic situations require an immediate fast reaction. A typical situation is an accident, obstacles on road due to storm or load from other vehicles or unexpected people on the road. The active vehicle management needs the information immediately, but without the information about position, traffic situation, weather, etc., it could be difficult to initiate the right reaction to the sudden event. A publisher could be implemented which provides the highest priority information and would lead to an immediate reaction at the query managers for the active vehicle management. The query manager could provide the necessary information, so that an adequate reaction of the controlled vehicles could be initiated. Automated controlled vehicles could get an auxiliary lane, and the traffic control management could slow down the other traffic. With such a query management, such scenarios could be defined and implemented in advance, so that in case of such sudden events even the degradation scenarios are controlled as a normal operating situation, just with a higher priority.

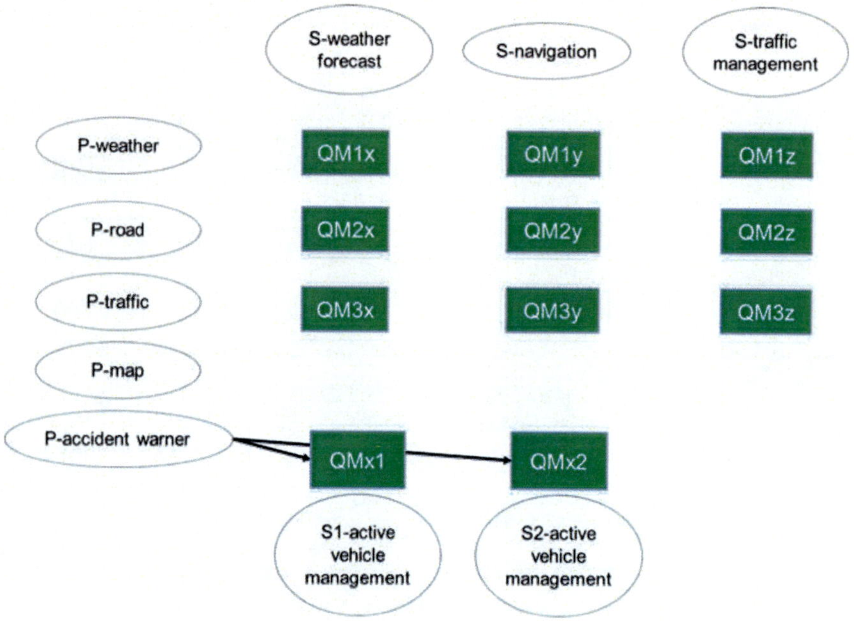

Fig. 6.30 Accident warner implemented as publisher

6.3.3 *Data Distribution Service (DDS)*

In the IOT world a new type of middleware established in the last years, it is suitable for real-time applications, and it supports several safety and security mechanisms. The Data Distribution Service (DDS) is a standard specified by the Object Management Group (OMG). DDS represents a middleware for data-centric communication in highly dynamic distributed systems. It is based on a publisher–subscriber concept that supports deterministic resource management.

The specification is divided into two areas:

- Data-Centric Publish–Subscribe (DCPS) describes the basic concepts for data distribution.
- Data Local Reconstruction Layer (DLRL) provides an abstraction layer for DCPS-based applications.

The DDS specification contains the following core concepts:

- an application-specific data type (e.g. defined in IDL) that defines types of information contained by a DDS package,
- a domain for logical structuring and containing sets of topics,
- a publisher is a participant in the DDS system that provides data (topics),
- a subscriber is a participant in the DDS system as the recipient for certain data (topics).

Using quality of service parameters, a subscriber can define its requirements for the quality of data transmission in a declarative manner. They can also create filters, for example, to receive only data in a certain value range of the topic. Depending on the implementation, the components (publisher and subscriber) are located either themselves or by means of a central server. There is a process of mutual discovery. Discovery without a central server is realised, for example, via multicast.

Commercial licences are available, which are providing a DO 178C certification, and there are solutions for air traffic control published.

There is also an open-source community "OpenDDS", and they provide support, tools and software packages for DDS. https://opendds.org/about/dds_overview.html.

They are providing detailed handling and application support for DDS.

Such a "publish and subscribe" approach allows distributed networking without any "central" server, which allows to maintain the communication in many possible fault conditions. With the containment mechanism, defective, corrupted or even attacked parts of the network could assure correct functioning of the remaining parts of the network.

Such a "publish and subscribe network" could be a solution for the communication network in a future automated road vehicle.

6.3.4 Communication Networks and Energy Transport

Any communication is useless unless the power for actuation is not available. If you command a brake or steering system, current motor vehicle uses just assistant system to reinforce the driver's power, usually provided by the legs or arms. If the driver is not available to initiate the actuation, a system must assure that energy and command information are at the actuator at the right moment in time available. This means that an energy management system is required which can ensure that sufficient energy is available at the actuator at any time it is needed. A significant advantage of a motor vehicle is that unless it drives slowly it also does not need too much energy to brake. As vehicle drives as more kinetic energy is in the vehicle to control the vehicle. As faster a vehicle drives, as less energy is needed to steer the vehicle. Only the propulsion system requires more energy to accelerate or drive faster. Illuminations and control systems require not that much energy, but in sum, they are also a significant factor as well as heating and air conditioning.

As long a vehicle moves, there is enough energy in the vehicle to control the vehicle as long as the vehicle EE architecture is capable to assure availability of energy for actuation and control. In order to assure degraded operation or limp home modes, sufficient independent networks are needed inside of the vehicle. In our days, the basic idea is to load a battery, which is good for a driving distance of more or less 500 km. Since many years, it is known that the public energy net needs to be improved. The basic idea is a smart grid in order to provide energy wherever needed, as much as necessary at any moment in time.

There are various activities around and many solutions standardised in order to assure the availability of energy. It is necessary to provide the energy and to supply the billing systems with data to compensate costs. The energy distribution needs to be continuously optimised, so that the valuable electrical energy is not unnecessarily burned or even used for unprofitable purposes.

Another important aspect is the safe availability of energy, without risk of power failure. A sudden power failure could be compensated with distributed energy storage systems also inside vehicles as well as inside living facilities up to family houses, but any failure means energy losses. Touch protection in the household area is more or less established as well as for the energy distribution, but smart solutions are more and more applied for protection mechanism. Major protection sectors are in the domain of safety, privacy and authenticity. In the various applications, the number of safety domains immediately grows depending on how people use the energy.

There are typical systems available to perform the following functions:

- Fault detection, isolation and restoration (FDIR) systems are focused on improving the reliability of the distribution networks.
- Integrated volt-VAR control (IVVC) systems are designed to reduce electric feeder losses improving the overall voltage distribution and conservation in both peak and non-peak time periods.
- Organic flash cycle (OFC) systems can potentially increase power generation from thermal energy reservoir.

Many standards have been established around the smart grid and their communication systems in order to make the energy for the users available.

There are a series of relevant protocols that can also be used on these networks, in conjunction with the previously listed technologies:

- Distributed Network Protocol (DNP): It is a set of communication protocols designed specifically for the communication of device part of automated systems. It is used mainly in SCADA systems to interconnect the master station with the RTUs and IEDS. It has been designed to be reliable, but lacks security measures.
- IEC 61850: It is a standard designed to be used to control the automation processes on electrical substations. It defines data exchange between control systems and the substations themselves. It includes many required features to carry out these functions, including data modelling, data storage (substation configuration language or SCL), fast transfer events and reporting schemes. It also includes variations of this protocol for different installations, such as Hydroelectric Power Plants (IEC 61850-7-410) or Distribution Automation Systems (IEC 61850-7-420).
- IEC 60870: It defines the systems needed for the supervision and data acquisition of power automation systems. It has been divided into five parts: transmission frame formats, data link transmission services, general application data structure, coding definitions for information elements and basic application functionalities.

Therefore, not all the technologies and protocols used for the communications through the Internet support protection mechanism, safety and security measures by default, and as such it becomes necessary to use additional measures in order to protect them. Some of the most common security technologies that can be used to protect communications include:

- Transport Layer Security (TLS): It is a cryptographic protocol designed to protect communications over a network. This is achieved by using asymmetric cryptography and client certificates to authenticate both the sender and the recipient and symmetric protocols to carry out data encryption. Its predecessor, SSL v3, is also commonly used, although its use is not recommended due to several security flaws that have been detected on it.
- IEC 62351: It is a standard designed to handle the security of several protocols including IEC 60870, IETC 61850, IEC 61970 and IEC 61968. Among its features, it includes TLS encryption, node authentication, message authentication and several other specific security profiles.
- IEV IEC 61850-90-12: It provides definitions, guidelines and recommendations for the engineering of WANs, especially regarding their protection, control and monitoring. It is based on IEC 61850 and several related protocol standards. It is mostly used for communications between substations and the control centre. Internet Protocol Security (IPSec): It comprises a set of protocols designed specifically to protect IP communications by applying authentication and encryption to them. It supports mutual authentication and works on the Internet layer (while TLS works on a higher level, on the application layer).

- Secure Shell (SSH): It is a protocol that provides a secure connection to remote machines, by applying encryption to protect the data. For this communication, the remote machine must have an operational SSH server to which the client will connect to.
- DNP3 Secure: It is an upgrade to the standard DNP3 protocol designed to provide additional security measures, including authentication and data encryption. It is compliant with IEC 62351-5 standard, and in some cases Virtual Private Networks (VPNs) are also used to secure IP networks.
- Virtual Private Network (VPN): It is not a protocol per se, and it is more of a concept, but it can be considered as a relevant security measure in order to protect communications. In a VPN, the use of point-to-point private network over a public network or the Internet is specified. A VPN uses tunnelling protocols to make available private communication and to make use of encryption protocols to protect the confidentiality of the data transmitted.

These are just a collection of standards and measures, which provide risk reduction solutions. The challenge of future mobility is how to identify all the risk and finding the most effective measure to reduce the risk below a level tolerable for the society.

6.4 Hazard and Risk in Road Traffic

Key of any road traffic safety is a common understanding of all road traffic participants of the common basic safety principles. Mutual consideration and conservative behaviour towards the weaker road users are the principles of all traffic laws. Even in the case of mistakes by others, it is necessary to avert damage and to avoid accidents or minimise the extent of damages as good as possible. Misunderstanding, misleading traffic rules, disregard, ignorance or misjudged situations and skills often mean a significant traffic safety risk. In particular, the mobilised road users, such as e-scooters, cyclists and motorists, have to control their means of transportation adequately in addition to observing the traffic area. They have to know how the transportation system

- has to be controlled in the respective traffic environment,
- where are the rules documented and what signs mean for their behaviour,
- and for what they are accounted, liable or responsible while using the transportation system in public road traffic.

While a definition is just way to understand road traffic safety, it is always important to remember that safety based on continuous care.

There are objects or obstacles on or near the road that have a right to be in the place and those that have no right to be there. For a safe detection of the objects, all objects that represent a safety risk for the vehicle under control or for other vehicles must be perceived.

Rules and definitions matter because they are the filter with which we assess our environment.

Such rules need to be defined in advance for the entire possible road traffic scenarios. Unfortunately, the legal regulations do not address all the risk and the relevant scenarios. Without such a common understanding, it is not possible to define degradation or any other kind of accident scenarios. Implementing the rules based on just one viewpoint would lead to violate the society rights since any social group claims other protestations. In case of an accident, the viewpoint of the judge is different, from the viewpoint of a driver in an emergency case.

What is a Hazard Based on This Context?

Hazards are one of the core concerns of road traffic safety. A hazard:

- is a condition, such as an object, situation, circumstance, that poses an unacceptable level of danger,
- occurs once in the safety mishap lifecycle,
- can lead directly to risk occurrence (i.e. safety mishap, accident, etc.) if not mitigated and
- arises from hazard mechanisms, such as initiating actions and hazardous sources.

The Federal Aviation Administration (FAA) defines a hazard as, “A condition that could foreseeably cause or contribute to an aircraft accident ...” (14 CFR *§ 5.5*).

Understanding what a hazard is also involves identifying what a hazard is ***not***:

- benign objects (birds, mountains, people), which are hazardous sources,
- safety mishaps, which are another way of saying risk occurrences,
- damages, which are a product of risk occurrence and
- dangerous actions, which are associated with initiating mechanisms.

In the actual operational environment, with real situations, people may disagree as to what point was the dangerous condition.

How could we convert this to road traffic safety?

Benign objects are objects, which have to be taken into account, which we have to expect. Animals are frequently on urban roads and on motorways, and consequently they are no hazards. People are frequently close to road traffic in urban environment, in industrial areas and on motorways, and they are less frequent. However, even on motorways, where pedestrians are forbidden, people are sometimes on the road for example if they had an accident. Mountain, hills, lost transportation goods, and obstacles like stones, wood etc. have to be considered in road traffic. Weather conditions, even like rain, ice and thunderstorm, are usual situations, which need to be considered.

Misshapes and damages of road infrastructure could lead to hazardous driving scenarios similar like road damages due to weather impacts or accidents due to damages at a vehicle. A horn is a safety device, failure of a horn could lead to accidents as well as flat tyres or missing light in the dark. Drunken drivers are considered worldwide as a risk also for vehicle occupancies and other road traffic participants.

The most frequent cause of road accidents is too high speed. It is a dangerous activity as well as overtaking on a winding country road.

Accident prevention and road traffic control, for example, by the police are accident prevention measures as well as road and vehicle maintenance.

All road traffic actors like all traffic participants, the police and maintenance staff etc., should understand the basic framework of the lifecycle of safety incidents. It helps the hazard identification process and builds the necessary discipline due to safety awareness, hazard reporting and penalties.

The following aspects could be identified as safety mishaps:

- Hazardous sources exist, which are the starting point for safety concerns and root causes for accidents.
- Initiating mechanisms interact with hazardous sources and increase threat level to targets any road traffic participant.
- Lack of identification of the increasing danger makes a danger appear always anew.
- Hazard occurs, whereby the operational environment reaches an unacceptable level of safety.
- Breakdown of risk controls/lack of response to mitigate hazard into an acceptable range.
- Risk occurrence happens, incurring damages.

Risk management in road traffic is a must, and during the entire lifecycle of vehicle development a continuous process accompanies the development of new function and features.

Safety concerns can be largely mitigated at any point leading up to risk occurrence. However, hazard occurrence will be most obvious "sign" that something bad is about to happen. For this reason, it is essential that all road traffic participants and actors like police and maintenance staff, etc., have a solid and common understanding of risk and hazards.

Examples of Hazards

The best way to demonstrate what is a hazard is to simply provide some examples. Here are some examples of hazard occurrence, as they all demonstrate unacceptable level of danger:

- road damages,
- operating or driving in close proximity to areas with people needing particularly protection, such as children, or irrational creatures such as wild animals,
- lack of proper traffic guidance (i.e. weather, traffic density, high speed, etc.),
- incorrectly behaving technical systems (e.g. traffic lights, sensor, but also vehicles and their feature),
- road traffic regulation mechanism (traffic management, but also built-in vehicle functions (e.g. lane keeping, ACC),
- unreliable safety data (e.g. data to the vehicle, but also data within the vehicle),

- any kind of common cause effect, which could lead to failure of any kind of redundancy,
- possible unintended use, misuse, gross negligence and so on.

These are just examples, and details require a systematic hard and risk analysis, but the primary similarity between all of these points is that they are a condition of unacceptable level of safety.

How to identify hazards in road traffic environment?

Hazard identification is actually quite a bit more flexible than it is often given credit for. It does not simply mean identifying a hazard when it occurs. To be sure, this is one element of hazard identification.

However, hazard identification also includes:

- identifying growing threat of operational environment,
- identifying unsafe interactions (i.e. initiating mechanisms) with hazardous sources and
- identifying an unsafe situation and understanding how to mitigate it.

The formal process of hazard identification also includes hazard communication. This simply includes reporting of safety concerns and best practice how to identify or prevent by any kind of road traffic participants by television or other media. Product, road traffic and driver's observations are even an obligation by law, and this is the purpose of the driver's register. Continuous update of legislation by crowing road traffic is necessary to keep the current level of risk.

6.5 Horizontal and Vertical Engineering

In order to evaluate all the impacts of mobility to safety, many influence factors are to be considered. Those influences could affect a horizontal signal chain from

- physical effect to a sensor,
- transmission and processing of data to the actuator,
- and physical effects directly at the actuator.

The vertical influence from top down depends strongly on the vehicle and system design. Here, any design decision could lead to new hazards.

ISO 26262 also addresses the case as follows:

> 6.4.4.7 Hazards newly identified during safety analyses or during the system architectural design that are not already covered by a safety goal shall be included in an updated hazard analysis and risk assessment in accordance with ISO 26262 3:2018.
>
> NOTE Hazards not already covered by a safety goal may be non-functional hazards. Non-functional hazards are outside the scope of ISO 26262, but they can be annotated in the hazard analysis and risk assessment; e.g. by annotating the hazard with the following statement, "No ASIL is assigned to this hazard as it is not within the scope of ISO 26262".
>
> *Source* ISO 26262:2018, Part 4.

The scope of ISO 26262 is very limited to malfunctions and failure of EE systems. Driver's impact, impact due to other technology than no EE and all external influences from the ITEM (see definition from ISO 26262:2018) which includes all environmental impacts could influence the system behaviour. If the unintended system behaviour due to EMC or whatever could have the potential to lead to harm, could be only evaluated by proper safety analysis beyond the scope of ISO 26262.

One very common approach in many industries is the idea of a digital twin. It is an artwork to find for the twin in the reality the right level of abstraction.

In order to find the right level of abstraction, the following questions need to be taken into account:

- What is the system of interest (SOI)? What part of the systems lifecycle needs to be considered with what level of granularity?
- What is the context for the system? What stakeholder needs to be considered? What kind of relation have the different actors with the system?
- What is the intended use? What is the purpose behind the system or the product?
- What technology should be applied?
- How do virtual effects affect people?

In order to evaluate all the impact on the system under development, a huge matrix is required. The entire development plan needs to be tailored as a systematic knowledge building process. Systematically with any iteration, any element of such a matrix needs to be elaborated (Fig. 6.31).

The digital twin requires a continuous improvement process or a continuous build-up of experience values, which includes the following for the operational layer:

- influences or stimuli for the vehicle under consideration,

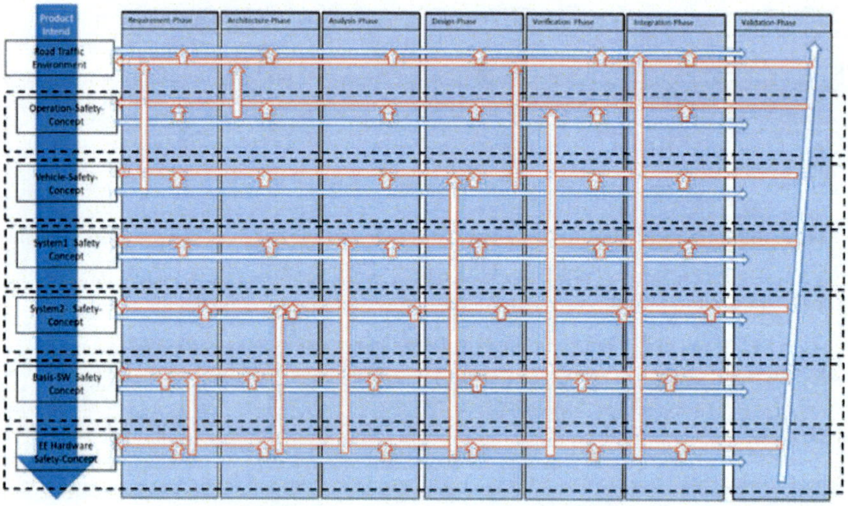

Fig. 6.31 Vertical and horizontal activity landscape

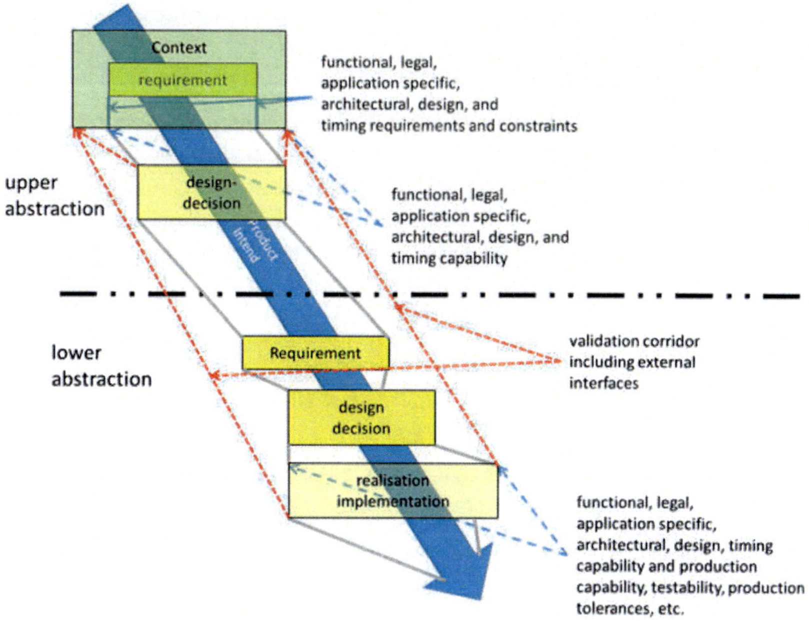

Fig. 6.32 Breakdown from higher to lower level of horizontal abstraction

– reactions caused by the stimuli and due to vehicle internal factors like control systems or the driver inside the vehicle, etc.

The vehicle represents a black box with certain degree of freedom.

The next horizontal layer describes all the basic dependencies within the vehicle. Theoretically, it is the EE architecture of the vehicle. On this layer, you could see the boundary of the vehicle including the outer and inner interfaces on a specific level of abstraction. Inside the vehicle boundary, you see at least the energy distribution, information exchange and physical interfaces. Even the steering column is just a media that transfers the physical movement at the steering wheel to the steering knuckle. In that consequence, it provides at the same time information and energy from the drivers' hand to the front axle, which affected a lateral movement of the vehicle, unless the vehicle itself is moving. During standstill, the steering movement only causes heat and mechanical wear on the tires.

On that layer, the vehicle becomes more than a black box and the following observable entities could be considered:

– energy management and the distribution in the vehicle and outer interfaces like charging interfaces,
– the communication facilities within the vehicle,
– the signal chains including their inputs at sensors, control devices and their final elements like output at actuators,

– the physical effects inside the vehicle like mechanical movement, physical touching, material exchange or transfer, but also physical effects like pressure flow in hydraulic or pneumatic systems or electrolysis in battery or fuel cells, etc.

This layer and the internal structure and interdependencies provide in the first step new requirements and constraints for the layer above (Fig. 6.32).

Any architecture or design decision in the lower level of abstraction could lead to new constraints and requirements in the higher level of abstraction. Since Turing explains us the black box problem, we know how to identify memory effects inside, but we still have no completeness argument for any methods.

6.5.1 Closed-Loop Control and Signal Chains

Specifying a top-down approach, the requirements for a system element need still the inner aspects for sufficiency arguments (Fig. 6.33).

– In a pure piece of wire, we could have temperature-dependent resistor, which could delay the signal and even lead to no reaction or complete other reaction at the output. If temperature, EMC, noise etc. as an external impact or an internal effect could of course also change the observable reaction at an output.
– Even a piece of wire could have already capacitive effects, which provides already very simple memory effects. Any further and more advance memory capability could significantly change the IO relation.
– If there is already a simple switch inside, it is possible to use the one or other way of the switch. Any other device which provides based on certain condition different transitions could change the observable behaviour. Any internal feedback or feedback from outside means of course also possible new states and transition with potential observable reaction at outputs of the element.
– If there are optional transitions in the observed system, not only the time of producing an observable reaction is questionable, also at what output the reaction

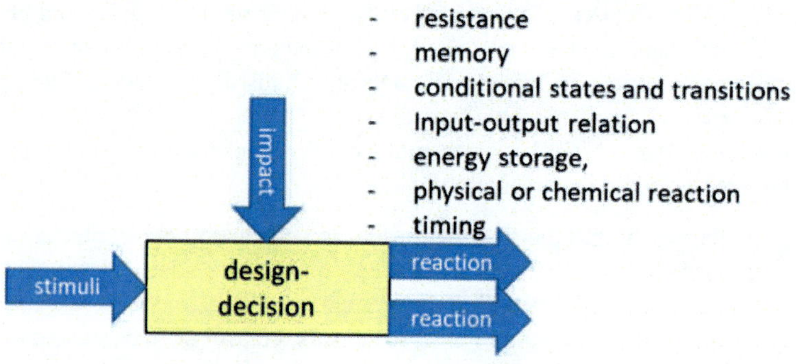

Fig. 6.33 Black box and its inner effects

could be seen could be questioned. In case of an explosion, the reaction is easily observable, but the assignment to a specific output is not traceable anymore.

- Any energy storage could release or store energy depending on the design. The observable output always depends on inner and outer effects.
- In case of battery, fuel cell and unintended physical and chemical effects like corrosion provide changes in the observable reaction.
- All the effects above could also lead to accelerated, delayed, sporadic reaction or any change of the output characteristics over time.

A pure top-down specification needs to be questioned by any design decision, which could lead to intended effects in the system element and unintended reaction of the element under consideration. Any combination of the effects above makes transparent that analysis is necessary. It is imperative to question the design decisions (Fig. 6.34).

In case of typical control cascades, we easily see that the number of states and transition easily could lead to very complex and heterogeneous matrix of input–output relation.

A typical safety case according to ISO 26262 provides metrics and table in ISO 26262: part5 for electronic control systems, but not for closed-loop control circuits (Fig. 6.35).

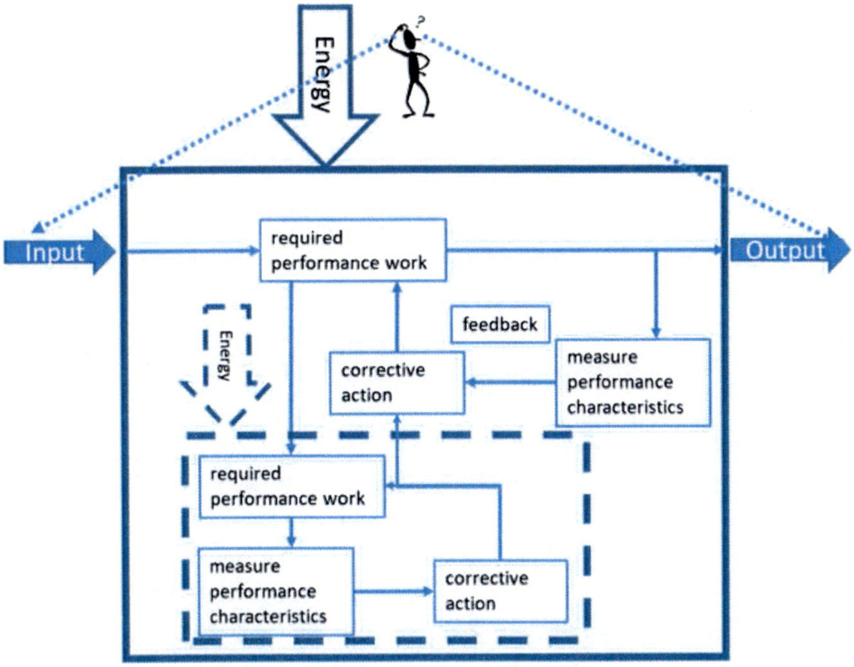

Fig. 6.34 Observer of control cascades

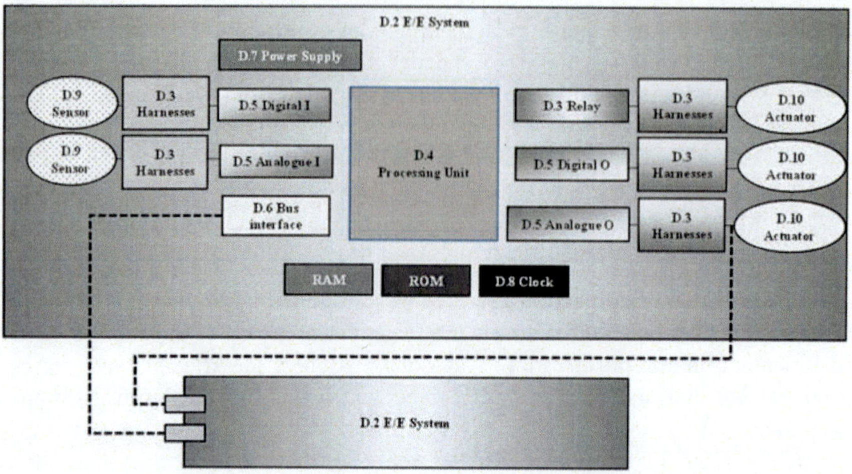

Fig. 6.35 Figure D.1—generic hardware of a system. *Source* ISO 26262:Part5, Annex D, Figure D1

The typical closed-loop control circuit provides continuous feedback and the failure in a signal chain affecting very differently the actuator (Fig. 6.36).

In case of human driving, the driver closes the loop between the reference sensors, feedback sensors and actuators of the vehicle. The driver controls the actuator by control devices such as pedals, steering wheel and gear shift lever. Driver uses besides optical perception by the eyes, touch sensing at the control devices and the balance sensing as reference and feedback sensor. During driving school, the driver learns

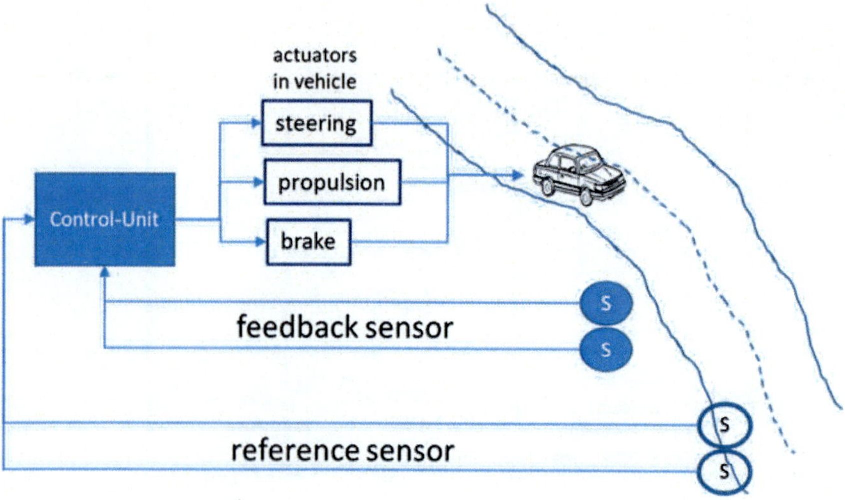

Fig. 6.36 Main control loops for vehicle control

to work as part in such multiple closed-loop control systems. The physical training teaches how to react on the various actuator devices.

In a system control architecture, it is necessary to provide a unique closed-loop control architecture so that traceable control functions could be implemented.

The reference sensors need to detect at least the maximum drivable speed and the adequate lane for the vehicle. The reference sensors must perceive all necessary deviations to remain in an adequate limit within the driving corridor.

Usually, the control loop for the driving direction is a static loop unless we ignore recuperation in an electric drive motor.

The brake control system is already a closed-loop system, which in modern vehicles includes wheel speed sensors, yaw rate, lateral and longitudinal acceleration sensors for brake force distribution and stabile braking.

The propulsion system uses closed-loop control to compensate for the system's own moments of inertia, torque and external effects such as the vehicle-specific CW value as well as friction to the road surface, etc.

An electronic power steering system (EPS) compensates with internal control loops many short-time effects which could lead to lateral movement of the vehicle.

In case of cooperative steering between brake-functions like (ESC, electronic stability control) and by-wire-steering systems, there are several control cascades which consist of internal and external (e.g., the road feedback) close-loop controls.

In a typical motor control system, the output-rectifier uses the current feedback for temperature, but also for reacting on torque or phase position. A resolver in the motor could provide further feedback inside the motor control unit. The reaction and the feedback of the motor could also derive from the actuation of the motor. However, the output of the power rectifier depends on the external torque of the actuator driven by the motor. In case of strong headwinds, the drive motor must apply more torque. Depending on the motor and the power rectifier design, it could be slip or phase position at a resolver, current or voltage feedback in the electrical circuit (Fig. 6.37).

For classical assistance systems where the driver was responsible to control all the vehicle external effects, but for automated driving the ability of the driver to control such heterogeneous control loops is nearly impossible. The driver has no ability to cooperatively control the vehicle with such automation system. In case of system failure or even system malfunctions, the driver completely fails as a fallback.

In cooperative driving scenarios with more than one controlled vehicle, it is nearly impossible to synchronise the vehicle control loops without understanding the vehicle control loops and their respective timely behaviour.

The key challenge is that the control loops for the different sensors do not just compensate functional failure, and particular delays in timing in combination with random effects from the environmental impact lead to different controller characteristics. Besides functional and timing failure, also the wrong control characteristics could lead to wrong vehicle behaviour.

Typical control characteristic errors like:

- too much or too short feedback time,
- too high or too low proportional average,
- too low or too high integral and differential average,

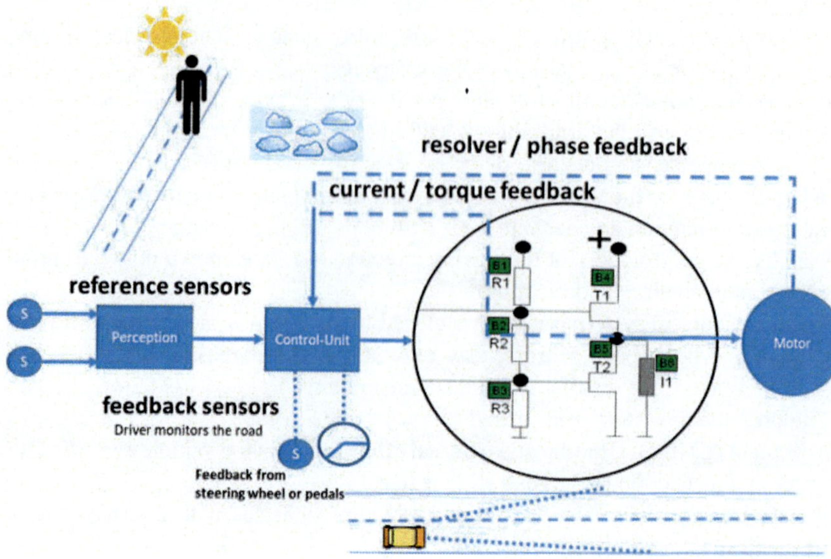

Fig. 6.37 Closed-loop control with system internal and vehicle external control loops

lead to different effects depending on the controller configuration.

Since steering and propulsion are continuously operating systems, the impact on typical on-demand functions like electronic brake functions, ESC, brake force distribution, etc., operates on a complete other control principle. Using the same sensors for multiple control loops could work in a correct designed control corridor and the same time basis.

6.5.2 *Closed-Loop Control on Various Layers of Abstraction*

A typical signal chain is in comparison with closed-loop control cascade a far higher complexity. The traceability from a higher level of abstraction to lower level of abstractions is only possible if the control cascades are transparent for the higher level (Fig. 6.38).

Since the vehicle is the ultimate boundary, all control loops based on the vehicle design and a systematic breakdown of the level abstractions could be applied since the lower control circuit architecture is available (Fig. 6.39).

In case of a typical IPO system, the elements and the interfaces between the elements inherent form the higher level of abstraction to the lower level of abstraction (Fig. 6.40).

The output of the controller in a standard control circuit reacts continuously based on the feedback, and the commanded reference value remains usually steady. Abstracted the control loop to a steering system, the commanded control value is

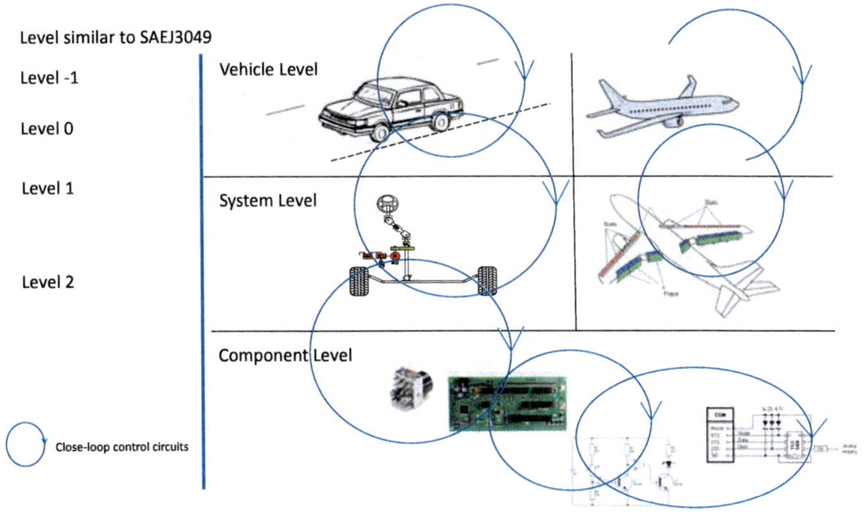

Fig. 6.38 Typical comparison of vehicle layer of abstraction with airplane levels

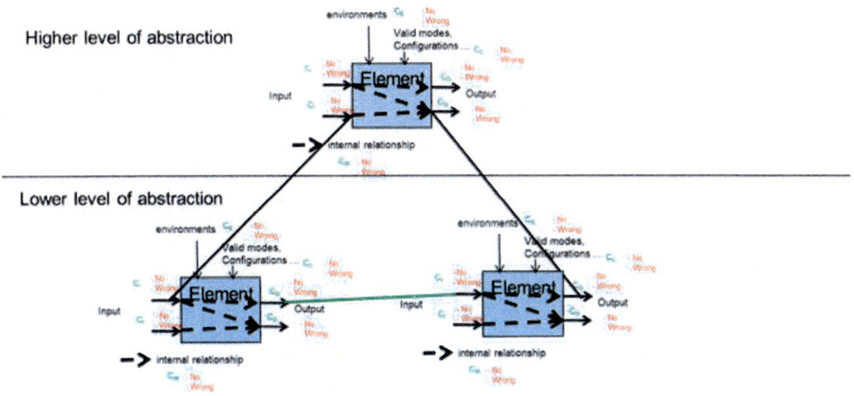

Fig. 6.39 Hierarchical breakdown of a typical IPO system

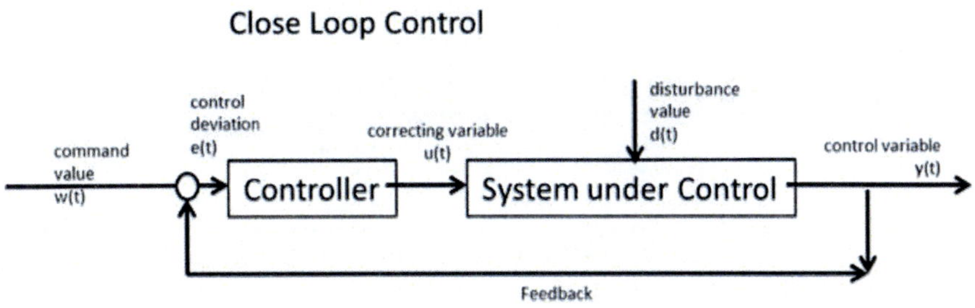

Fig. 6.40 Typical closed-loop control circuit with only one feedback and no cascades

derived from the road lane characteristics. Usually, the control target is to keep the vehicle on the lane centreline. The measured correcting value is the positive or negative distance from the ideal centre line (Fig. 6.41).

From higher level of abstraction, the element breakdown and the interfaces are inherent as well as even the fault of the elements and the interfaces. The faults in the feedback and possible impacts from the controlled system or even from road or environment are not traceable. It leads to the typical fault negative versus fault negative dilemma. You do not know if it is an error in the feedback or a real disturbance from the real world or the system under control (Fig. 6.42).

In the higher level, the control loop consists of a command input and a current feedback loop. The output for the valve is a normal output, which provides input from two sources. Both levels of abstraction need to be considered to develop an adequate control loop design (Fig. 6.43).

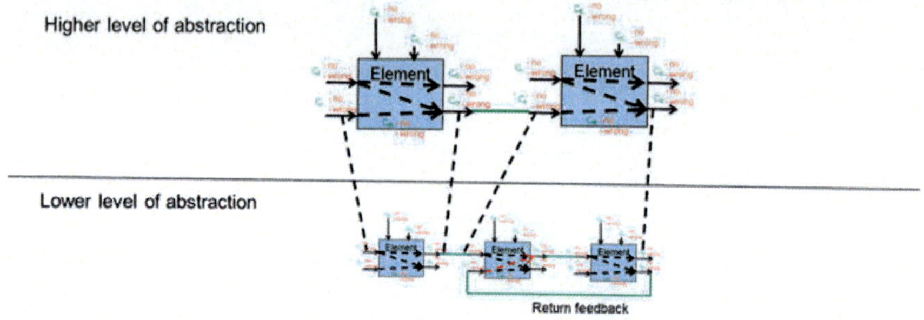

Fig. 6.41 Perspective of a control circuit from a higher level of abstraction

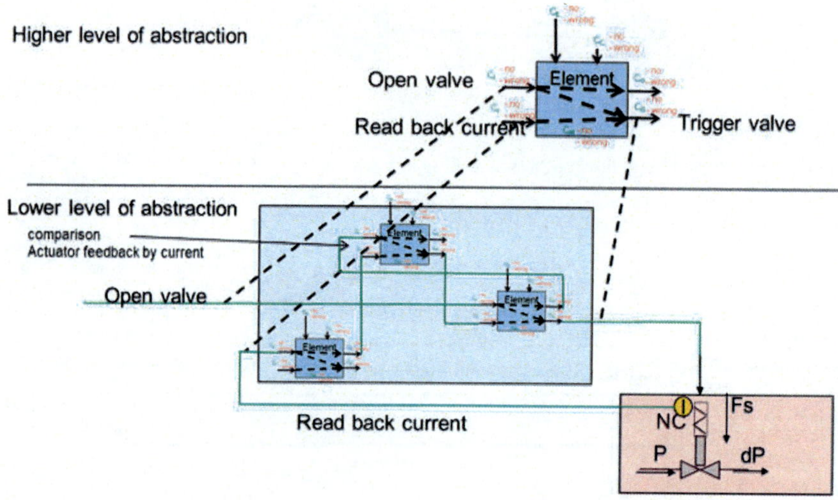

Fig. 6.42 Control of a valve by a pressure sensor

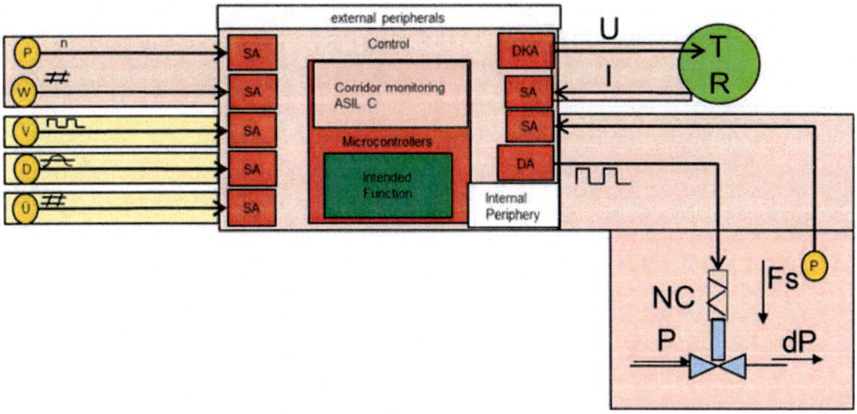

Fig. 6.43 Typical corridor controlled closed-loop system

For closed-loop control circuits, it is very common to control the permitted upper and lower limits for the actuator. Since there are design-related limits, higher or lower output values are negligible for a propulsion system. For a steering system, the thresholds are much closer and the timing constraints how long a failure could tolerate are very small (Fig. 6.44).

The current read back on the electronic design level reacts very fast, since a feedback in the power transistors affecting directly a higher current at the transistor gate. Those current feedbacks could lead to higher torque from the drive motor due to wind. The microcontroller providing the command reference for the motor power rectifier is not even involved in such a control loop. Such control loops usually could react in microseconds to expected disturbances. The reliability of the selected electronic hardware parts is usually not relevant unless the electronic circuits operate within the

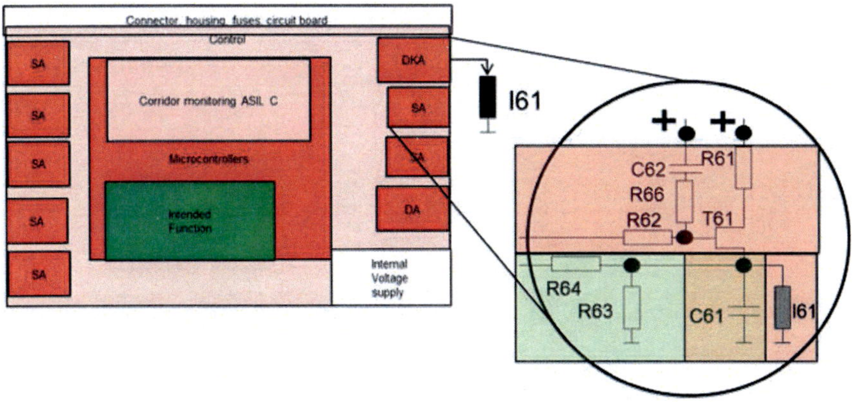

Fig. 6.44 Breakdown of a valve control circuit, the electronic design layer

design limits. The design limits could usually be controlled by the microcontroller via the current feedback (Fig. 6.45).

If we see the both layers of abstraction, you see typically two closed-loop control circuits: the current feedback on electronic layer and the pressure feedback on a system level of abstraction.

In case of a motor management system, the two control loops usually run in several hundred millisecond ranges. In a steering system, such control loops operate in a ten millisecond order of magnitude (Fig. 6.46).

A typical Level 3 AD function needs to consider various feedbacks. The figure above considers the following feedbacks:

- Current feedback on EE hardware level from the power electronics, which gives an equivalent to the current torque demand of the electric motor. Sometimes, there are already filters just in the power electronics to the gates of the power rectifier. This provides a feedback loop in microseconds. Reading the current back into a microcontroller depends already on the cycle time of the microcontroller.
- System feedback via a resolver to read the slip or the phase delay in a motor operates typically in a time interval which is at least double of the microcontroller cycle time.
- The driver closes the loop as long as he touches the steering wheel or operates with the accelerator or brake pedal by a physical damping or even active control during driving.

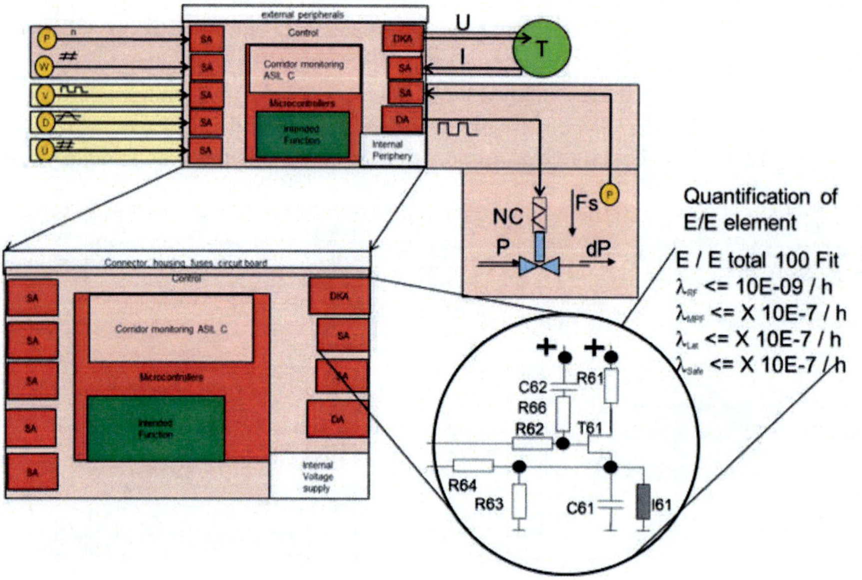

Fig. 6.45 Closed-loop cascade via two levels of abstraction

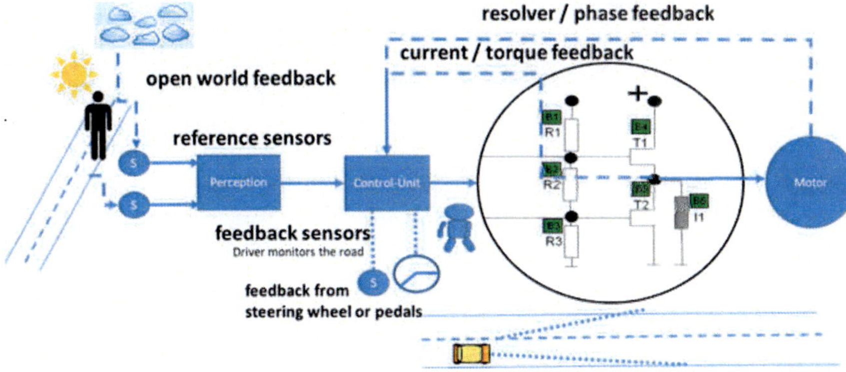

Fig. 6.46 Multiple feedback in an AD function

- The sensors at the steering wheel or at the pedal measuring the drivers influence and provide feedback to several control units by communication bus or to the specific control unit, which is the dedicated control device for the sensor.
- Usually, the environmental sensor does not only provide the reference values for the lateral or longitudinal sensor, but also provide the feedback about the quality of the brake or steering control loop. For example, they provide the information, how good a vehicle follows the lane or if the vehicle keeps the lane during braking. For a typical lateral control, there is already overlapping with the ABS brake or ESC sensors, such as wheel speed, yaw rate or lateral or longitudinal sensors within the ESC sensor cluster.

There are several overlapping closed-loop control circuits in a typical actuator control unit. The cascades do not overlap continuously but depend on the following:

- various operations or driving situations,
- weather conditions,
- vehicle or system states or conditions,
- vehicle or system operating modes, etc.

Usually inside a brake or a steering controller, an arbitration controls all the state and modes for the various situations and conditions.

The control and feedback issues are also addressed in methods like STAMP/STPA, etc. (Fig. 6.47).

A systematic analysis of failure modes does not address the required feedback for the control systems of transportation systems. In Nancy C. Leveson's approach, STPA a comparison is given which makes the missing context (legal or regulations, etc.) requirements and the issue with the feedback of closed-loop controller transparent (Fig. 6.48).

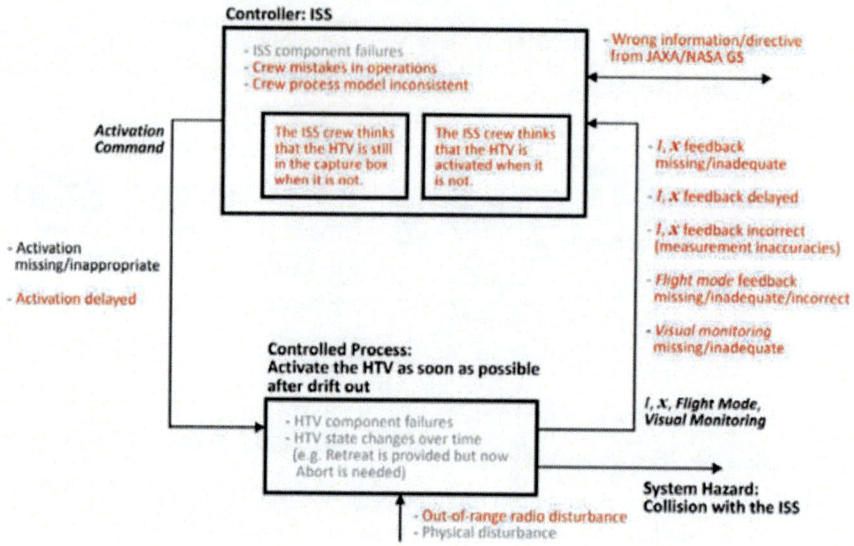

Fig. 6.47 Example from STAMP, controller in a controlled process

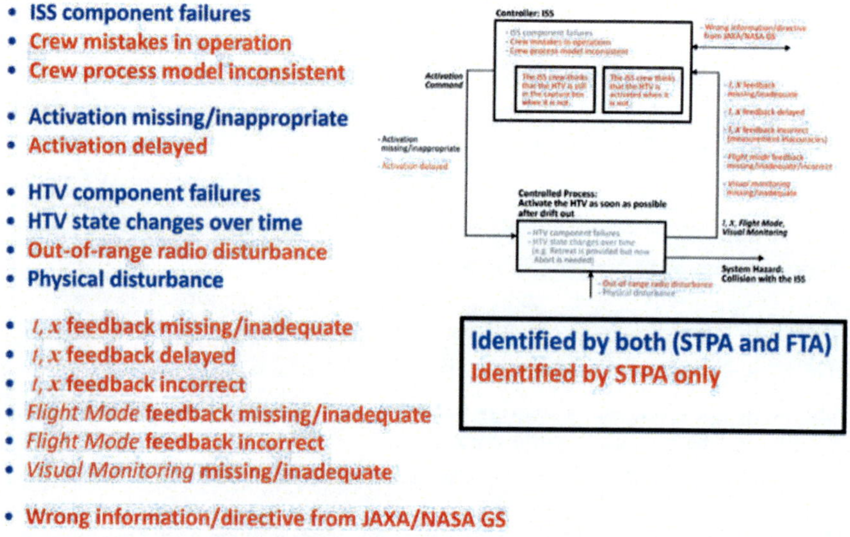

Fig. 6.48 FTA versus STAMP. *Source* Nancy C. Leveson

6.5.3 Analyse Methodology

In any level of abstraction, the following questions need to be addressed:

– What are safe behaviour corridors within the dedicated context?

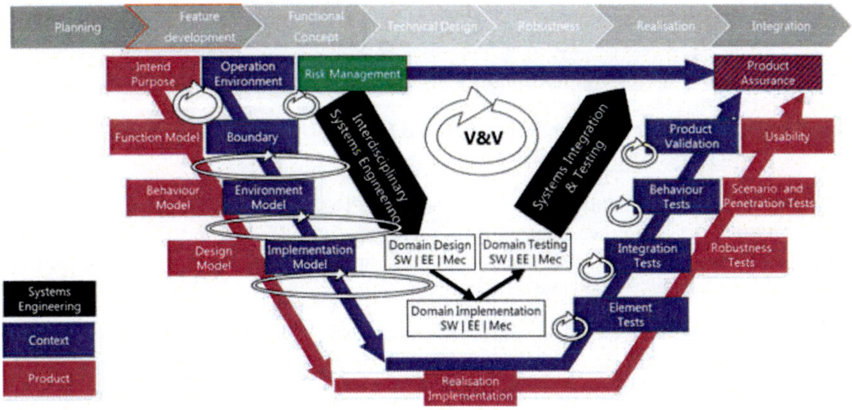

Fig. 6.49 Model maturity as V-model

- What degradation scenarios are acceptable?
- What transitions are possible, and which of them require additional safety mechanism?
- In what timing corridor the behaviour could be considered as sufficient safe?

Consequently, it is necessary to break down the product requirements from the system of interest (SOI) down, and requirements and constraints from the context (Fig. 6.49).

The digital twin is a continuous growth of the twin during development.

6.5.4 *Layered System Analysis*

Since fault propagation in dynamic system depends on external events and internal states or conditions, an overall safety concept would lock the system. It is in the nature of a dynamic system, which in one state, a specific activity, function or feature is demanded, and in another state, it is undesired and could even lead to failure or malfunctions.

IEC 61508 considers already two layers of protection, which are extended in IEC 61511 for the process industry.

IEC 61508, Part1:1998 defines the requirements as follows:

> 7.5.2.4 Where failures of the EUC control system place a demand on one or more E/E/PE or other technology safety-related systems and/or external risk reduction facilities, and where the intention is not to designate the EUC control system as a safety-related system, the following requirements shall apply:
>
> a) the dangerous failure rate claimed for the EUC control system shall be supported by data acquired through one of the following:

- actual operating experience of the EUC control system in a similar application;
- a reliability analysis carried out to a recognised procedure;
- an industry database of reliability of generic equipment;

b) the dangerous failure rate that can be claimed for the EUC control system shall be not lower than 10–5 dangerous failures per hour;

NOTE 1 – The rationale of this requirement is that if the EUC control system is not designated as a safety-related system, then the failure rate that can be claimed for the EUC control system shall not be lower than the higher target failure measure for safety integrity level 1 (which is 10–5 dangerous failures per hour; see Table 3).

c) all reasonably foreseeable dangerous failure modes of the EUC control system shall be determined and taken into account in developing the specification for the overall safety requirements;

d) the EUC control system shall be separate and independent from the E/E/PE safety-related systems, other technology safety-related systems and external risk reduction facilities.

NOTE 2 – Providing the safety-related systems have been designed to provide adequate safety integrity, taking into account the normal demand rate from the EUC control system, it will not be necessary to designate the EUC control system as a safety-related system (and, therefore, its functions will not be designated as safety functions within the context of this standard). In some applications, particularly where very high safety integrity is required, it may be appropriate to reduce the demand rate by designing the EUC control system to have a lower than normal failure rate. In such cases, if the failure rate is less than the higher limit target safety integrity for safety integrity level 1 (see Table 3), then the control system will become safety-related and the requirements in this standard will apply.

Source IEC 61508, Part1:1998, 7.5.2.4.

The two systems are:

- the basic control system of the equipment under control (EUC, which is in AD the ego-vehicle) and
- the safety instrument system (as they call the EE-PES system) that is developed according to a safety integrity standard (Fig. 6.50).

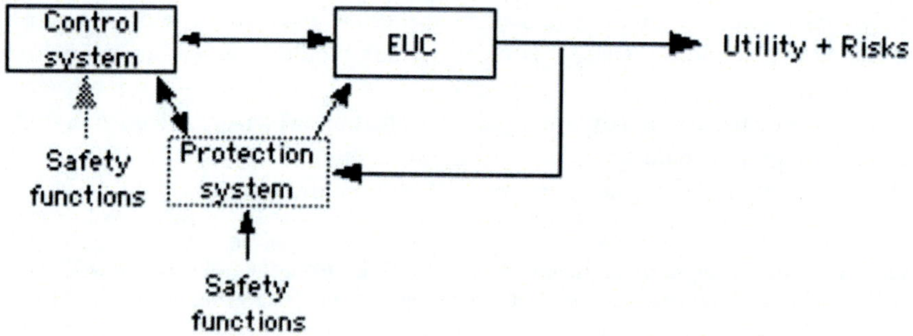

Fig. 6.50 Basic safety concept of the IEC 61508:1998

The E/E/PE safety-related systems, as they called it in IEC 61508, are part of a protection system. The basic control system provides already a certain level of risk reduction, where the standard defines a limit of 10E-2 for the maximum risk reduction. The protection system could provide according to IEC 61508 a risk reduction of 10E-7 if the system is designed according to that standard (Fig. 6.51).

Considering a frequency of 10E-9 of occurrence of the unwanted event, according to IEC 61508 at least two layers of protection are necessary. The problem of the fault tree is that it can only deal with independent faults, therefore a qualitative analysis is necessary in addition to the quantitative analysis.

IEC 61508 provides another method in the original standard to make the need of independent safety mechanisms, or as the standard called, "safety functions" are necessary (Fig. 6.52).

a) The following requirements underpin the matrix and each one is necessary for the method to be valid: the safety-related systems (E/E/PE and other technology) together with the external risk reduction facilities are independent;

b) each safety-related system (E/E/PE and other technology) and external risk reduction facilities are considered as protection layers which provide, in their own right, partial risk reductions as indicated in figure A.1;

NOTE 1 – This assumption is valid only if regular proof tests of the protection layers are carried out.

c) when one protection layer (see b) above) is added, then one order of magnitude improvement in safety integrity is achieved;

NOTE 2 – This assumption is valid only if the safety-related systems and external risk reduction facilities achieve an adequate level of independence.

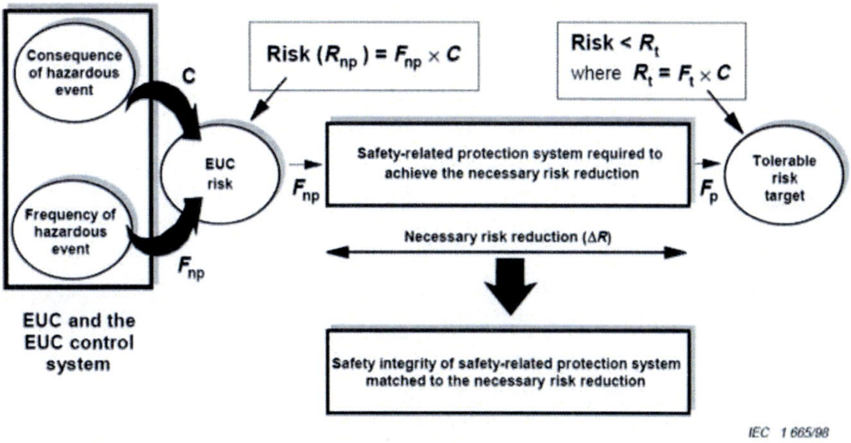

Fig. 6.51 Protection system consisting of risk reduction, by basic control system and protection system including E/E/PES developed according to IEC 61508

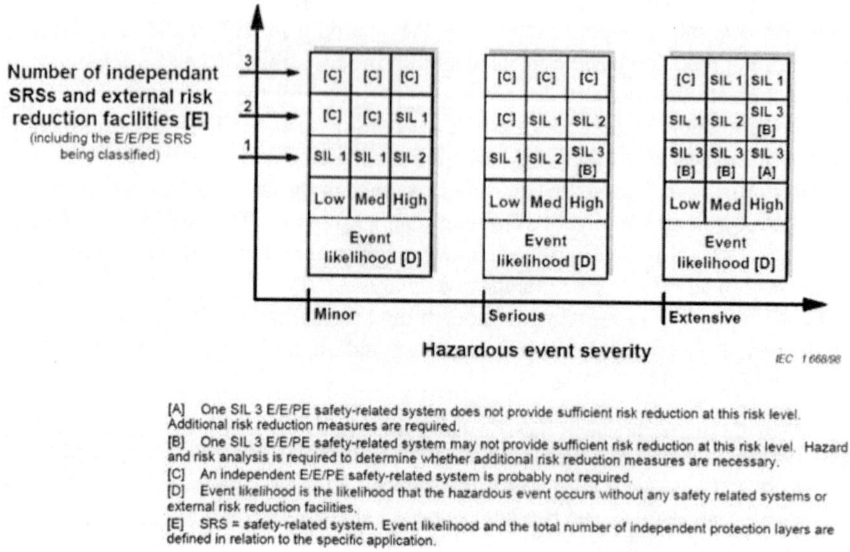

Fig. 6.52 Hazardous event matrix. *Source* IEC61508, 1998

d) only one E/E/PE safety-related system is used (but this may be in combination with another technology safety-related system and/or external risk reduction facilities), for which this method establishes the necessary safety integrity level.

The above considerations lead to the hazardous event severity matrix shown in figure E.1. It should be noted that the matrix has been populated with example data to illustrate the general principles. For each specific situation, or sector comparable industries, a matrix similar to figure E.1 would be developed.

Source IEC61508, part5 1998.

There are three hazardous events considered, which require a different set of measures. The overall risk reduction for the EUC is the sum of three types of events:

- a minor event
- a serious and
- an extensive event.

The overall protection system based on three independent protection layers is:

- external measures and measures of other technology,
- safety functions provided by the basic control system with some safety functions which claim an safety integrity with SIL 1 (difficult to implement, since the independence would lead to some challenges),
- E/E/PES system developed according to IEC 61508 up to SIL 3.

Since in plant processes at that time, not only full electronic process control systems or safety-related PESs (safety PLCs) were installed, most of the safety mechanism had been implemented by independent control loops (separate sensor, controller and actuator). The approach became forgotten, since today's EE design implements multiple control and safety functions. The important experience of that approach was the need of the independence also of the safety functions.

Another major change in the later standardisation is the approach of multiple system level approaches, based on

- protection layer for the EUC and its basic control system (which provide the intended function of the entire system),
- the safety functions which provide the necessary integrity to fulfil the requirements for the risk reduction provided by the E/E/PES safety-related system.

6.5.5 *Quantitative Aspects of Layer of Protection Analysis*

The example given in the previous chapter based on an example from IEC 61508 is derived from principles from plant engineering, e.g. in the chemical industry. Especially by analysing the system capability, we need a bottom-up approach, rather than a top-down approach as we need it to break down the user requirements (Fig. 6.53).

A quartz is a major part or utility of a microcontroller system. It provides the time base for clocking the microcontroller and the cycles for the implemented software. Furthermore, it provides the basis for the synchronisation of all tasks and processes inside of the controller. The output of the quartz are just pulses, but those pulses could have the following failure modes:

- time drift between individual pulses,
- different lengths of the pulses,
- missing impulses,
- pulses stretch and compress on the time axis,
- delay and acceleration on the time axis, etc.

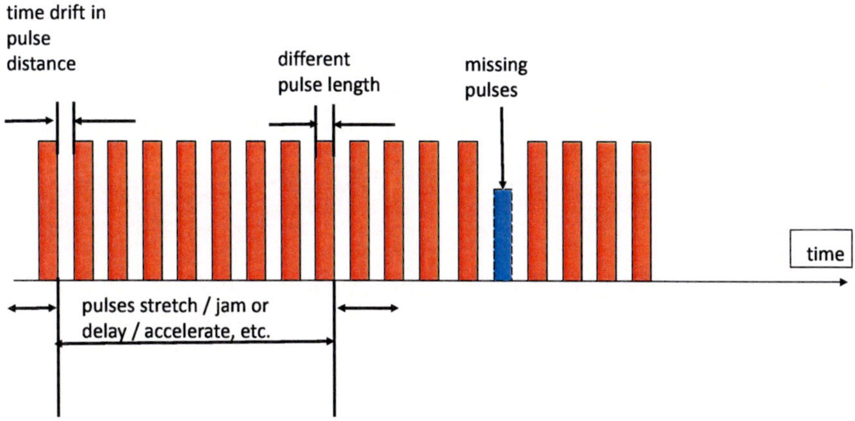

Fig. 6.53 Error conditions of pulses from a quartz

The cause of such error conditions could be

- too less power or power drops in the system,
- functional faults due to design, or ageing effects of the physical elements,
- EMC impacts, etc.

which also leads to errors in the pulse form, pulse height, etc., depending on how you use the quartz output the fault effects the system in many different ways (Fig. 6.54).

Very typical approach to improve the reliability of a quartz is a redundant implementation of the quartz. A comparison depends on the use of a quartz.

- Using the quartz as a pulse provider in a dedicated time frame, it might be sufficient to measure a reference trigger point of any pulse and compare the trigger events between each other. In this way, you only know how good the quartzes are in comparison with each others.
- In order to make qualitative statements about the quartz, any of the pulses must be analysed during runtime and their accuracy against the specification monitored.

For a giga-hertz quartz, it is nearly impossible to measure and analyse all properties during runtime. The only way is to make a reliable design of a quartz. But the design just confirms a statistical correctness of the specified design characteristics. Total correctness not possible, we must assure by an adequate integration strategy to assure that the possible faults could be controlled.

For a redundant quartz, it is not the best approach to overlap the pulses simply. Some solutions using a passive filtering system but trusting one quartz unless it is correct seem the fastest principle. Whenever certain deviations are detected in the active quartz, it could switch to the other quartz, and you can do it by passive and very reliable elements. Such a devise we could understand as a gate-keeper, which opens the door for the most healthiest pulses. In such a hot standby concept, the silent quartz is also continuously running, and the gatekeeper avoids just the error propagation in case of their occurrence (Fig. 6.55).

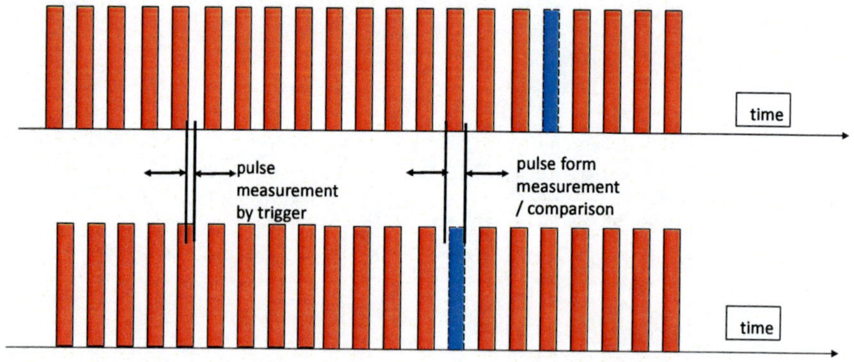

Fig. 6.54 Redundant quartz and their comparison

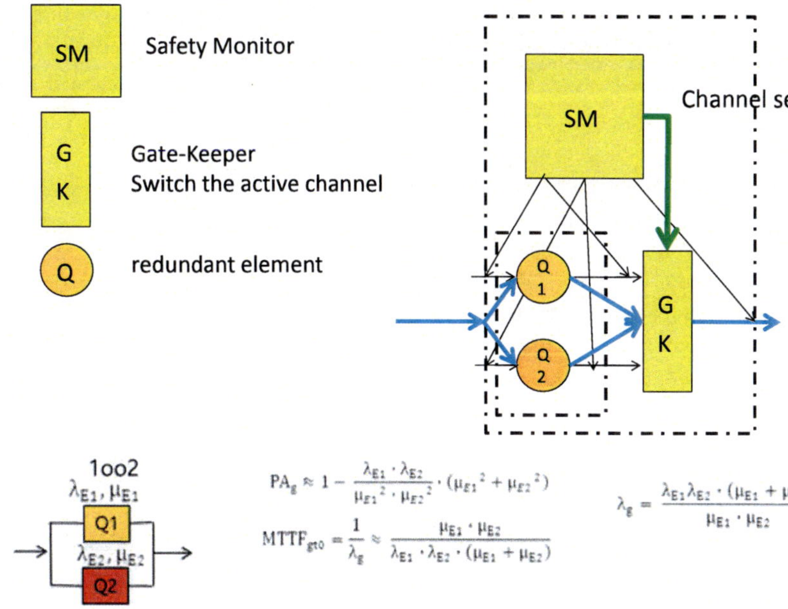

Fig. 6.55 Gatekeeper approach to control a redundant quartz

Since we need to consider faults of a quartz (Qx) during runtime, we need all technical possible faults. By adequate design, we could improve of course the frequency of the occurrence of the faults.

The probability of the occurrence follows the following formula:

$$PA_g \approx 1 - \frac{\lambda_{E1} \cdot \lambda_{E2}}{\mu_{E1}^2 \cdot \mu_{E2}^2} \cdot \left(\mu_{E1}^2 + \mu_{E2}^2\right)$$

where

- λ_{E1} is the failure rate of the element and
- μ_{Ex} is the repair rate or recovery rate.

For the redundancy of $Q1$ and $Q2$, we get the overall failure rate by the following formula:

$$\lambda_g = \frac{\lambda_{E1}\lambda_{E2} \cdot (\mu_{E1} + \mu_{E2})}{\mu_{E1} \cdot \mu_{E2}}$$

which leads to a "mean time between failures (MTBF)" for the redundancy as follows:

$$\mathrm{MTTF}_{gt0} = \frac{1}{\lambda_g} \approx \frac{\mu_{E1} \cdot \mu_{E2}}{\lambda_{E1} \cdot \lambda_{E2} \cdot (\mu_{E1} + \mu_{E2})}$$

Such an architecture could be mapped on any functionality. We need to evaluate the following characteristics of the architectural elements by typical questions like the following:

- What are the redundant parts?
- What are the design-related reliability measures and their effectiveness?
- What are the possible failure modes for the redundant elements as a black box; since they are using memory and require feedback from outside (e.g. a closed loop control circuit), implemented state machines, etc., we have to understand the internal possible error states and possible additional failure modes. In a redundancy concept, these are all causes of dependant failure.
- Internal reliability and the ability or possibilities to improve the reliability.
- Which error conditions are controlled, by which measures and what is the effectiveness of those measures, etc.

That means the qualification of the element for a specific safety-related applications means intensive systematic analysis of the integration candidate. All that mechanism needs to be implemented into the monitoring function. The monitoring function does not influence in any kind the information flow from the sensors to the actuators. The monitoring should be usually designed in a way that monitoring provides only information to the gatekeeper, but it should be free from any interference to the intended functionally or the purpose of the basic functionality. Of course is the sufficient independence a major design criterion. Failure in the monitoring function shall lead only to a switching of the gatekeeper, but not of a failure of the intended function on the considered architecture.

The gatekeeper needs to be designed in a way that a hot standby is possible by means of a tolerable switch over time and no functional impact for the output of the unit (Fig. 6.56).

Such an architectural approach could be also applied on a higher system level, for example as a part of a sensor fusion.

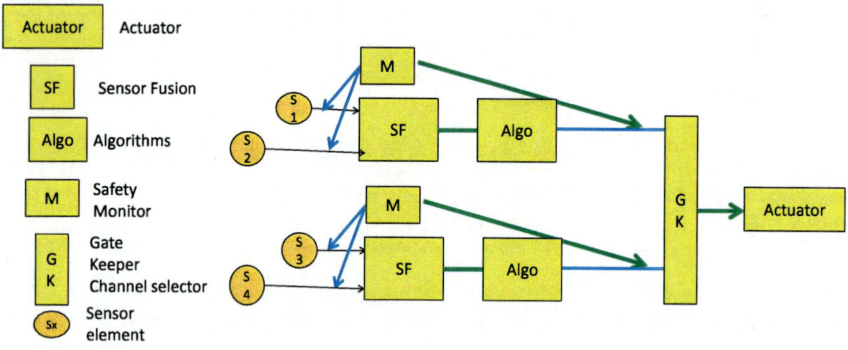

Fig. 6.56 Application of redundant gatekeeper architecture

Considering a redundant sensor fusion and a related algorithm (algo) with 2 sets of redundant sensors (here 4 sensors), such an architecture could be applied as well including a gatekeeper which switches conditionally from one redundant path to the other. As closer the gatekeeper is to the actuator, as more beneficial the architecture supports also the availability of the entire system. The sensor fusion and the various algorithms could be also of different logical and design-related principles. Such an architecture is very typical for a dissimilar implementation. By using the same sensors but on different logical principles, we could come to the same consequence for the actuator. The brake system should brake in front of obstacles, or the steering system should steer according the resulting lane and trajectory commands. For example, a triple modular system could be implemented with the following logical principles:

- object detection and static localisation,
- optical flow or dynamic behaviour model,
- free space detection, etc.

Since we need to understand only the internal error condition, such application with dissimilar approaches is a typical safety application for artificial intelligence such as neuronal networks and machine learning (Fig. 6.57).

Using the example from the sensor fusion, the redundant path could be as long and could consist of as many elements as possible. The principle for the architecture and their qualification methodology for safety and availability considerations remain the same (Fig. 6.58).

Considering the redundant element with a failure rate: $E_{\text{rxtotal}} = 100$ fit, the safety monitoring with SM = 100 fit and the gatekeeper with GK = 10 fit, we reach with diagnostic coverage for latent and single point failure a total failure rate of $E_{\text{Rtotal}} = 310$ fit (Fig. 6.59).

Such an application could be also implemented in parallel, so that we get a cascade of such gatekeeper architects. By such a consideration, we see the influence of the

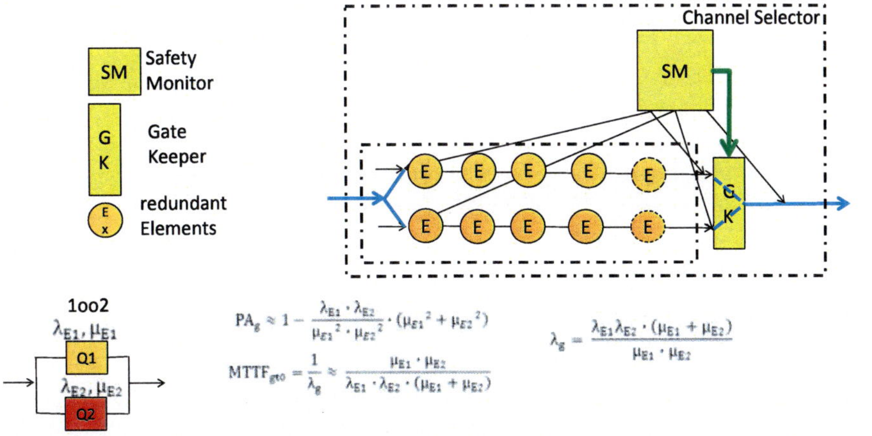

Fig. 6.57 Redundant elements controlled by a safety monitor

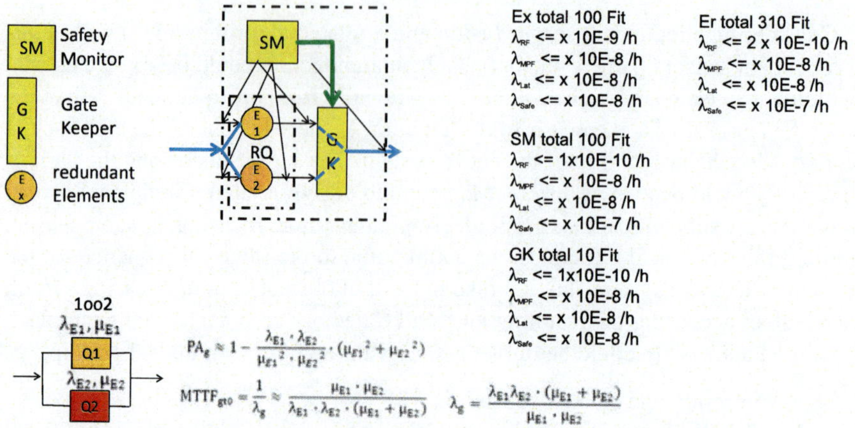

Fig. 6.58 Quantitative analysis of the gatekeeper architecture

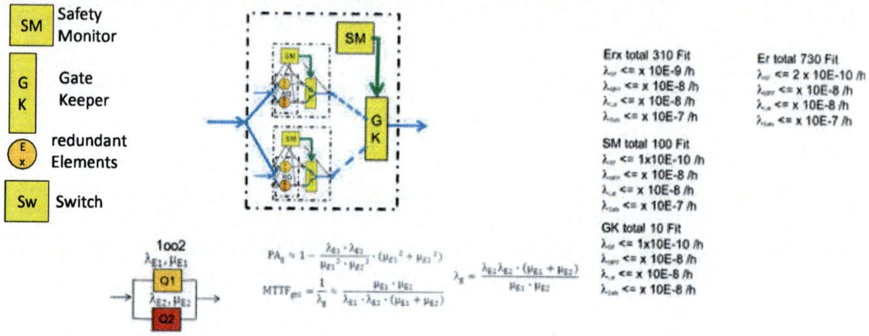

Fig. 6.59 Quantitative analysis of a redundant gatekeeper architecture

safety mechanism is negligible, and the major failure rate is dominated by the gatekeeper. In case of a high available design, the gatekeeper should be replaced by a cross-comparison unit, where any of the monitoring units monitored the redundant intended functionality (Fig. 6.60).

The gatekeeper works very similar to a voting principle. But it assures better control principles, higher degree on diagnostics and resulting higher availability, by following adequate design and redundancy principles.

6.6 Methods for Risk Management

The methods for risk management need to be adapted to the use case. Since the use cases in mobility industries, the methods need to be adapted. On the other hand, experiences from other industries become more and more important, in order to

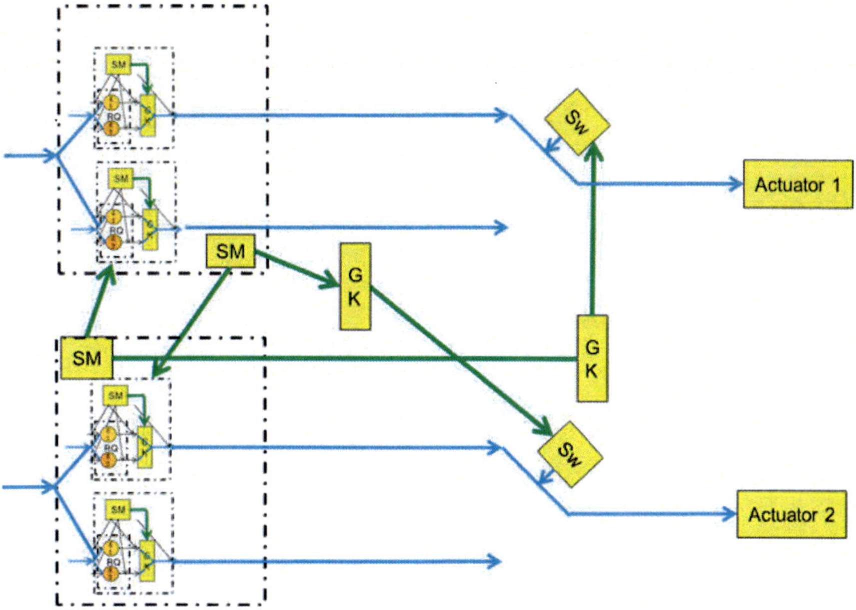

Fig. 6.60 Cross-monitoring approach based on a gatekeeper architecture

understand the risk scenarios. Since we are not bringing the system under consideration in the right context and in an entire lifecycle approach, we will miss many risk and hazardous scenarios to be analysed.

Example for such changes is methods like STAMP or STPA, which had been described in the early 2000 by Nancy C. Leveson. Many proven risk methods had been compiled and explained on avionics and space application. Today, we often use those methods to analyse automation scenarios with the use cases for automated driving in automotive road traffic. In a paper, we have a good overview on what are typical gaps of a traditional fault tree analysis (Fig. 6.61).

The gaps are in the following direction

- Closed-loop control requires further failure scenarios.
- Open world control requires further architecture and design methods, which consider not just the system it needs to include also the operating environment and the integration environment, e.g. into the motor vehicle.
- Legal compliance becomes a more important aspect of product development, since many legal stands could be relevant to the system design, the usability and their constraints.

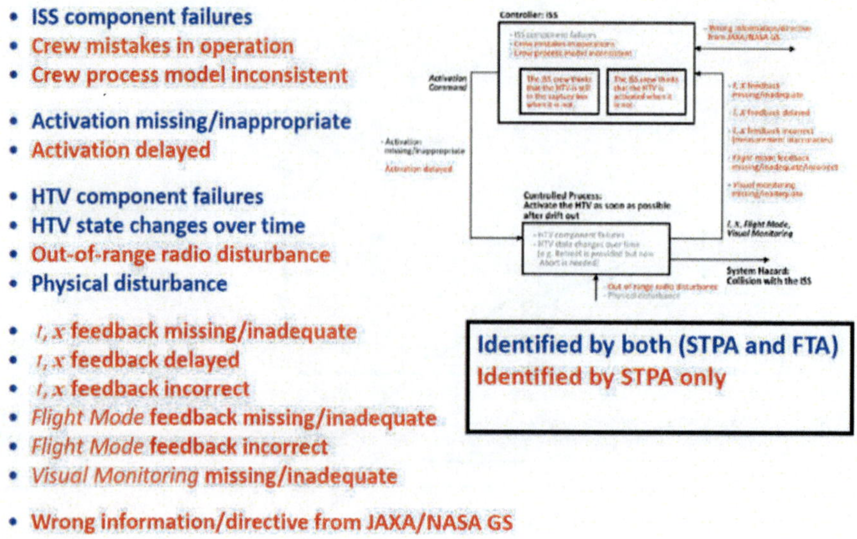

Fig. 6.61 FTA comparison to STAMP. *Source* Nancy C. Leveson

6.6.1 Failure Mode and Effect Analysis (FMEA)

The classical design FMEA considers mechanical components, and its aim was the adequate design and resulting characteristics. From an exploded view of a component, the detailed characteristics had been evaluated and the consequences of certain deviation from nominal or defined tolerance had been analysed (Fig. 6.62).

The automobile associations like VDA and AIAG have described the essential methods in this context. Standards had been improved in other industries based on their requirements. FMEA is according to ISO 26262 an inductive method for the safety analysis. However, all FMEA methods in the automobile industry are widely based on the sequence of failure cause, failure and failure effect. The kind of measures to improve the product or avoid, mitigate errors as well as their propagation had been

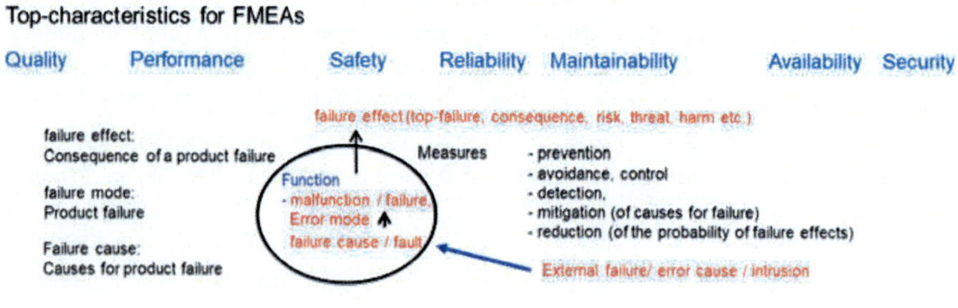

Fig. 6.62 Basic principle of FMEAs

defined and applied differently in the standards. The risk priority number based on the multiplication of the following factors is:

- severity of damages (S),
- probability of the error occurring (O),
- probability of the error detection (D).

The severity class of a FMEA is generally determined by the failure effect. This severity (S) is defined differently than the severity of a hazard and risk analysis. Generally, it can be said that the severity of the hazard and risk analysis refers to human impacts or harm (and environment damages in other standards). The severity class in a FMEA refers more to the vehicle itself. This is why the vehicle is often used as the root element of a VDA-FMEA. In a hazard and risk analysis, also the driving and operating conditions are considered, which would lead to very heterogenic structures for a FMEA. This does not mean that it could not be applied for simple systems. The probability of the class for fault occurrence (O) and the class of detection of faults (E) are generally based on the assessment of the failure cause. Those three factors build the **risk priority number (RPN)**. The two factors S and O (SxO) are often combined to the so-called criticality. However, it is often useful in the different FMEA methods, to assess the factors separately. The probability of error propagation is often not considered in the classical FMEAs.

The first step defines the scope and boundary of the system under consideration. The typical failure analysis itself only happens in the fourth step. Steps 2 and 3 are analyses and information, which are needed in FMEA in order to present the

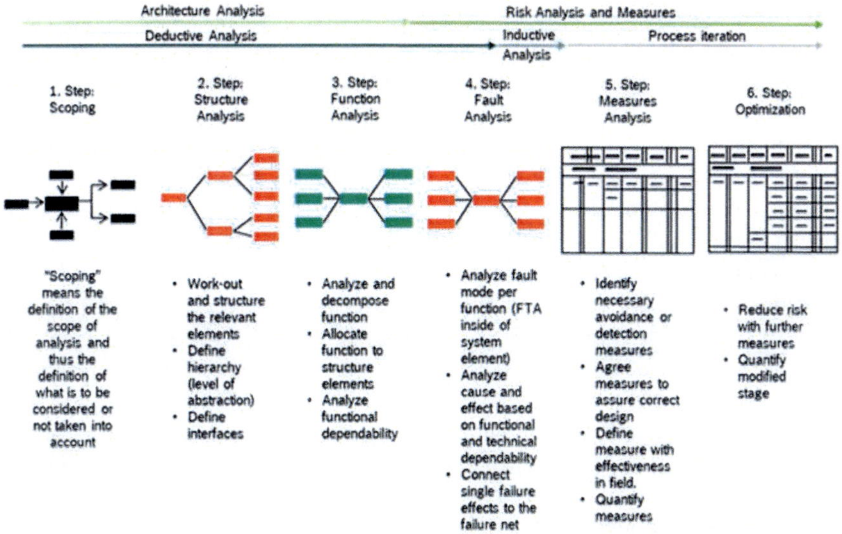

Fig. 6.63 FMEA in 6 steps similar to joint development group of VDA 4 and AIAG. *Source* FMEA, joint development group of VDA 4 and AIAG, 2018

object of analysis or are other analysis by themselves. Steps 1–4 could be seen as the illustration of a deductive analysis, since functions and structures are broken down or decomposed (Fig. 6.63).

There are many types of forms available, which cover at least the following columns:

- functions (intended function),
- failure type,
- failure cause,
- failure effect,
- occurrence, control and/or detection measures,
- risk factors of the risk priority number.

6.6.2 Fault Tree Analysis (FTA)

The fault tree analysis is a key method for the development and analysis of safety-relevant systems in almost all industries and for nuclear power plants or the aerospace sector. With the help of the fault tree analysis, it can be examined, in which combination, by using Boolean logic, relevant elements could fail and cause undesired states or events, like a failure of an engine. The aim of the fault tree analysis is to determine the minimal amount of events, which can cause such a top event, and therefore to detect specific weaknesses and unintended states in the system.

The historical background of the fault tree analysis comes from the military sector. At the beginning of the 1960s, this technology was first used by the U.S. Air Force and then spread to other areas of the aerospace as well as the nuclear energy sector.

Efforts are made to illustrate and analyse more and more comprehensive and complex systems as fault trees. Fault trees are based on Boolean logic, which can be investigated on minimal cut-sets with various algorithms with different targets. Special forms of the cut-set analysis are Quine–McCluskey for the minimisation of the Boolean logic, the MOCUS algorithm, the algorithm of Rauzy on binary decision diagrams, the algorithm of Madre and Coudert with meta-products and the search strategy CAMP DEUSTO. Those algorithms are based on the different data structures and procedures for the determination of minimal cuts.

The Boolean algebra and graphic fault trees generally build the foundation for such analyses. The fault tree analysis is seen in ISO 26262 as deductive analysis, whereas this cannot be said for the higher analyses of the cut-sets. These analyses are also not necessary for the requirement or architecture development at the descending branch of the V-model; they would more support the analyses of ISO 26262, part 5, Chap. 9. This is the ascending branch of the electronic V-model considering already a first realisation of the product. Here, required analysis and the related metric like (PMHF, probabilistic metric of random hardware failure) target to identify error propagations based on failure rates for random hardware faults and their potential to violate given safety goals.

6.6.3 Markov Analysis

The Markov analysis is primarily used to assess the transition from one condition to another. The formula for the safety architectures in the informative part 6 of IEC 61508 was derived from such models. Those formulas are gladly used for the EE safety architectures. However, the basic principles and assumptions, under which the models were designed and the formulas derived, are often not known or not applicable for the realised automotive architectures. Often only one failure at the time is assumed; therefore, ageing affects, error combinations and dependent, transient or latent failures are not derivable from this formula. For approximations or as help for the quantification, these formulas are applicable if corresponding further analyses are applied.

Markov Analysis in Road Traffic Scenarios

Markov analysis will get more importance in cooperative road traffic scenarios, since the behaviour of the various road traffic participants could not be considered as deterministic.

The prediction of road traffic accidents is of importance in locating the most hazardous sites in order to improve the safety management level. Through the comparative analysis of traditional hazard and risk analysis methods, Markov prediction models using the grey system theory and Markov chain theory are constructed. The results indicate that the prediction model can be well applied in road traffic accident prediction with strong engineering practicability.

For road traffic density, Markov models are often used in combination with GPS, particularly to estimate the road traffic density. The following steps could be applied:

1. For navigation part, the time-dependent network routing problem and the derived navigation path can be updated timely by referring the travelling time arrangement.
2. In the prediction part, the equilibrium Markov chain handles the traffic congestion condition by the traffic state computation. That is, the congestion alleviation strategy is designed to balance the equilibrium states via the computation of Markov transferring probability matrix.

Markov Chain Monte Carlo Technique

In large-scale deployments of traffic flow models, estimation of the model parameters is a critical but cumbersome task. A poorly calibrated model leads to erroneous estimates in data-poor environments and limited forecasting ability. There are methods for calibrating flow model parameters for a discretised scalar conservation law using only velocity measurements. Such methods are based on a Markov chain Monte Carlo technique, which is used to approximate statistics of the posterior distribution of the model parameters. Numerical experiments highlight the difficulty in estimating jam densities and provide a new approach to improve performance of the sampling through re-parameterisation of the model.

6.6.4 HazOp (Hazard and Operability Studies or Analysis)

Hazard and operability study (HAZOP) is a structured and systematic technique for system examination and risk management. In particular, HAZOP is often used as a technique for identifying potential hazards in a system and identifying operability problems likely to lead to non-conforming products. HAZOP is based on a theory that assumes risk events are caused by deviations from design or operating intentions. Identification of such deviations is facilitated by using sets of "guidewords" as a systematic list of deviation perspectives. This approach is a unique feature of the HAZOP methodology that helps stimulate the imagination of team members when exploring potential deviations.

IEC 16882 provides a methodology for HAZOP for the plant automation. IEC 61882:2016 provides a guide for HAZOP studies of systems using guidewords. It gives guidance on application of the technique and on the HAZOP study procedure, including definition, preparation, examination sessions and resulting documentation and follow-up. Documentation examples, as well as a broad set of examples encompassing various applications, illustrating HAZOP studies are also provided. The focus parts are:

- clarification of terminology as well as alignment with terms and definitions within ISO 31000:2009 and ISO Guide 73:2009,
- provision of an improved case study of a procedural HAZOP. Keywords: HAZOP, risks and operability problems.

HAZOP is a qualitative analysis of the hazard potential of failure conditions or malfunctions of individual technical elements. In interdisciplinary teams, i.e. architect, system analyst or tester, the target function (also the design appropriate functionality) is inferred from the detailed description of the analysed object as well as the possible failure functions or failure behaviour and measurements from a structured questioning of the target function.

Similar questions, just like those for HAZOP, are also considered in the fault tree analysis or in variants of FMEAs in order to find causes of known failure (Fig. 6.64).

SAE had founded a working group and published a standard concerning detailing of the HA&RA according to ISO 26262, considerations for ISO 26262, ASIL Hazard Classification, SAE J2980 Prop Draft F2011ff. In this standard, the key or guidewords from other industry HAZOP had been derived to automotive applications (Fig. 6.65).

This standard provides further very important background information, which helps to perform hazard and risk analysis for automotive applications.

Guidewords	Importance	Aspect
NO, NOT, NOT	negation the goal function	no part of the goal function is performed, but it also happens nothing
MORE	quantitative growth	Physical size: weight, speed temporal aspects: too late, too early Covers: too late, too short, too high, too low
LESS	quantitative decrease	Feature: Material, dynamic, thermal conductivity Dynamics: heating, pressure build, move, rotate
AS WELL ... AS	qualitative growth	Function object is achieved, side effects occur as - Dynamic effects: heat resistance increase capacity reduction, overshoot - Tangible effects: contamination, wear, Corrosion, fire Contact
PARTIALLY	qualitative decline	Partial feature - Performance is not reached - Vibrations (signal always breaks) - Information or incomplete signal - Sub-functions or sub-elements without function
REVERSAL	Negation of goal function	Direction, sign, action principles
DIFFERENT TO	Operating states	Ignition cycle, sequences, state machines, Memory organization in the microcontroller, Generation of data fields

Fig. 6.64 HAZOP—keywords, meanings and aspects. *Source* SAE J2980, Considerations for ISO 26262 ASIL Hazard Classification, 2015

→ GUIDEWORDS

System Function Vs. Guidewords	No Activation	Incorrect Activation 1 (More than requested)	Incorrect Activation 2 (Less than requested)	Incorrect Activation 3 (Wrong direction)	Autonomous activation (Request is zero)	Locked Function (Failure to release)
Electric Steering Assist Function	Loss of Steering Assist	Excessive Steering Assist	Reduced Steering Assist	Steering in the opposite direction	Unintended Steering	Locked Steering
Brake by Wire (base brake functionality)	Loss of Brakes	Excessive Brake Apply	Reduced Brake apply	-	Unintended Brakes	Stuck Brakes
Stability Control Function (ESC with brakes)	Loss of ESC	Excessive Yaw Moment Correction	Inadequate Yaw Moment Correction	Incorrect Yaw moment Correction	Unintended ESC Apply	Stuck ESC

Fig. 6.65 HAZOP—guidewords adapted to automotive applications. *Source* SAE, J2980 Revised Proposed Draft F 2011

HAZOP Tailoring to Specific Analyses Targets

The basic framework for the HAZOP methodology could be very good tailored by target-oriented guidewords. Adequate tailoring of guidewords could be applied for the following examples:

– hazard and risk analysis (see example SAE, J2980),

- support of design and system FMEA or FMEA with focus during development and in-use analysis,
- security, privacy or authenticity analysis,
- software safety or software architectural analysis, etc.,
- operational scenarios,
- misuse scenarios.

The documentation of the HAZOP-based tables was causes and measures to avoid hazardous situations that are similarly elaborated.

HAZOP is a method for assessing hazards in facilities, equipment and processes and is capable of assessing systems from multiple perspectives like:

Design:

- assessing system design capability to meet user specifications and safety standards,
- identifying weaknesses in systems.

Physical and operational environments:

- assessing environment to ensure system is appropriately situated, supported, serviced, contained, etc.

Operational and procedural controls:

- assessing engineered controls (e.g. automation), sequences of operations, procedural controls (e.g. human interactions), etc.,
- assessing different operational modes, like commissioning, standby, normal operation, steady and unsteady states, normal shutdown, emergency shutdown, etc.

6.6.5 Preliminary Hazard and Risk Analysis (PHA, PRA)

Preliminary hazard analysis is performed to identify areas of the system, which will have an effect on safety by evaluating the major hazards associated with the system. It provides an initial assessment of the identified hazards. PHA typically involves:

- determining hazards that might exist and possible effects,
- determining a clear set of guidelines and objectives to be used during a design,
- creating plans to deal with critical hazards,
- assigning responsibility for hazard control (management and technical),
- allocating time and resources to deal with hazards.

Brainstorming techniques are used during which the design or operation of the system is discussed on the basis of the experience of the people involved in the brainstorming activity. Checklists are commonly used to assist in identifying hazards. A classical preliminary hazard analysis (PHA) is based on the following main artefacts, which are often represented in the form of a table:

A brief description of the system and its domain.

- a brief description of any subsystems identified at this phase and the boundaries between them,
- a list of identified hazards applicable to the system, including a description and unique reference,
- a list of identified accidents applicable to the system including a description, a unique reference and a description of the associated hazards and accident sequences,
- the accident risk classification,
- preliminary probability targets for each accident,
- preliminary predicted probabilities for each accident sequence,
- preliminary probability targets for each hazard,
- a description of the system functions and safety features,
- a description of human error, which could create or contribute to accidents.

The advantages of using the PHA method include:

- It identifies the potential for major hazards at a very early stage of project development.
- It provides basis for design decisions.
- It helps to ensure plant to plant and plant to environment compatibility.
- It facilitates a full hazard analysis later.

The disadvantage of PHA is that it is not comprehensive and must be followed by a full hazard and operability (HAZOP) study.

Preliminary Risk Assessment (PRA)

The PRA is a first line analysis (hence the name "preliminary") that you would use to determine at a glance the effects of identified risks. It is often considered as a scaled down version of a failure mode and effects analysis (FMEA) that may be very beneficial in the risk-based thinking.

- Using a cross-functional team, brainstorm "things that could go wrong". If necessary, you can process all of these potential things that go wrong through the PRA process. You may decide to only manage the top 5 or 10 depending on how the PRA results look.
- Determine the potential impact for each of the items in the first column. You might have multiple impacts for one thing gone wrong. See the sample below on how this may look in your PRA.
- Determine the potential causes for each "thing gone wrong". There may be multiple potential causes. See the sample below for insights on how this may look in your PRA.
- Define a metric for likelihood, and use a cross-functional team to assign a quantitative value to the likelihood of this event occurring.
- Define a metric for severity, and use a cross-functional team to assign a value for the severity of the consequence if this event were to occur.

- Simply multiply your likelihood and your severity columns to create a risk score.
- Consider what possible additional controls could be effectively implemented to mitigate or control the risk.

PHA and PRA follow both a systematic brainstorming techniques, which are supported, by scaling or metrics to make the result assessable. Both methods follow a process, where the detailing is supported by means of a HAZOP or a FMEA.

FHA According to FAA

The Federal Aviation Association (FAA) provides a very similar approach, and they call it functional hazard analysis (FHA).

In the FAA System Safety Handbook, they define the steps as follows:

> 8.2 Sources for detailed safety design criteria include Occupational Safety and Health Administration (OSHA) standards, MIL-STD-454, Requirement 1, and MIL-STD-882. Design review is typically a continual process using hazard analyses. Active participation at internal and customer design reviews is also necessary to capture critical hazards and their characteristics. ff..
>
> 8.2.1 What is the Role of the Hazard Analysis? Hazard analyses are performed to identify and define hazardous conditions/risks for the purpose of their elimination or control. Analyses examine the system, subsystems, components, and interrelationships. They also examine and provide inputs to the following National Airspace Integrated Logistics Support (NAILS) elements: ·
>
> - Training
> - Maintenance
> - Operational
> - and maintenance environments ·
>
> System/component disposal Steps in performing a hazard analysis:
>
> 1. Describe and bound the system in accordance with system description instructions in Chap. 3.
> 2. Perform functional analysis if appropriate to the system under study.
> 3. Develop a Preliminary Hazard List.
> 4. Identify contributory hazards, initiators, or any other causes.
> 5. Establish hazard control baseline by identifying existing controls when appropriate.
> 6. Determine potential outcomes, effects, or harm.
> 7. Perform a risk assessment of the severity of consequence and likelihood of occurrence.
> 8. Rank hazards according to risk.
> 9. Develop a set of recommendations and requirements to eliminate or control risks
> 10. Provide managers, designers, test planners, and other affected decision makers with the information and data needed to permit effective trade-offs
> 11. Conduct hazard tracking and risk resolution of medium and high risks. Verify that recommendations and requirements identified in Step 9 have been implemented.
> 12. Demonstrate compliance with given safety related technical specifications, operational requirements, and design criteria.

ff..

8.2.3 What is the Relationship between Safety and Reliability? Reliability and system safety analyses complement each other. They can each provide the other more information than obtained individually. Neither rarely can be substituted for the other but, when performed in collaboration, can lead to better and more efficient products.

Two reliability analyses (one a subset of the other) are often compared to hazard analyses. Performance of a Failure Modes and Effects Analysis (FMEA) is the first step in generating the Failure Modes, Effects, and Criticality Analysis (FMECA). Both types of analyses can serve as a final product depending on the situation. An FMECA is generated from a FMEA by adding a criticality figure of merit. These analyses are performed for reliability, and supportability information.

A hazard analysis uses a top-down methodology that first identifies risks and then isolates all possible (or probable) causes. For an operational system, it is performed for specific suspect hazards. In the case of the hazard analysis, failures, operating procedures, human factors, and transient conditions are included in the list of hazard causes.

The FMECA is limited even further in that it only considers hardware failures. It may be performed either top-down or bottom-up, usually the latter. It is generated by asking questions such as "If this fails, what is the impact on the system? Can I detect it? Will it cause anything else to fail?" If so, the induced failure is called a secondary failure. Reliability predictions establish either a failure rate for an assembly (or component) or a probability of failure. This quantitative data, at both the component and assembly level, is a major source of data for quantitative reliability analysis. This understanding is necessary to use it correctly. In summary, however, hazard analyses are first performed in a qualitative manner identifying risks, their causes, and the significance of hazards associated with the risk.

Source FAA System Safety Handbook, Chap. 8: Safety Analysis/Hazard Analysis Tasks December 30, 2000.

Common Approach of the Safety Analysis

In some standards, the typical hazard or risk identification and risk assessment processes are separated from the methods for the systematic safety analysis. Since the Risk Identification and its classification based on experiences and is a cross-functional approach, since the systematic methods like

- HAZOP
- FMEA
- FTA, etc.

Those methods built more a secondary process that details and verifies the quality of the hazard and risk analysis. They support the specification of measures on various levels within the organisation, the process and development team and for implementation into the product as well as evaluation of their necessary effectiveness.

6.6.6 Operational Safety Assessment (OSA)

The aviation industry defines on top of the risk management and functional risk assessment an OSA "operational safety assessment".

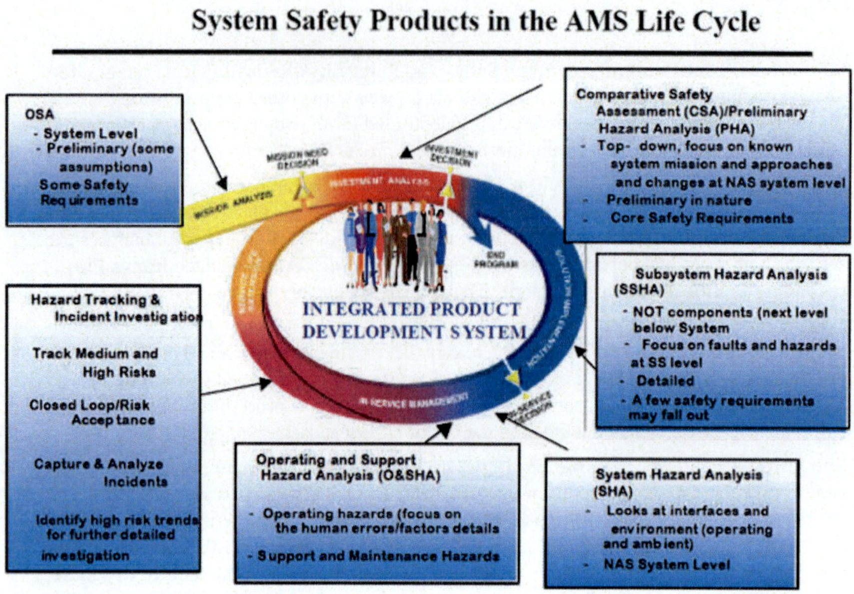

Fig. 6.66 Integrated product development system. *Source* FAA System Safety Handbook, Chap. 4: Pre-Investment Decision Safety Assessments, 30 December 2000

> An Operational Safety Assessment (OSA) has been designed to provide a disciplined, and internationally developed (RTCA SC189) method of objectively assessing the safety requirements of aerospace systems. In the FAA, the OSA is used to evaluate Communication, Navigation, Surveillance (CNS) and Air Traffic Management (ATM) systems. The OSA identifies and provides an assessment of the hazards in a system, defines safety requirements, and builds a foundation for follow-on institutional safety analyses related to Investment Analysis, Solution Implementation, In-Service Management, and Service Life Extension. The OSA is composed of two fundamental elements: (1) the Operational Services & Environment Description (OSED), and (2) an Operational Hazard Assessment (OHA). The OSED is a description of the system physical and functional characteristics, the environment's physical and functional characteristics, air traffic services, and operational procedures. This description includes both the ground and air elements of the system to be analyzed. The OHA is a qualitative safety assessment of the operational hazards associated with the OSED. Each hazard is classified according to its potential severity. Each classified hazard is then mapped to a safety objective based on probability of occurrence. In general, as severity increases, the safety objective is to decrease probability of occurrence (Fig. 6.66).

The operational safety assessment (OSA) includes the operational hazard assessment (OHA). The OHA is a qualitative assessment of the hazards associated with the system described in the OSED.

Once the system has been bounded and described, and the functions determined in the OSED, the analyst is ready to determine the hazards associated with the system. For these types of assessments, the best method is to assess scenarios containing a set of hazardous conditions. Therefore, the following definition can be used to define the hazards in a Preliminary Hazard List (PHL):

Hazard - The potential for harm. Unsafe acts or unsafe conditions that could result in an accident. (A hazard is not an accident).

Hazard or hazardous condition. - Anything, real or potential, that could make possible, or contribute to making possible, an accident.

Hazard. - A condition that is prerequisite to an accident

Source FAA System Safety Handbook, Chap. 4: Pre-Investment Decision Safety Assessments, December 30, 2000.

The OHA and OSA focus on the operational aspects and the interrelation of the involved development parties with focus on the requirements derived from the operational intend.

OSA addresses the following aspects:

- operational hazards,
- harm and severity
- safety objectives (authorities, operator, aircraft, etc.),
- functional failure,
- operational safety requirements.

Usually, the OSA provides the input for the functional hazard assessment (FHA).

6.7 Application from Avionics

Most of the applications are derived from experiences in military, space and aviation industry. Unmanned space application is famous, and since 1961 cosmonaut, Yuri Gagarin became the first human in space. The first animals in space were a group of fruit flies, launched to an altitude of 42 miles by American military scientists on 20 February 1947. Neither the group of fruit flies nor Yuri Gagarin did fly or drive the space ship or the rocket. All the data had been pre-developed and implemented into the rockets. Thanks to Norbert Wiener, we learnt how to control rockets and of course the principle was also used for later spaceships.

The principle of the rocket was of course also transferred to jet aircraft. The pilot does not control the turbine directly, and he controls the physical thrust of the turbine with a joystick equivalent to force feedback in today's by-wire aircraft. In many commercial aircraft, there is a cable that still connects directly to the throttle valve of the turbine, so that in the event of an electrical power failure, the turbine can be controlled manually. This also means that the pilot feels the forces on the throttle valve as feedback directly on his hands. The auto-pilot is just an extension of the by-wire technology, which allows the pilot to fly hands-free in case of released air traffic situations. The pilot or his co-pilot must continuously monitor the instruments and press a button to indicate that he still controls the aircraft (Fig. 6.67).

The autopilot in an aircraft consists of three major control devices. The mission control works close together with the navigation system. The mission of an aircraft is usually a flight from one airport to another airport. A typical navigation system

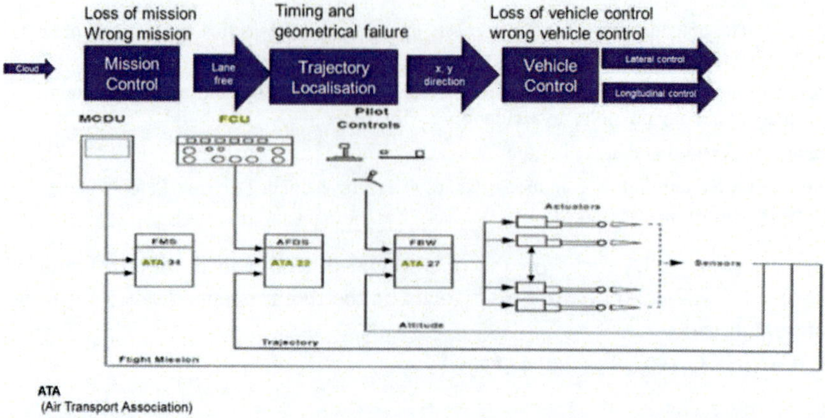

Fig. 6.67 Typical autopilot of a commercial aircraft.
Source Aircraft Systems, Ian Moir and Allan Seabridge, 2001

consists of much more details, and functional units as navigation systems are known for motor vehicles. The map includes, for examples

- detailed maps of the airports including the local regulations at the airports,
- information about the electronic support equipment for weather forecast, flight control and navigation,
- flight corridor details including regulations for their usage,
- critical sectors such as forbidden areas near the flight corridors, etc.

The mission control system is usually based just on a dual control device. A higher redundancy usually is not necessary, since the pilot could switch to manual operation mode. The pilot has other options to get support from air traffic control to find the nearest airport. In general, a pilot is educated to fly an aircraft without the mission control within the given flight corridors. If an aircraft flies to the wrong airport in compliance with the applicable safety regulations, it leads to the lost of its mission, but not to any safety-critical situation. The Air Transportation Association (ATA) provides design rules for aircrafts. The so-called ATA chapters are a framework for the individual systems of an aircraft. All technical systems of the aircraft are divided into corresponding groups and subgroups; e.g. the tanks belong to Chap. 28-10: Group 28 Fuel Systems, Subgroup 10 Storage. This system can be found on every aircraft which is certified according to FAR 25 and EASA CS-25. This means that every civil aircraft which is used for the transport of passengers and/or freight and has a max. take-off weight of more than 5.6 t is certified according to these regulations. The ATA chapter is often part of the part numbers. This enables maintenance personnel, logistics personnel or even design engineers to work in a uniform manner and easily identify aircraft systems and parts. ATA34 considers a dual control system for the navigation.

The second control device is autoplaner according to ATA22, and it consists of the trajectory planner and the localisation unit. Trajectory planning without localisation is impossible. Therefore, both functions need to be available during the ration of the autopilot until the safe takeover of the manual control by the pilot. A sudden failure of the trajectory planning and the localisation could not be immediately controlled by the pilot. The pilot or the co-pilot has the obligation to monitor all functions and instruments during the flight with the autopilot, but in case of failure or significant malfunctions the probability of entering critical flight situation is significantly present in various failure scenarios. Therefore, the control device is designed by a dual–dual redundant architecture. That means a redundant control device is implemented twice.

The third control device of the autopilot is the flight control; the specific design regulation is ATA27, and it controls the turbines. A turbine is a complex system, which however technically can deliver only too less or too much thrust. If the pilot or the system has decided to increase thrust, then at some point, there will be a point of no return and the turbine will irrevocably increase thrust. Only in the next control cycle, an error could be corrected. Therefore, all sources affecting the thrust control must lead to a distinct reaction at the turbine.

There are three typical feedbacks from the turbine:

- The flight control gets via the feedback the current attitude of the turbine.
- The autoplaner gets via the feedback the current trajectory of the turbine in combination with the tail, etc.
- The navigation gets via the feedback the current flight mission.

All control parameter could be compared between the intended parameter and the measurements from the feedback (Fig. 6.68).

The pilot gets on the display the comparison of the data within the design limits of the aircraft. In case of manual flight, the pilot can supervise the stabile flight conditions of the systems, which are indicated that the current operations are within

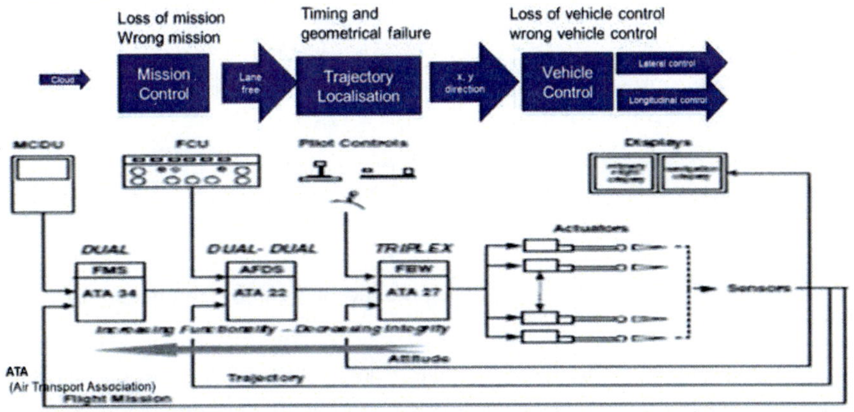

Fig. 6.68 HMI and override for autopilot

the permitted limits of the aircraft. In case of takeover, he always could monitor the robustness margins of the entire manoeuvre. In case of system-operated flight, the pilot can monitor the robustness margin and he always knows if the aircraft is in stabile flight conditions. As more stabile the aircraft is, as more performance or functionality the system could demand from the aircraft. As closer the aircraft operates to its design limits, as lower is the integrity of the aircraft systems. All control algorithms, the sensors and the actuators operate closer to their design limits. For an example, if the airplane flies in good weather conditions, as faster it can fly, since the entire flight and the performance of the control systems are more stable. Nevertheless, increasing functionality means decreasing integrity for a dedicated aircraft or their systems.

6.7.1 Flight Envelope

There are various definitions and different systems in an aircraft that are called a flight envelope. One definition is based on the technical capability of an aircraft or a missile. It is the envelope of the possible performance of a missile in an altitude–velocity diagram or load–multiple velocity diagram. Flight envelope graphs, for example, indicate the permissible operating range of a missile. It is determined by a matrix of mutually influencing parameters, which in turn may only assume certain values. All conceivable combinations of speed, altitude, angle of attack, angle of sideslip, weight, centre of gravity and aircraft configuration (e.g. landing gear, flaps, slats, afterburner) are examined to determine the limits of an aircraft when these variables are changed. The manoeuvrability, structural loading and engine behaviour must remain within acceptable limits. Engines also have envelopes for their permissible power range. At the HMI to the pilot, a flight envelope protects against erroneous actions of the pilot. The flight envelope protection is an independent, background protection against dangers exceeding of the permitted and technical feasible control corridor by the pilot. A computer checks the pilot's control commands before they are executed.

One of the key causes of aircraft accidents and incidents is derived from "loss of control" (LoC). The following main types classify LoC:

- airplane stall during manual speed control; typical causes are
  - pilot mishandling (e.g. excessive bank angle at low speed) or neglect,
  - loss of spatial orientation leading to stall and spin.
- airplane stall during operations with automatic or partially automatic control, caused by
  - partial automation—flight crew manually controlling trust/airspeed,
  - full automation—flight crew selecting excessive climb rate command.

- lateral directional control upsets due to
  - atmospheric conditions (e.g. wake turbulence),
  - partial automation—control asymmetry due to icing, engine out, fuel imbalance.
- overstressing the airplane structure, caused by
  - mishandling during atmospheric upset (e.g. rudder control wake vortex encounter),
  - mishandling during stall recovery,
  - excessive normal load factor (NLF) manoeuvres, e.g. firefighting flight operations.
- exceeding of the high-speed airplane limits caused by
  - partial automation—flight crew manually controlling trust/airspeed,
  - full automation—flight crew selecting excessive descent rate command.
- failure of critical flight control function(s) due to
  - design deficiencies,
  - hardware failure.

While the flight envelope could be a typical characterisation of an aircraft or a missile, a flight envelope protection system is a mechanism that prevents for failure of the pilot. In both cases during the design process, the capability and design limits of the aircrafts are elaborated and validated for the permitted system modes. The flight envelope is also the basis for the design of the control circuits. Such an engineering approach requires a complete forecast of the aircraft in all permitted flight situations. Based on the cascaded closed-loop control circuits, all design parameter of the entire aircraft including the position within a given flight corridor could be continuously controlled and monitored. In case of failure of the pilot or of technical devices, the available margin for the degradation is always available for the system. Bad weather conditions or critical flight manoeuvre is always monitored for the design limits of the aircraft, in case of violating any limits, adequate degradation of performance or systems could be planed, and the resulting behaviour of the aircraft could be forecasted.

6.7.2 Adaption to Automated Driving

The principle of the flight envelope could be also applied for the control of automated vehicles. An automated vehicle also has a design typical drive envelope. The degree of freedom of a motor vehicle is more or less limited to the longitudinal and the lateral axis. Unless a motor vehicle does not brake too fast, deceleration leads to lower risk level. Usually in an aircraft, the safe flight corridor is provided by the air traffic control. The aircraft must always be able to operate within the given flight corridor. That means the aircraft could plan the trajectories and the necessary thrust of the turbine. Depending on the aircraft design and the weather conditions, particularly the wind, the aircraft must be able to provide the minimum thrust for the intended flight manoeuvre. For a motor vehicle, loss of power control is not a critical condition, since vehicles usually drive stabile on a paved road. However, a vehicle must also maintain its ability to steer and brake in all relevant situations, weather and traffic conditions. Railway systems usually need to control only the longitudinal axis, since rails and switches guide and limit the train for lateral movements. The railway authorities usually plan and control the operation for the railways, switches and signalling systems. The railway infrastructure usually should be independent from the operator of the rolling equipment, the trains.

The ultimate safe state of an automated motor vehicle is stopping within the given lane. It is not the desired condition to block the traffic, but for motor vehicle, it means the condition of lowest risk. As we learnt from safe road design, we also consider a safe lane for motor vehicles, since the vehicle operates within the lane markings with sufficient margin to the left and right side. This default lane is usually given by the road design for the dedicated roads. A human driver learns during driving school to drive in the centre of the lane, and he should adapt the speed depending on weather, traffic and road conditions. A system could get the default lane from the navigation system, similar to an aircraft the default flight corridor including the information of the considered limitations. Today maps have all the road markings and traffic signs already implemented including the given tolerances (Fig. 6.69).

For a given driving speed, a vehicle requires a permissible tolerance corridor in which certain deviations from the prescribed lane are permitted. As higher the speed, as wider the tolerance corridors. In case the vehicle could not anymore maintain the tolerance limits, the speed needs to decrease. Similar to the aircraft, the integrity decreases with higher speed, since timing gets closer to the given deadlines for the control algorithms. Because of this, as better the real-time capability of an automated road vehicle, as higher the possible performance and functionality of automated vehicles (Fig. 6.70).

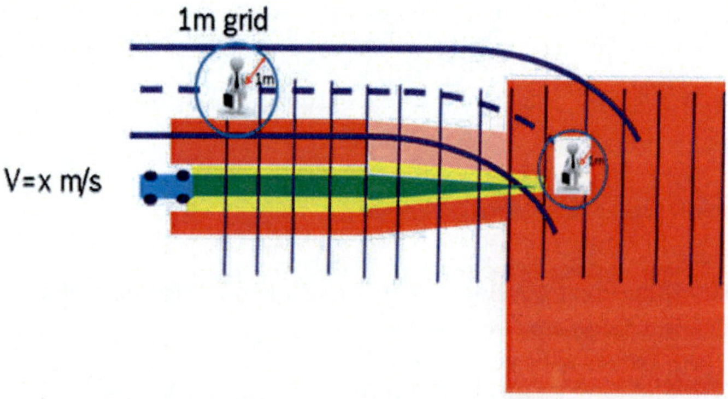

Fig. 6.69 Safe driving corridor and operational limits

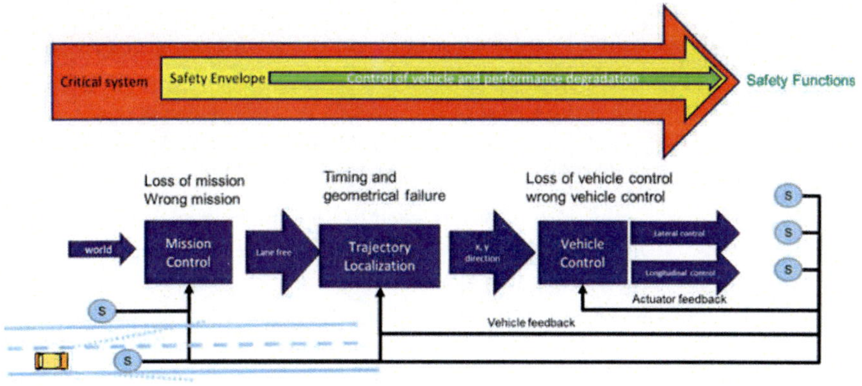

Fig. 6.70 AD vehicle signal chain derived from autopilot

The control chain of the autopilot could be also transferred to a vehicle EE architecture.

- The final element is the actuator control loop. The actuators get their feedback directly from the actuator output, so that deviations could be corrected within the actuator control device.
- The trajectory planer and the vehicle localisation need to monitor the entire vehicle behaviour and compare it with the lane. The feedback should provide information how good the vehicle follows the trajectory within the lane. A continuous comparison of the position of the vehicle and the lane provides the quality of the directional stability.
- The navigation system continuously compares the current vehicle vector with the mission.

Unless the vehicle is on a free lane, it does not lead to accidents. Therefore, the failure in the mission does not ultimately lead to hazardous situations. At the actuator control, there is a similar issue, and the vehicle can only drive faster or slower and could keep the direction or steer to the left or right. Failure of the trajectory is tolerable, unless the vehicle remains in its lane. In the “real-world monitoring”, a continuous observation of the road, weather and traffic conditions is necessary, so that by performance degradation the speed could be adapted to the current driving situation.

There are various independence criterions in the entire signal chain, which are provided by the aircraft regulations, which are as well necessary to apply such a safety concept for road vehicles.

- The control units for navigation, vehicle control and actuator control need to be separated because any common cause could not be controlled in the final control device.

- The feedback loops require sufficient independence, any sensor failure would propagate into the closed-loop control, and it could not be detected in the dedicated control unit.
- The timing in all control units is different. Many final elements operate within a milliseconds time-frame, even faster as an automotive steering system. Delay in feedback would lead to many different impacts for the vehicle control.

As a consequence, most of the sensors need to be implemented redundantly, so that failure could be prevented before they propagate into the next control unit.

This requires a very systematic engineering approach, since the violation of the requirements of independence has at any interface the potential to lead to accidents.

6.8 Outlook on Future Automated Mobility

The discussions about robot taxies get more and more quite, since the industry realised that automation of mobility requires another engineering discipline than electronic protection systems or driver assistance systems. As long as society sees no advantage in automating individual mobility and the modes of transport are shown to offer only minor safety benefits, it will be difficult to obtain funding for automating individual mobility. Consequently, automated protection systems will be increasingly demanded. The approaches for air taxis are becoming more and more mature. Even the aviation authorities such as EASA and FAA are already working on the necessary certification standards. The certification standards or advisory circulars are published for reviewing. The accepted functionality and the performance will be always balanced with the safety capability of such systems.

Outlook

Today's systems in vehicles still reflect the organisational structures of vehicle manufacturers. These are then broken down in the supply chains to the suppliers. This means that, of course, no functionalities can offer lifecycle support and value creation can no longer be offered at the initial sale but over the entire period of use of the vehicle. At the end of the lifecycle, we will of course offer customers the motivation to buy newer basic technology so that high-performance or more complex functions can then be offered. The challenge behind this is of course how to ensure the safety of the systems and their applications. Like the discussion about cloud solution, the private cloud obviously offers simpler solutions for cybersecurity than a private cloud. For a private cloud, the threats can be clearly identified and assessed, whereas in a public cloud, the possible threat scenarios mean different levels of threat for all users and dependents. In order to have a great benefit from a system platform, the maximum safety and security standards must be applied. The book showed which aspects have to be taken into consideration for the development of a reliable platform, which of course have to be put into the right context for the respective project. In order to ensure that such a platform is always secure for future new functionalities, the risks for new customer and additional functions must also be predictably managed. This means that the necessary safety and security measures must be provided in the platform architecture. This also means that the safety architecture must be modularly expandable so that additional functionalities, including their safety, security and protection functions, can be integrated into the platform architecture.

But safety will not be the only challenge for future mobility. A key task will be to be able to provide the necessary energy at the right time and place. Making vehicles over a ton heavier to achieve the normal range of a diesel car is not a sustainable solution. In many countries, living with the risk that significant amounts of energy are used for heating or air conditioning is also questionable. But also, the timely availability of energy and information within the vehicle or the means of transport must be solved in order to be able to implement the advantages of digitalisation safely. The energy availability for a human driver in an automobile was previously given because the driver served as a fallback level for function, energy or information for many safety

H.-L. Ross, *Safety for Future Transport and Mobility*,
https://doi.org/10.1007/978-3-030-54883-4

functions. In the event that the human system failed, it was not the system manufacturer who was liable but the driver who had to pay for most of the damage, or even an administrative offence or even a penalty was imposed on him. In the context of safety in use, the question was not whether system functions were suitable for everyday use, but only the ability of the driver, passengers and maintenance personnel to handle the vehicle and its functions without danger. The availability of energy and its environmental compatibility is currently the subject of very controversial discussions, the combustion engine is no longer accepted and nuclear energy is completely excluded as a solution. Converting solar energy directly into kinetic energy will keep science busy for some years to come, so that we will have to apply a low-loss energy conversion through various measures in order to implement the individual driving wishes of the vehicle users in an environmentally friendly way. How to implement highly available control systems for the propulsion has been known for many years from the aerospace industry known. Within the book, the adaption of the redundancy principles and the control technology for motor vehicle steering systems and lateral control for motor vehicles was described. However, the technical implementation probably requires some extended processes, methods and corresponding organisational structures, especially if such functions are to be changed during their operation time. The adaption of the closed-loop control circuits and the cascades lead to different control strategy and consequently to other implementations. The book can introduce other control strategies and explain the related protection and safety mechanisms. With the herein described process and method principles, the necessary activities could be developed and could provide insights for a development plan for safety-related automated control systems. Since all function and system development depends on the context, the relevant stakeholder and the intended purpose of the products or their functionality, the analysis of the context is always the key success factor.

In the same way that one has to look several times when comparing the perception of the system with the real world, the development processes are of course still circular. However, the cycles do not only refer to the sample construction for components, which have been known for over 50 years as A, B, C, D samples, or the verification and validation loops, which are also regarded as iteration loops in the V-model. These typical spiral processes must also now be applied in the development within the operational concept for function development. Whether it is the intended function, a protective function or a safety mechanism, these development processes will have to be adapted to each other in a maturity model, and thus, stable architectures for the new mobility functions will have to be developed. Of course, the agile world will not allow you to have production cycles of six years, as was previously the case for a vehicle platform. Therefore, functionalities such as software or firmware over the air will be an essential means to adapt the vehicles to the new conditions and the extended experience. Safety-relevant functions will not be allowed on public roads without laboratory validation. These precursors are already anchored in the new homologation regulations.

To develop one's own consciousness is a characteristic of humans, probably even the characteristic that distinguishes them significantly from all beings on earth. When we talk about big data and fast data today, it is important to understand which of the many information are actually the most important for acting correctly and how to prioritise them. In all this information, the automated car must also find its own "I" in order to make the right decision. Nevertheless, we cannot leave this to a self-learning algorithm; the development team must bring all the necessary experience to the vehicle so that it is able to do the right thing at the right time. In the end, a reflex-like movement such as braking, steering or accelerating abruptly is only a yes or no decision; it refers to "do I do it or don't I do it". Throughout evolution, one characteristic has emerged that is essential for the survival of the species and that is adaptability. However, it is not the best adaptability that brings the greatest success, but the ability to adapt and act continuously in different situations.

The book tried to make transparent that automation of mobility does not mean an autonomous automobile. The mission and the purpose of the entire mobility concept requires deep analysis of the intended operation environment and the risk to any human in the environment and related to any role a human could have within the lifecycle of the mobility concept and their systems. A mobility concept in the intended operation environment requires a multi-layer engineering approach. For the operation domain, a human centric engineering approach is recommended, so that any actor could identify their benefits and the related risks. From organisation view point, a close cooperation with the relevant authorities is necessary, since they represent the minimum tolerable risk level of the society. The requirements for any operation domain independent in what area, nation or climate zone are for any transportation system the same. If they follow a pure manufacturer strategy, any manufacturer would try to develop their own marketing strategy and would try to manage the risk depending on the needs derived from their marketing strategy. A coexistence of various different risk strategies end in an endless complexity, since every party tries to find their benefits for their marketing strategy. Unfortunately, many vehicle manufacturers announced already their automation strategy, but the legislation is just under development. For the transportation system similar as in the aviation industry, UN ECE provides now basic rules and regulations harmonised with the Vienna Convention, which based on the principle of road-worthiness for automated road vehicles, which lead to harmonised expected vehicle behaviour in critical operation environment and requirements for the specific control equipment for the automated vehicles. The general idea of autonomous vehicles could not anymore develop. The vehicle could not be more intelligent than the rule makers and the developers that develop the automated vehicles according that given rules. Any behaviour in the intended use and also in case of failure or degradations must operate within the limits of the given regulations. A huge challenge will be the adaption of the road traffic regulations for the different nations and the differ operation domains like the urban traffic. Here, any nation need to find an acceptable compromise between the mobility aspects and the burden for other road traffic participants. If you are in a country that promotes individual mobility, you will find the advantages for private vehicles more pronounced. Automated vehicles will be allowed to drive faster and will be able to claim more space for themselves. It is important, however, that every road user is aware of the local situation and knows that a vehicle can legitimately drive at 50 km/h in the traffic area in which he or she is moving at any time. As an experienced road user, you know that at this speed, you must be prepared for a stopping distance of up to 30 m. A small child will not be allowed to move freely in the traffic area; school children will receive appropriate traffic education so that they can learn to move in the traffic area and assess the dangers. This means, however, that the traffic area must be prepared for "safe traffic" and the laws must be adapted for this purpose and also be discussed transparently in public. On the other hand, the necessary balance between private and public local and long-distance transport will have to be discussed for a long time in the post-Corona period. The main argument for private individual transport is still flexibility. If one wants to right now, one can get into one's car and drive from A to B. At the moment, society has to accept cars in the city centre, motorcyclists on country roads and e-bike riders on cycle paths. Parties that want to give more space to cyclists or pedestrians are becoming more and more popular and their arguments are getting more and more heard.

It is a fundamental necessity for a state authority to lay down the rules on what to do with each other on public roads. It is also the task of this state authority to define the minimum requirements for the basic functions, as well as the safety and protective functions. Defects, such as design and production errors, as well as insufficient instructions for the use of the various means of transport, are the responsibility of the manufacturers and are regulated worldwide by product liability. However, instruction errors in the education of drivers, the way to communicate correct behaviour to road users in public transport, are again the task of the traffic authorities. If they do it by road traffic regulations or by education, if they do it by traffic signs, road markings or general rules depend on the historical experiences and tradition. Fortunately, the worldwide traffic rules base on the Vienna Convention, so that we have a common root for the road regulations.

The same basic rules are also available for air traffic. In aviation, a common understanding of air traffic law and air traffic safety has been established over almost 100 years. Even if today an Airbus pilot is not allowed to fly a Boeing without additional training, the basic safety principles originate from the same roots worldwide. Traditionally, different risk scales have been applied to military and space applications, but the risks for investors, users, operators, developers and all stakeholders are comparable.

This book does not address solutions for air taxies or even air robots; the examples from the aviation industry should only show solutions and proven safety principles how to avoid, mitigate or control risk. Proven design principles and safety mechanisms have been borrowed from the aerospace industry. Even if these mechanisms are used to reduce comparable risks, their effectiveness and applicability always depends on the given context. The context and risk scales of mobility and space systems are very different. Thus, mobile transport systems already form a certain safety barrier against cybersecurity as long as they cannot be directly influenced in their behaviour by digital communication. In any case, as long as a pilot or driver is entrusted with the command and control, a plausible control instance must continue to be overcome. However, from a certain age on, a vehicle is also gladly maintained in a cheaper repair shop. In the case of a civil aircraft, it would be incredible for a nonprofessional to be allowed to repair the aircraft. Of course, one expects also from a certified workshop that the maintenance processes are as such that sabotage, like implementation of Trojans or the installation of defective parts, etc. are avoided. In the safety society, there is no agreement at all on what is meant by external influence. Sabotage is clearly an offence, which is intended harm by the system, to users of the system or people who are in the environment of the system. Misuse in a workshop can also be understood as a threat to the safe operation of the automated system. A suicide can often even result in many imitators. In railway accidents with suspected suicide, the course of events is often not published. There is a consensus among experts that rare events, foreseeable misuse and all conceivable external influences, which could interact with the automated systems, must be regarded as risks. The assessment and determination of appropriate measures is of course a challenge.

The general development process for an aviation system is similar to the process now being introduced by the UN ECE for the new automated driving functions. The three pillars of continuing to rely on classic product tests in the vehicle context, auditing the development processes at the manufacturer and validating the systems and products in the real-world context are comparable with the typical processes at EASA or FAA. The methods for aviation certification start in the operating environment to develop the requirements for the systems that significantly influence the behaviour of the aircraft. The operational risks are analysed in parallel to the design and feature development of these relevant systems. The aviation process considers the aircraft as a black-box with standardised system domains, as they are defined in the ATA-Regulations. Such system domains are today not defined in the automobile industry and all vehicle manufacture see the interfaces inside a vehicle as their specific domain competence. Some companies see their core competences and their key-selling points in the chassis design, others more in the vehicle body or in the interior design. Some of the mayor supplier already offer horizontal integrated signal chains (from sensor to actuator), but the entire vertical integration from the vehicle feature down to the software implementation remains at the vehicle manufacturers side. There are vehicle manufacturers which intend to implement now their own software, up to the idea to develop their own operating system for the control devices but over the history from OSEK to Adaptive AUTOSAR that would be an ambitioned approach. Especially, the lifecycle maintenance for the software and their adaption to new functions and features could be a challenge.

In recent years, all major vehicle manufacturers have begun to set up modular construction kits that are standardised across platforms in terms of mechanical system structure, functional design and electrical/electronic interfaces. Unfortunately, the interfaces to the main actuators, such as steering, brakes and propulsion, have remained largely unchanged. This means that one still has proven interfaces to the key actuators that significantly influence vehicle behaviour, but new interface requirements mean a completely different EE architecture for future vehicles. This is an easy transition from the internal combustion engine to the electric motor, but the influences on the steering and braking systems also require revolutionary adaptation. Previous attempts to develop braking and steering systems as an evolutionary advancement based on ESP and steering assistance systems have brought only limited progress towards automated driving. These systems were previously the domain of the major suppliers, but even the approaches of steer-by-wire and brake-by-wire have only been operated with the aim of improving assistance systems. The new interface requirements, such as short reaction times, simultaneous information and energy availability, deterministic, time-synchronous and timely behaviour of the three decisive actuators cannot be implemented with today's vehicle architectures. Communication networks and power distribution in vehicles are still in their infancy. Some new players are already communicating their initial successes very aggressively, but both established and new players with their current platforms are lagging behind in terms of large-scale production. In addition, there are other aspects that also require a significant degree of user confidence in this technology. The approaches with the charging infrastructure already show typical problems in this respect. An enthusiastic Tesla driver naturally praises the good handling of his vehicle during charging. Other vehicles at the charging points of the various suppliers, different plug concepts, billing systems, etc. still overtax most users immensely. Even if fancy navigation systems and mobile phone apps give you many good tips, one cannot yet speak of socially accepted processes. The complete handling of hydrogen technology is currently neither communicated nor do field tests with prototypes or first small series show that a technology for broad sections of the population is establishing itself. Enthusiastic drivers are gaining initial experience with the abominated assistance systems, which are designed to make lane keeping and distance control easier for them. All these systems must be monitored by the driver, but taking the driver out of responsibility is a much greater hurdle. There are enormous outcries about the publication of the new homologation regulation automatic lane keeping systems (ALKS) and the methods for its validation. The main advantage for the industry is that automation functions could now be released and certified independent from a vehicle type. There is of course an additional process required for the implementation and integration into the basic vehicle. Up to now, it was only possible to release such functions as part of a vehicle type approval. Today, the established vehicle manufacturer see their benefits that they only could manage the horizontal and vertical system integration and maintenance is preferred in certified repair workshops, at least for the premium vehicles. All processes and actions where humans must or should interact with machines must be proven so effective that it can be intuitively and safely implemented or used, or the human must learn the processes and actions. As long as the human being will only be a passenger in his own car, he will no longer build up a bond to his car. In other words, it will not matter to him which vehicle he can drive from A to B; the main thing is that it is safe, sufficiently flexible and economical. This raises the question of whether today's motorist really wants to pay for the ownership of a vehicle or only for its use. The burden of parking, maintenance, servicing, insurance, taxes, fees, etc. is usually borne reluctantly by the driver, and leasing is already becoming an alternative and car-sharing services are growing in the cities. With all the pay-by-use vehicles, another challenge is emerging. The vehicles must follow uniform standards; otherwise, it will not be possible for users to handle them safely. A driving licence for passengers in pay-by-use vehicles would be just as paradoxical as a driving licence for taxi customers. Just as with propulsion hybrids, there will probably be automated driving hybrids for a long time to come. In other words, on the way to work, in speed-limit zones or on a long holiday trips, the system should drive, and the owner drives conventionally if he wants to have fun driving. However, cost-conscious users are considering whether it would not make more sense to rent a vehicle. Sooner or later, the delivery service will bring the purchase, and for health reasons, reason walking and cycling at least during sunny days might be better.

Having a parking space or parking space for two vehicles per household is also a burden that not everyone is happy to accept. There would be the idea that after driving fun, like in valet parking, the vehicle drives alone to a large parking lot and that you can then call up the vehicle repeatedly from there if necessary. Nevertheless, do you have to own the vehicle for such a service? Probably not if you are not dependent on the vehicle and you only need the service if you want to experience driving pleasure.

The legislator will define the minimum requirements that will serve as a basis for the acceptable risks. However, industry will have to agree on binding standards and norms because the legislator does not want to restrict the freedom for innovation. Incentives to buy existing technology will not be supported by state funds; support for infrastructure for more environmentally conscious or environmentally friendly mobility or a higher level of safety will always remain a social interest and will be supported accordingly by the state authorities. The solution or the methodology how the safety requirements are implemented is the responsibility of the standardisation bodies of the industrial associations. The decisive factor is the recognition that the solution space will continue to be very large, even if it currently appears very complex for industry. Which solutions are actually sustainable will again depend extremely on which interest groups develop the most effective standards. If every manufacturer or every interest group tries to implement its own standard, the systems in mutual cooperation will lead to new dangers that we cannot assess today. What does the mobility of the future actually look like? The railways are hoping to become a serious competitor to air travel in the medium distance range. For urban transport routes, more and more tunnel solutions or structurally separate express roads are emerging, which are very well and safely suited for automated vehicle solutions. Such solutions can be monitored very well, cyclists and pedestrians are effectively protected by the structural measures. Alan Musk also relies on this solution with his "Boring Company". The further the distance the more the ideas of the Transrapid and the Hyperloop come into question. Here, too, personal protection and cyclist protection is guaranteed by structural measures. The driver or operator does not necessarily have to sit in the means of transport; he can also overlook the entire traffic area from outside and release the safe routes. The entry and exit to and from such systems must of course be ensured safely, but otherwise, the users of the means of transport are only passengers. Hybrid systems are once again considered ideal for this purpose. Why not sit in the same cabin as in a railway coach, the long-distance route is transported on a train, plane, Hyperloop or Transrapid, and for the last mile to the settlements and the individual houses, the cabin gets wheels, brakes, steering and a propulsion system, so that you can come home with dry feet. In all mobility systems, the primary focus will be on the quality of the interconnected systems and their reliability in getting from A to B. Every change of the means of transport is associated with risks. The transport planned in advance with the specific means of transport over the designated route can be safeguarded very well and effectively with traditional means. These examples are intended to show that the subject of safety and security need not present a particular hurdle, provided that the applicable rules are respected. It is important that there are clear governmental regulations which ensure that the various parties involved in traffic can coexist. The split between safeguarding the route, the means of transport and how the systems interact with each other and with other systems must be clearly defined and comprehensible for all parties and stakeholders. Selfish politics, loud polemics or populism only lead to hyper-motivation and thus to uncontrollable risks. Whenever this transparency does not exist, this lack of transparency gives rise to new risks that are not predictable and therefore not controllable. Of course, individual mobility is also conceivable in the air. Anyone who attended a Formula 1 race in the pre-Corona era knows the helicopters circling over the race tracks, flying the stars to the grandstands at the race track. In some countries, air taxis will also become an acceptable solution, so here, too, we should look for environmentally friendly, effective and efficient propulsion systems and reliable energy concepts and develop these into safe means of transport suitable for series production.

Cybernetics offers a solution for all conceivable systems. These methods have already created the precondition for people to come to the moon and for us to be able to explain space to ourselves. Television, radio technology and also the Transrapid could never have been invented or even built without the basics of cybernetics, because we would have lacked the ideas for such a thing. Of course, cybernetics is not a construction kit from which one can help oneself and then build the next means of transport. But we as the "Lego" generation like to think in such modular structures. Many stages of maturity had been necessary, before Lego developed from a basic structure and architecture to be applied to highly complex solutions with microprocessors and controllers for machine learning. Cybernetics offers the possibility of explaining, perceiving and understanding the dependencies between humans, systems and their context. As soon as one is able to describe dependencies and problems, the basis for dealing with such systems has been laid and the problems and arising risks could be mastered. The safety term for safe perception cannot be reduced to the error-free nature of the sensors used. Safe perception in real-world applications means sufficient perception to be able to perform the planned automation function within the scope of a defined tolerable risk. From cybernetics, an approach is emerging that systematically deals with predictability and its uncertainty. In this way, one quickly realises that the real world is not as it needs to be for a technical system to function safely. Another aspect we recognise is that we never perceive the world from the different angles in the way we describe it to a third party. Therefore, the perspective from which the describer observed the space or behaviour must be known. Sometimes, it is necessary to observe twice to check the assumptions of the first observation. If these assumptions are not valid, the entire observation and its description may be invalid. In a simplified way, the principle of the Kalman filter is to be mentioned. The first look creates a hypothesis of the observed and the second look gives hints about the confidence of the hypothesis. Such principles are essential to create a consistent and communicable safety argumentation. It is a major challenge that an organisation or even the relevant supply chain has the ability to develop a safety-relevant cybernetic system. This is of course based, as proposed in ISO 26262, on the necessary safety culture including the processes and a common organisational understanding of methods. If you approach a problem of person recognition systematically, you will soon realise that the challenge must be understood by the entire development team. In the operation area of an automated vehicle, there should be primary no person, that could be endangered. Also, in the surrounding area, there should not be a person you should not expect to be there. The safety requirement for the intended function of perceiving people in the operating area of automated vehicles often means making sure that there are no people in the operating area. If persons are present there, an appropriate reaction of the safety system must be ensured. The typical negation of the intended function at traffic lights is simpler. You often hear that the intended function is called "recognise red lights". After a systematic technical risk analysis, including the legal context and consideration of misuse scenarios, as well as foreseeable uncertainties, one would formulate the following safety requirement: Recognise the traffic light status, and if the traffic light shows "not safe green", reduce the vehicle speed conservatively so that the vehicle comes to a standstill in time. Such negations effectively prevent systematic errors in the intended function during development and at the same time in the safety or protective function to be implemented. With each validation, a logical discrepancy would arise either in the intended function or in the safety function. Safety engineering is not the task of a safety manager but requires a well-organised development team.

Cybernetics also helps us to integrate artificial intelligence (AI) into technical systems in a meaningful way. If humans cannot use their limbs, an AI-based system will limit their access to actuators even if access is restricted. In simple words, AI is the attempt to transfer human learning and thinking to the computer and thus gives it intelligence. Instead of being programmed for every purpose, an AI can find answers and solve problems independently. However, if the system is not able to communicate the solution to the problem or to implement it, essential advantages for the use of AI are missing. Bionic has helped us to understand the human system and has given us the basis to build similar adaptive and efficient systems and computer architectures. Cybernetics and comparable mechanisms in humans also show how their functions can be implemented reliably, dependably, timely and thus safely. However, it is also shown that not every reflex reaction is always the best reaction to survive. To be able to control the reflexes systematically and to react quickly at the right moment is a task for a system that requires a structured development process. When which reflex is useful and with what intensity the reflex-like reaction should occur is an important learning success.

The bionic not only helps us to realise dependent functions in distributed networks or developing new controller architecture, it also gives us many hints how to build intelligent actuators. The human limbs know exactly which abilities they have trained and where their limits are. If the brain and its nervous system expects a perfectly trained footballer to be able to perform Swan Lake, then it will not work with the footballer's legs. Enormous data compressions are necessary in order to realise actually reflex-like functions. Such a reflex must be of course safeguarded during their activation, certain redundancies are unavoidable. In cybernetics, the term "redundancy" is used in a wider scope, especially when the context of a function is included in the redundancy. It is not just a matter of physical or temporal redundancy; there are also patterns that allow certain plausible relations or verifications. This happens in a human organism of course exactly at the time when such reflexes are necessary. Such redundancy as part of the reflexes has been learned by humans during their development and is passed on to their descendants. Of course, we as humans learn to control and manage our reflexes even after birth. There are processes that are more complex necessary to train a neuronal network as to find mechanisms to develop a system with functional safe reflexes.

Today's actuators in vehicles are still commanded slaves of the control systems. If you give the steering or the drive motor a torque command, it will slavishly execute it. A by-wire brake system, even if only slavishly, converts brake pressure to demands on the brake disc. In the human system, however, the actuation limits are stored, and the limbs only perform what they can do. This means that in an "intelligent vehicle" or the smart phone on wheels, we will need something like smart actuators that can request and subscribe the necessary data for the conversion themselves and then implement the necessary "safe" functions. This means, however, that the functions that the actuator can implement are already available in the actuator. They only have to be activated from the outside at the right time, for the right function in the right situation. Neurons only activate data and energy flows; muscles are responsible for the conversion of forces, but they are often no more intelligent than the steak on the footballer's plate. During such reflexes, the human being sometimes releases adrenaline, which then activates the rapid movements in the limbs or other organs of the body. But the adrenaline only activates functions that are stored in the actuators. Fortunately, today there are network technology solutions that can technically implement such functions. We also have great smart actuators in prosthesis engineering for disabled people that can even be controlled by the nervous system. We must now try to implement such actuators for smart vehicles as well, otherwise the whole AI in the vehicles is of limited use. How such data streams can be functionally safeguarded should not be unknown to a modern safety engineer. However, it should be clear to him that humans have two arms and two legs; this redundancy not only serves to improve the conversion of force but also has an essential task with the ability to synchronise within its limits. Nevertheless, we should not despair of the fact that all safety solutions can then again only be solved by redundancy. The car as it stands there today even has four wheels, which can also be seen as an opportunity.

Technical systems interacting with the real world do not always require machine learning algorithms. In many cases, closed-loop control circuits are already sufficient to be able to continuously adapt to changes in the real world through appropriate system reaction. Today, we can build quantum computers, but the vast majority of all computer systems are still based on a von Neumann software architecture. Even in the communication systems of today, mechanisms like those of a hundred years ago would still mean an improvement in communication capabilities. Semaphores, relays, switches used as in the human nervous system can make communication more reliable and faster. Especially, the ability to build adaptive systems can be copied from humans. Why do giga-bytes of video data have to be transmitted and road markings have to be recognised at the desired location by a camera system? A pure information, yes, the line is at the presumed location, with the assumed properties would be sufficient. The principle of putting the recipient of information into the right context in advance during development or at a much earlier point in time can contribute to a significant reduction in the amount of transmitted data. Such data reduction measures can of course be usefully applied to all monitoring mechanisms. Using the example of a remote-controlled car, it can mean that you only need one joystick, which can be set to move faster forward and slower backward, and the sideways movement can be set accordingly by the right–left movement. All other degrees of freedom in a remote-controlled vehicle cannot be influenced at all from a distance, as long as the actuator technology is not significantly changed. The extent to which remote control is possible and how the necessary traffic space can be safely observed is another challenge, but one that can be well regarded with the help of cybernetics.

If the risks to people and the environment are not analysed and evaluated consistently or incorrectly, such systems will always become a risk to society. If such systems are a risk for society, they will always be a risk for investors.